High Performance CMOS Range In

Devices, Circuits, and Systems

Series Editor
Krzysztof Iniewski
Emerging Technologies CMOS Inc.
Vancouver, British Columbia, Canada

PUBLISHED TITLES:

PUBLISHED TITLES:

PUBLISHED TITLES:

Reconfigurable Logic: Architecture, Tools, and Applications
Pierre-Emmanuel Gaillardon

Semiconductor Radiation Detection Systems
Krzysztof Iniewski

Smart Grids: Clouds, Communications, Open Source, and Automation
David Bakken

Smart Sensors for Industrial Applications
Krzysztof Iniewski

Soft Errors: From Particles to Circuits
Jean-Luc Autran and Daniela Munteanu

Solid-State Radiation Detectors: Technology and Applications
Salah Awadalla

Technologies for Smart Sensors and Sensor Fusion
Kevin Yallup and Krzysztof Iniewski

Telecommunication Networks
Eugenio Iannone

Testing for Small-Delay Defects in Nanoscale CMOS Integrated Circuits
Sandeep K. Goel and Krishnendu Chakrabarty

Tunable RF Components and Circuits: Applications in Mobile Handsets
Jeffrey L. Hilbert

VLSI: Circuits for Emerging Applications
Tomasz Wojcicki

Wireless Medical Systems and Algorithms: Design and Applications
Pietro Salvo and Miguel Hernandez-Silveira

Wireless Technologies: Circuits, Systems, and Devices
Krzysztof Iniewski

Wireless Transceiver Circuits: System Perspectives and Design Aspects
Woogeun Rhee

High Performance CMOS Range Imaging: Device Technology and Systems
Considerations
Andreas Süss

FORTHCOMING TITLES:

Advances in Imaging and Sensing
Shuo Tang and Daryoosh Saeedkia

Introduction to Smart eHealth and eCare Technologies
Sari Merilampi, Krzysztof Iniewski, and Andrew Sirkka

High Performance CMOS Range Imaging

Device Technology and Systems Considerations

Andreas Süss

CRC Press
Taylor & Francis Group
Boca Raton London New York

CRC Press is an imprint of the
Taylor & Francis Group, an **informa** business

A BALKEMA BOOK

Cover Information
No signal TV illustration. Scalable vector. Error concept
Copyright: kasha_malasha
Courtesy of Shutterstock

CRC Press
Taylor & Francis Group
6000 Broken Sound Parkway NW, Suite 300
Boca Raton, FL 33487-2742

First issued in paperback 2021
First issued in hardback 2020

Typeset by MPS Limited, Chennai, India

No claim to original U.S. Government works

ISBN 13: 978-1-138-61207-5 (pbk)
ISBN-13: 978-1-138-02912-5 (hbk)

This book contains information obtained from authentic and highly regarded sources. Reasonable efforts have been made to publish reliable data and information, but the author and publisher cannot assume responsibility for the validity of all materials or the consequences of their use. The authors and publishers have attempted to trace the copyright holders of all material reproduced in this publication and apologize to copyright holders if permission to publish in this form has not been obtained. If any copyright material has not been acknowledged please write and let us know so we may rectify in any future reprint.

Library of Congress Cataloging-in-Publication Data

Applied for

Visit the Taylor & Francis Web site at
http://www.taylorandfrancis.com

and the CRC Press Web site at
http://www.crcpress.com

Publisher's Note
The publisher has gone to great lengths to ensure the quality of this reprint but points out that some imperfections in the original copies may be apparent.

Table of contents

Foreword

Range imaging is a common name for techniques used for non-tactile image acquisition of spatial objects. These techniques can vary from one-dimensional acquisition (e.g. 1-D range finders) to three-dimensional acquisition (3-D imagers). They usually use either ultrasound waves or electromagnetic waves, e.g. microwave radar or infrared and visible radiation. With respect to the last technique, there is a variety of approaches involving e.g. active and passive triangulation, interferometry, use of structured light, light-field imaging, and time-of-flight methods. All of these approaches have their merits but also disadvantages – thus the decision which approach to use depends on costs and application.

The technology described in the dissertation of Andreas Süss is based on the time-of-flight method and employs scannerless silicon-based near-infrared (NIR) imaging. The use of silicon imaging appears to be particularly attractive, because it enables the co-integration of electronics on the same silicon chip as the photodetector array and has the potential to yield a compact, small size, low power, and low cost 3-D camera. This is possible due to silicon sensitivity in the NIR range.

The time-of-flight method relies on the measurement of the time which elapses when an emitted light signal travels from a light source to a distant object and is then reflected back to a photodetector collocated with the light source. The distance can be readily determined using the light velocity and measured travel time. The emitted light beam must be widened in order to obtain the desired divergence and thus illuminate the entire field-of-view. Also, the beam has to be modulated, either using a continuous-wave modulation or a pulse modulation, so that a signal travel time can be determined as a delay between the emitted and received signal. In order to be able to capture all signals from the scene a photodetector array is used. Typically, laser diodes or LEDs are employed as light sources.

It appears that a crucial component of the time-of-flight is the photodetector array which must exhibit high photosensitivity, high speed, and low noise. Andreas Süss introduces here a so called lateral drift photodiode (LDPD) which can be easily realized in a slightly modified CMOS technology. This is basically a buried photodiode with a top layer pinned to the substrate potential. This construction reduces significantly the dark current and thus the associated shot noise. The doping gradient of the photodiode generates a lateral drift field which speeds up the flow of photo-generated charge carriers into a floating diffusion. This diffusion serves as a sense node since it converts the charge into an output voltage. The resulting time-of-flight CMOS image sensor thus contains an LDPD array with all necessary control and readout electronics and is operated with pulse-modulated laser light. The sensor enables the evaluation of arrival times of laser pulse trains even with a superimposed ambient illumination.

Mr. Süss pays in his dissertation a particular attention to the noise since the noise essentially defines the detection limit of this approach. He describes the noise of electronic components used and introduces a new method for noise analysis of time-discrete analog circuits. Furthermore, he shows that the theoretical limit of the time-of-flight detection is imposed by photon shot noise. Besides noise, he investigates various concepts of the LDPD arrays that enable easy signal evaluation, a reduction of non-ideal effects, and a cancellation of ambient light effects. Among others, a novel readout approach is introduced that employs the above mentioned floating diffusion as a gate of a parasitic junction-field effect transistor (JFET). Unlike MOS transistors JFETs exhibit very low 1/f-noise and are thus predestined for implementation of low-noise readout circuits. This is corroborated by measurements of integrated devices. Several of the presented new concepts of time-of-flight sensors have been integrated in the CMOS technology available at the Fraunhofer Institute of Microelectronic Circuits and Systems in Duisburg and the measurements are shown and discussed. Also, non-ideal effects encountered are investigated and appropriate empirical models are developed that allow a proper operation of time-of-flight sensors using pulse modulation.

The dissertation of Andreas Süss contains an impressive variety of topics and new ideas related to the 3-D time-of-flight imaging based on pulse-modulation. In course of his work the author developed several CMOS chips in order to verify the underlying principle while incorporating his ideas. His excellent work represents a major scientific contribution in this area. It is highly innovative and it is certain to draw a lot of attention of the engineering community.

Duisburg

Prof. Bedrich Hosticka, Ph.D.

Professor emeritus of Microelectronics
and Micro Systems Technology

Former head of Microelectronics and
Micro Systems Technology department
University Duisburg-Essen

Former head of Signal Processing
and Systems department

Fraunhofer Institute of
Microelectronic Circuits and Systems
Duisburg

Preface

This work is dedicated to CMOS based imaging with the emphasis on the noise modeling, characterization and optimization in order to contribute to the design of high performance imagers in general and range imagers in particular.

CMOS is known to be superior to CCD due to its flexibility in terms of integration capabilities, but typically has to be enhanced to compete at parameters as for instance noise, dynamic range or spectral response. Temporal noise is an important topic, since it is one of the most crucial parameters that ultimately limits the performance and cannot be corrected. This work gathers the widespread theory on noise and extends the theory by a non-rigorous but potentially computing efficient algorithm to estimate noise in time sampled systems. The available devices of the $0.35\,\mu$m 2P4M CMOS process were characterized for their low-frequency noise performance and mutually compared by heuristic observations and a comparison to the state of research. These investigations set the foundation for a more rigorous treatment of noise exhibition and are thus believed to improve the predictability of the performance of e.g. image sensors.

Many noise sources of CMOS APS have been investigated in the past and most of them can be minimized by usage of a PPD as a photodetector. Remaining dominant noise sources typically are the reset noise and the noise from the readout circuitry. In order to improve the latter, an alternative JFET based readout structure is proposed that was designed, manufactured and measured, proving the superior low-frequency noise performance of approximately a factor of 100 compared to standard MOSFETs.

ToF is one key technology to enable new applications in e.g. machine vision, automotive, surveillance or entertainment. The competing CW principle is known to be prone to errors introduced by e.g. high ambient illuminance levels. The PM ToF principle is considered to be a promising method to supply the need for depth-map perception in harsh environmental conditions, but requires a high-speed photodetector. This work contributed to two generations of LDPD based ToF range image sensors and proposed a new approach to implement the MSI PM ToF principle. This was verified to yield a significantly faster charge transfer, better linearity, dark current and matching performance. A non-linear and time-variant model is provided that takes into account undesired phenomena such as finite charge transfer speed and a parasitic sensitivity to light when the shutters should remain OFF, to allow for investigations of large-signal characteristics, sensitivity and precision. It was demonstrated that the model converges to a standard photodetector model and properly resembles the measurements. Finally the impact of these undesired phenomena on the range measurement performance is demonstrated.

Acknowledgements

This work was developed as part of the activities at the chair *Electronic Devices and Circuits/Elektronische Bauelemente und Schaltungen* (EBS) at the *University Duisburg-Essen* in collaboration with the *Optical Sensor Systems* department at the *Fraunhofer Institute for Microelectronic Circuits and Systems* (Fraunhofer IMS) in Duisburg, Germany.

I want to acknowledge Prof. Bedrich J. Hosticka, Ph.D. for having had me at the department of Optical Sensor Systems, for having accepted the supervision of my doctoral studies and for having offered the interesting topic. I am further grateful for the willingness of Prof. Dr.-Ing. Holger Vogt for having been a second supervisor of this work and for his continuous supportiveness.

Also, I hereby acknowledge Werner Brockherde, Daniel Durini and Andreas Spickermann for contributing in the definition of parts of the topics discussed in this work.

I want to thank the Professors at my Alma mater – the *University of Applied Sciences, Düsseldorf* – for helping me defining the basis skill set to perform scientific work. Most important for me, here, is to thank Prof. Dr. rer. nat. Hans-Günter Meier whom I consider as a mentor and friend and who was invaluable to me during my studies, and still is. I am also glad to have studied under Prof. Dr. rer. nat. Wolfgang Schnitker, Prof. Dr.-Ing. Rainer Lackmann, Prof. Dr.-Ing. Hans-Georg Lauffs, Prof. Dr.-Ing. Gregor Gronau and Prof. Dr. Peter Pogatzki whom I learned a lot from. I am also thankful for having been exposed to all the kind and supportive professors I could not list here and educational support staff members as for instance Rainer Schulze or Jürgen Brieger.

During my studies I met many exceptional people from whom I can now proudly tell that some of them became great friends and colleagues. I am enormously glad to call Gabor Varga, my former colleague at the Optical Sensors Systems department my friend. I am thankful for all the countless enriching discussions which were truly important to me. I am equivalently proud to name Wolfgang Betz my friend who was always there for me. I hope I can be a friend as partially as good as these guys are. I am furthermore thankful for all the other colleagues of the Optical Sensor Systems department from whom Melanie Jung, Stefan Gläsener, Adrian Driewer, Christian Nitta and Wiebke Ulfig are especially important for me to mention and the remaining staff of the Fraunhofer IMS and especially Stefan Dreiner, Karsten Göpel, Uwe Paschen, Janusz Pieczynski, Dieter Greifendorf, Reneé Lerch and Kay Gorontzi and all the others that simply don't fit in this page but still supported me one way or another. I also want to take the opportunity to acknowledge all the dedicated students I met who were willing to work under my supervision and thus supported this work – without their efforts this work would have been much less interesting to read. Most importantly I have to name and hereby grant the enormous efforts Andrey Kravchenko, Iyappan Subbiah and Tan Yenn Leng spent during their time at the Optical Sensor Systems department. Finally, I want to thank for the nice atmosphere at the office due to my former colleagues Stefan Bröcker, who also supported this work by his great hands-on knowledge and Rana Mahdi.

Of course, I reserved the most important people for the end of this substantially reduced list of people I am grateful to know – my family. I am extremely happy, proud and thankful to have such kind, supportive and caring parents – Helene Sabine Süss and Frank-Olaf Süss – my brother – Stefan Süss and my remaining family of whom I particularly want to emphasize my aunt Irene Thiel, whom I always looked up to – Ich liebe Euch! Finally, I want to thank for the support of my significant other Natascha Alexandra Brandt Rodriguez who brings joy to my life, every day.

About the Author

Andreas Süss received the B.Sc. from the Univ. of Applied Sciences Düsseldorf in 2008 and Ph.D. degree from Univ. Duisburg-Essen in 2014. From 2007 until 2014 he was with Fraunhofer Institute IMS where he was mainly working on high-speed, low-noise imagers for e.g. ToF applications. From 2014 until 2015 he had a scholarship from the KU Leuven and worked as a postdoctoral researcher on global shutter imaging at the MICAS department in collaboration with IMEC, Leuven. As of 2015 he is hired as an R&D engineer in the IMEC imaging division, where he is currently responsible for the pixel development for global shutter and high-speed applications. His research interests include modeling, temporal noise, optimization, compressed sensing and depth imaging.

List of Figures

List of Tables

Chapter 1

Introduction

With the invention of the *charge-coupled device* (CCD) in 1970 by Willard S. Boyle and George E. Smith, which has been rewarded with a Nobel Prize in 2009, the first step towards *image sensors* based on *integrated circuits* (IC) being fabricated in semiconductor processes was taken [BS70]. Since then the total demand for image sensors has been continuously growing until now [Fro08]. Since the 1990s, research has began to focus on image sensors based on *complementary metal-oxide-semiconductor* (CMOS) processes that were capable of competing against the well established CCD solutions [Fos97], which are especially designed to serve the needs of high performance imaging and have virtues in parameters such as linearity, spectral responsivity, fill factor, possibilities of miniaturization, noise performance and dynamic range. Major drawbacks, however, are the need for nearly perfect charge transfer [Fos93], a lack of flexibility and the relatively high cost due to the need of an additional IC, which provides the readout circuitry. The flexibility of CMOS processes allows for on-chip signal conditioning circuitry or *in-pixel intelligence*. CMOS solutions are thus first of all often cost-efficient due to the possible *camera on-a-chip approach* [Fos93], which allows for instance in-pixel signal amplification, e.g. with *common-drain/source follower amplifiers*.

Such implementations – also referred to as *active pixel sensors* (APS) – do not suffer from the need of perfect charge transfer, and are thus applicable e.g. for high frame rate imaging. In order to achieve market potential in classical CCD domains, the challenges of CMOS have to be solved. These are for instance noise performance, dynamic range, fill factor or spectral responsivity. Due to the improvements in CMOS process quality and innovations such as the concept of the *pinned photodiode* (PPD) by Teranishi et al. in 1982 [TKI82], the market for CMOS image sensors faces a steady growth (c.f. Figure 1.1) [IC 12; Res11; Fro08].

As can be seen in Figure 1.2, CMOS image sensors find applications predominantly in consumer markets, for instance in products such as camera phones & PDAs, digital cameras, PC cameras et cetera. On the other hand, markets like industrial imaging, medical imaging, scientific imaging or space imaging are sharing a comparably moderate margin [IC 10; Res11; Fro08; Yol10]. The automotive market, however, seems to be a promising sector to develop future CMOS solutions.

In the past 30 years, plenty of research has been undertaken to complement the acquisition of two-dimensional images with range information [HSS06; Bla04; SHM00]. In general, the range detection of objects should be robust against the shape

Figure 1.1   Forecast for worldwide CMOS image sensor market [IC 12] (Reprinted with permission of ICInsights).

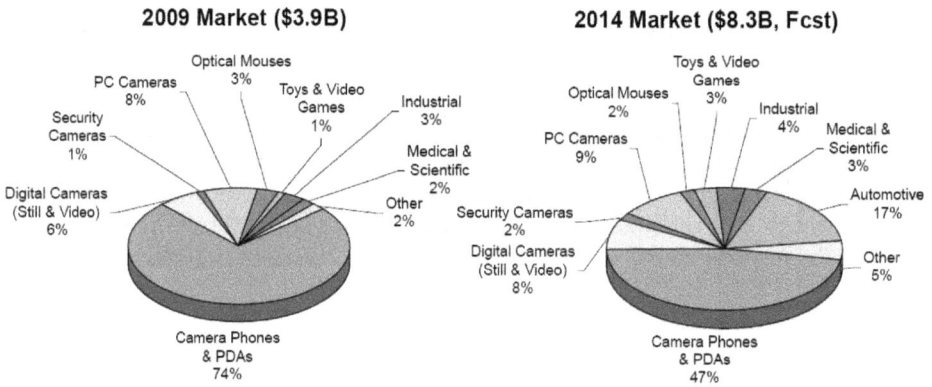

Figure 1.2   Forecast for CMOS image sensor market sections [IC 10] (Reprinted with permission of ICInsights).

and surface properties of the actual objects under observation, illumination conditions and the distance from the sensor [SHM00]. Range detecting imagers should provide a high flexibility, a high lateral and depth resolution, wide field-of-view (FOV) and high depth-of-field (DOF) and frame rate. In addition, they should be cost efficient, exhibit low-power dissipation and be easy to calibrate. Another issue is the eye-safety, which has to be maintained for certain applications, whereby the interaction with human beings is desired or cannot be avoided. Contactless 3-D shape measurements can be realized by the usage of sound (SONAR) or light wave packets (LIDAR/LADAR) as probing elements. The spatial resolution, however, is limited by diffraction. Furthermore, the propagation of waves is not independent of environmental conditions such

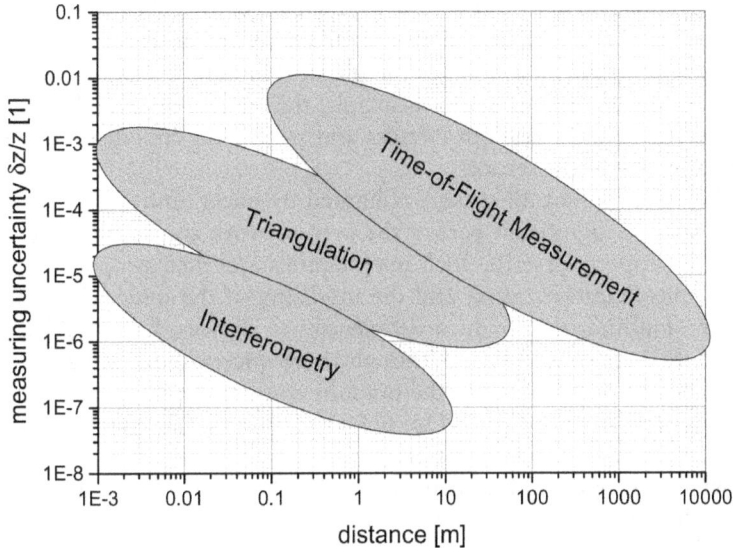

Figure 1.3 Comparison of methods for range detection (according to [SHM00]).

as humidity or air pressure. This is an issue – especially in case of sound waves. These restrictions predestine light waves as probing elements for range detection, since they allow a more robust detection and provide a higher resolution [SHM00; HSS06]. The actual methods for range detection can be subdivided into *time-of-flight measurements* (ToF), *triangulation* and *interferometry* [SHM00]. As depicted in Figure 1.3, these can be compared e.g. with respect to the achievable depth resolution [SHM00].

Based on the time-of-flight method, range imagers can be realized to detect an entire depth map of a given field of view with only one frame. Nevertheless, for the improvement of the depth resolution, acquisition of several frames may be necessary. Since the mechanics, which become necessary for scanning methods are not needed, this is a cost-efficient method for high frame rate range detection. In the time-of-flight method, a probing wave packet is emitted, which may be diffracted by optics, reflected by the actual object, focused by optics and transduced by the actual sensor. The time which elapses between the emission and detection of the wave packet is proportional to the range between the object and the sensor, and thus provides the range information.

The department *Optical Sensor Systems* at the *Fraunhofer Institute for Microelectronic Circuits and Systems*, at which the present work was developed, is contributing to the research on range image detection since 2001 [Jer01]. With the invention of a novel photon-detection device, referred to as *lateral drift-field detector* (LDPD) [Du10; Du09] and its application within a demodulator for the *pulse-modulated time-of-flight principle* [Sp11a; Sp11b], the foundation was set for the development of range imagers, that provide millimetre resolution from decimetres up to metres operating range and harsh environmental conditions such as high background illumination and varying object reflectance and shape.

The scientific contribution of the present work is thus focused on the actual realization of range imagers in a CMOS process, which comprise the proposed LDPD-ToF pixel structures and are based on the pulse-modulated ToF principle. Major issues covered in this work are the estimation of the physical limitations and achievable overall performance of the underlying sensor principle, the realization of high speed and high precision pixel-structures, readout circuits and peripheral circuitries that enable the desired performance of the sensor.

Since the resolution of all sensors is limited by uncertainties that are referred to as *temporal noise*, a significant part of the present work concentrates on probabilistic methods to properly describe such uncertainties and their propagation in circuits and systems, the characterization and the modeling of the underlying physical processes and the minimization of such deficiencies in devices, CMOS APS, peripheral circuitry and ToF-sensor systems. Although these uncertainties are an issue at least since Robert Brown 1827 observed the random motion of pollen grains [Lin08], they are still a major research topic [ITR11a; ITR11b] and thus large portions of this work are devoted to them.

To provide an insight in range detection methods and their basic limitations and to demonstrate the actual state of the art and recent research topics, Chapter 2 presents an overview of these. Chapter 3 is presenting existing mathematical methods that allow for a proper characterization and prediction of temporal noise in electronic devices, circuits and systems. After the presentation of the well-known calculus of the prediction of well-behaved *wide-sense stationary, ergodic processes* in *linear, time-invariant systems*, a short introduction to more general methods is given. Since general methods for the prediction of diverse processes in e.g. *switched capacitor circuits*, that can be used for signal conditioning, may be rather time-consuming, a simple method is presented that circumvents such insufficiencies. In Chapter 4, results of a characterization of the noise performance of available devices in a $0.35\,\mu m$ CMOS process are given. This is necessary, since it provides the foundation of the modeling process, which can later on be used to predict uncertainties in combination with proper simulation tools – so that optimization can be undertaken. Additionally it enables the designer to pick the devices, that offer the best performance. Noise sources and the reduction of such uncertainties in CMOS APS are discussed in Chapter 5. Here, a novel readout structure is presented that aims for a low-noise performance while omitting too many additional process modifications, so that low-cost design possibilities are preserved. Concepts and verification for the design of CMOS based ToF image sensors, necessary peripheral circuitry and the physical limitations of such solutions are discussed in Chapter 6. Finally, the obtained results are summarized in Chapter 7 and topics for further research and improvements are described.

Chapter 2

State of the art range imaging

In the past 30 years, plenty of research has been undertaken to complement the acquisition of two-dimensional images with range information [HSS06; Bla04; SHM00]. The contact-less measurement of entire depth maps enables diverse applications, which, e.g., rely on surface characterization or object recognition. These can cover, for instance, object recognition for automation, machine vision for robotics or safety and security solutions for automotive solutions.

Contactless 3-D shape measurements can be realized by the use of sound (SONAR), millimeter (RADAR) or light wave packets (LIDAR/LADAR) as probing elements. The spatial resolution, however, is limited by diffraction[1]. Furthermore, the propagation of waves is not independent of environmental conditions, such as humidity or air pressure. This is an issue – especially in case of sound waves. These restrictions predestine light waves as probing elements for range detection, since they allow a more robust detection and provide a higher resolution [SHM00; HSS06]. The actual methods for range detection can be subdivided in *time-of-flight measurements* (ToF), *triangulation* and *interferometry* [SHM00]. In this chapter the fundamentals of these methods are explained and an insight in their limitations is given. As this work concentrates on the development of time-of-flight sensors, a detailed comparison of different approaches based on the ToF principle is given and the state of the art is discussed later on.

2.1 TRIANGULATION

Triangulation has been used since the Greeks introduced it for navigation [Bla04]. Triangulation is based on the measurement of an angle between two known objects and the actual object under observation. Several realizations of this principle were

---

[1]Diffraction is a phenomenon, which is caused by the wave properties of e.g. light. If a wave packet encounters an obstacle such as a circular aperture, it is bended around the edges. This causes a varying intensity distribution on a detector located behind the aperture. In case of the circular aperture, the occurring pattern is referred to as *Airy disk*. A measure for the minimum density of the distribution, to which a spot light can be focused is defined by the *Rayleigh criterion*. It describes the smallest angular separation θ_p for which two spot light sources can be distinguished: $\sin \theta_p = 1.22 \lambda / d_{\text{aperture}}$. Here λ is the wavelength of the light wave and d_{aperture} is the diameter of the circular aperture. The angle defined by this criterion is connected to the case, where the maximum of one spot image coincides with the first minimum of the other image [PPBS08].

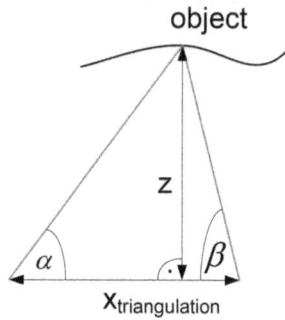

Figure 2.1   Schematic of passive triangulation according to [Lan00].

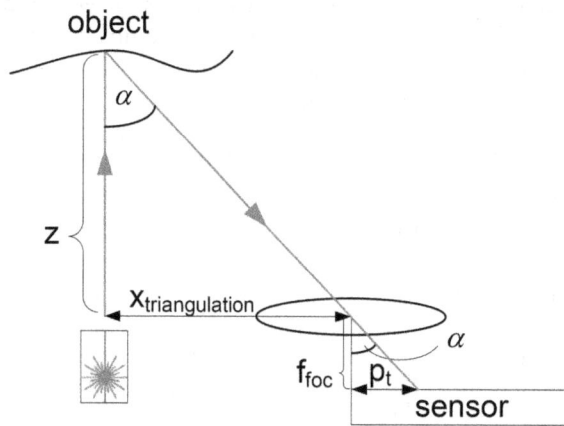

Figure 2.2   Schematic of active triangulation according to [Lan00].

demonstrated in the past [Bla04; SHM00]. A widely employed method is the so-called *passive triangulation*. An example for this is *stereoscopy*. For passive triangulation, measurements of an object or a scene are done from different perspectives, as is demonstrated in Figure 2.1. The distance z of the object from the triangulation base can be determined by

$$z = \frac{x_{\text{triangulation}}}{\frac{1}{\tan(\alpha)} + \frac{1}{\tan(\beta)}}, \tag{2.1}$$

where $x_{\text{triangulation}}$ is the length of the triangulation base. The measurements of the angles α and β, however, rely on proper ambient illumination conditions. In some applications this may not be tolerable. Active techniques allow to reduce these insufficiencies by exploitation of a well-defined active illumination. For this, an active light source may be employed, as depicted in Figure 2.2.

(a)   (b)   (c)

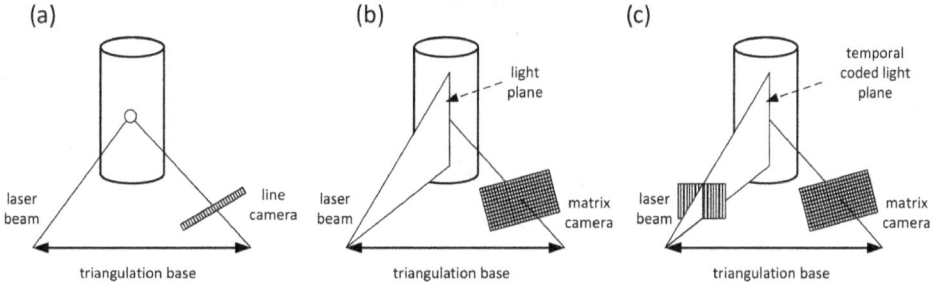

Figure 2.3 (a) point triangulation sensor, (b) light sectioning with a light plane and (c) fringe projection triangulation (according to [SHM00]).

The distance here can be computed by

$$z = \frac{f_{foc} \cdot x_{triangulation}}{p_t},$$ (2.2)

where f_{foc} is the focal length and p_t the position on the sensor that detects the impinging light beam. The sensor can, for instance, be a CCD or CMOS linesensor. The depth resolution can be defined by the absolute value of the differential dz:

$$|dz| = \left| \frac{\partial z(p_t)}{\partial p_t} \right| \cdot dp_t = \frac{z^2}{f_{foc} \cdot x_{triangulation}} \cdot dp_t.$$ (2.3)

The differential dp_t is determined by the lateral resolution of the sensor which is limited by the pixel pitch and the diffraction limit that is defined by the wavelength of the laser. More significant though for this method is the relation $|dz| \propto z^2$. Furthermore, for a good resolution a long triangulation base $x_{triangulation}$ is needed that results in an expensive and bulky system, which is also more prone to errors caused by shading effects.

Triangulation can also be used for the detection of entire 3-D maps. For this application, scanning mirrors are used that extend the point range measurements to lines or areas (cf. Figure 2.3, part a). The light beam can also be diffracted to simultaneously illuminate a line on the surface of the object under observation, what can be detected by an area sensor (b). Alternatively, structured light can be used to illuminate an entire surface. Patterned filters can be located in front of the light source. To extract the distance phase-unwrapping has to be done, which unfortunately involves high computational effort (c). In general, none of the above mentioned methods is feasible for cost-efficient, real-time depth measurements.

2.2 INTERFEROMETRY

In general, an interferometer is an instrument that splits a monochromatic and coherent light wave into parts by usage of a beam splitter, delays them by guidance along paths

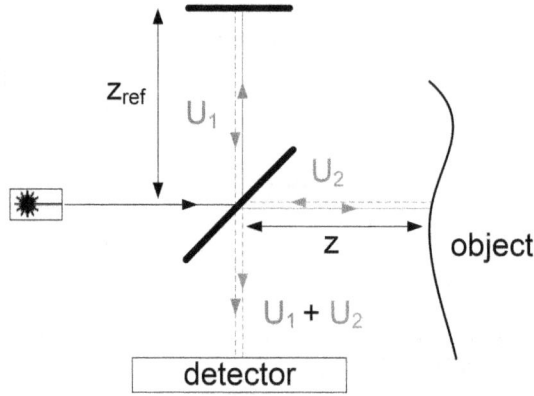

Figure 2.4 Schematic of the Michelson interferometer.

that may differ in length, leads them back together by usage of mirrors, superimposes them and detects their intensity [ST07]. Phase differences of the waves result in interference pattern that can be evaluated to yield e.g. a corresponding length difference. This principle can be utilized in metrological measurement setups. Exemplarily, in Figure 2.4 the basic operating principle of range measurements based on the *Michelson interferometer* is depicted. Here, after splitting the light beam one part is traveling along a reference distance z_{ref}, while the other is traveling along the unknown distance z. Both beams are reflected and then superimposed at the beam splitter, from where they are guided to an integrating detector [Lan00]. From the perspective of signal theory this method corresponds to a homodyne detection or a correlation where signal and reference share the same sinusoidal shape but differ in phase.

For explanation of the principle, a complex wave function $U(x, t) \propto \exp(j2\pi(t/T_{per} + x/\lambda))$ is defined, where x is the space coordinate and t equals the time, $j := \sqrt{-1}$ is the imaginary unit, T_{per} is the period and λ is the wavelength of the plane wave. A real-valued wave function can be defined[2] as $u(x, t) := \Re[U(x, t)]$. If then $I(x) := 2(1/T_{per}) \int_0^{T_{per}} u^2(x, t) dt$ is introduced as a measure of the intensity, the superposition of the waves in the Michelson interferometer can be evaluated to [ST07]

$$I = I_1 + I_2 + 2\sqrt{I_1 I_2} \cos\left(\frac{4\pi(z_{ref} - z)}{\lambda}\right). \tag{2.4}$$

Assuming $I_1 = I_2 = I_0$, this further simplifies to $I = 4I_0 \cos^2(2\pi(z_{ref} - z)/\lambda)$. Interferometry thus allows not for an exact absolute measure, due to it's ambiguity caused by the periodicity of the cosine. With the usage of multiple wavelengths, however, this ambiguity can be resolved [SHM00]. It can be derived, that the following relation

---

[2]$\Re : \mathbb{C} \rightarrow \mathbb{R}$ yields the real part of a complex number.

holds true for the resolution of classical interferometry with smooth surfaces[3] [Hä97; HE11]:

$$dz \propto z^{-1}, \tag{2.5}$$

while rough surfaces can be characterized by coherence scanning interferometry, which results in a distance independent resolution [HE11]. As well as triangulation setups, interferometric measurement systems depend on high precision optics and mechanics and are thus rather bulky and cost-intensive.

2.3   TIME-OF-FLIGHT

In the time-of-flight method, a probing wave packet is emitted, which may be diffracted by optics, reflected by the actual object under observation, focused by optics and transduced by the actual sensor. The time τ_{ToF} which elapses between the emission and detection of the wave packet is proportional to the range z between the object and the sensor, and thus provides range information:

$$z = \frac{c \cdot \tau_{ToF}}{2}. \tag{2.6}$$

c equals the phase velocity of light. Here the assumption was made that the light emitting source is collocated aside the detector. This coaxial setup has the advantage, that there are no correspondence or shadowing effects, which negatively affect triangulation based methodologies. Methods to actually detect the elapsed time, are the implementation of a *"stopwatch"* [HSS06], *indirect integration methods* which provide signals that are in relation to the actual time-of-flight [Koe08; ESM04; Sp09] or *correlation or sampling methods*, that relate a modulated wave packet to a reference signal [Lan00][4]. In principle accuracy of the measurements of the elapsing time is independent of the objects distance. However, due to dispersion and varying reflectance dependent on the objects surface, the amount of light varies with the objects distance. This results in an accuracy that is distance dependent. If demodulators are realized in a pixel-wise manner, a matrix of such devices can be utilized to detect an entire depth map of a given field of view with only one frame. For the improvement of the depth resolution, nevertheless, multiple acquisitions may be undertaken. Since the mechanics, which become necessary for scanning methods, are not needed, this is a cost-efficient and fast method for high frame rate range detection.

Sensors do exist, that employ the ToF based on acoustic waves; these are called *SONAR*. As mentioned before, these are limited by diffraction, the dependence of the acoustic velocity on air pressure and humidity and are subject to multiple reflections. Light waves though, are much more robust. Exploiting wavelengths ranging from the visible range to the infrared range, moreover, has the advantage of a lateral resolution,

---

[3]Here, surfaces with a roughness much smaller than $\lambda/4$ are considered as smooth [Hä97; HE11].
[4]Indirect integration methods, correlation and sampling methods are also referred to as *indirect time-of-flight methods*.

which is 10^2–10^3 times higher than *ultrasonic waves* allow for [SHM00]. A challenge of ToF based on light (LIDAR/LADAR), however, is the velocity of light waves, that equals approximately $30 \, \text{cm ns}^{-1}$ (in vacuum). So, ToF measurements rely on high precision and high speed devices and circuits. For a depth resolution in the millimetres range, time differences in the picoseconds range have to be resolved. In the recent past innovative optoelectronic devices have been invented and realized that provide in-pixel transduction so that this resolution was achieved. Nevertheless, so far research is still ongoing, since real-time, high resolution range image acquisition, which is robust against environmental conditions and still provides eye-safety has not yet been realized. In this section, range detection methods based on the ToF principle are discussed and an overview of the state of the art is given.

2.3.1   Direct time-of-flight

The most obvious time-of-flight approach to measure the distance of an object is surely the *direct time-of-flight* method. Here, a pulse is emitted, which is reflected by the object under observation and detected by a sensor. In the sensor a "stop-watch" is implemented, that may be realized in form of an oscillator in combination with a pixel-wise counter[SHM00; HSS06]. The frequency of the oscillator has to be very high, since it determines the time step size. Moreover, the frequency has to be very stable to achieve a satisfying accuracy. In each pixel a transducer has to be implemented that converts the impinging photons into electronic signals like charges or voltages. The counters are incrementing their values each period, that is defined by the oscillator, until the pulse actually impinges. This is one major drawback of this method, since the actual threshold above which the pulse is considered to be arrived is somewhat arbitrary and, of course, it is subject to noise. Considering the dissipation of the emitted optical energy due to the distribution over the field of view and the reflectance at the scenery which is defined by its surface conditions, the intensity defined by the light source has to be assumed to be unknown. Moreover, ambient illumination introduces a parasitic signal increasing the probability to accidentally exceed the threshold.

A method to attenuate the impact of ambient illumination and to avoid the dependency on a threshold value is the so-called *time correlated single photon counting* (TCSPC) method [NFK08]. Here *photon multipliers* or *avalanche diodes* are used to amplify the signal introduced by an impinging photon. Avalanche diodes that operate in *Geiger mode* are referred to as *single photon avalanche diodes* (SPADs). In systems comprising SPADs to enable the TCSPC method, the arrival of a photon is directly forcing the detector to switch a binary signal that stops the counter. This experiment is repeated periodically. Several measurements of the arrival-times are done to construct a histogram from which the experienced value – corresponding to the objects distance – can be estimated. Sensors with a timing resolution in the range of tens of picoseconds are reported [NFK08]. For this method it has to be assumed that the probability of having more than one photon impinging on the detector within one period is negligible. To justify this assumption the speed of the counter and the oscillator have to be adjusted to the illumination. Nevertheless, even with a narrow optical filter in front of the sensor, the parasitic photon-rate introduced by ambient light can be much higher than possible corresponding operating frequencies that would become necessary. Exemplarily, a setup comprising an optical bandwidth filter of 50 nm around 850 nm wavelength and a pixel pitch of $40 \times 40\,\mu m^2$ would result in an impinging photon-rate of approximately $3.44 \times 10^{10}\,s^{-1}$ if the ambient illumination would equal $1\,kW\,\mu m^{-1}\,mm^{-2}$. This amount of background illumination is approximately corresponding to *Planck's law* evaluated at the top of the atmosphere[5].

The TCSPC method thus allows for millimetre resolution if ambient light is negligible. The drawbacks of this method are the need for high computational effort, caused by the evaluation of the histograms, high power dissipation and the need for high bandwidths and high precision components such as a low-noise and low dark count rate SPADs and highly accurate oscillators and – of course – the attribute of being prone to errors caused by ambient illumination [WRH11].

2.3.2 Continuous wave method

Continuously modulated waves (CW) are widely employed in range image sensor systems (e.g. [Lan00; ZDZ10; LLM11; KKK12; KYO12; PMP12; PSM10]). In such systems continuous, periodic waves, with e.g. a sinusoidal shape, are emitted by a light source and reflected by the object under observation. The task of the detector is then to measure the phase-shift introduced by the traveled distance of the wave. The periodicity of the probing signal results in an ambiguity that can be resolved by multiple measures with differing angular frequencies. This, however, increases the complexity of the sensor system and reduces the frame rate.

The generation of such light waves can for instance simply be done by direct modulation of *light emitting diodes* (LED) [Lan00]. In applications, where the probing waves are not supposed to disturb human beings in the vicinity of the setup, infrared (IR) sources can be employed – often LEDs that emit radiation at a wavelength of $850\,nm$ are used (c.f. Tables 2.1, 2.2 and 2.3). By realization of a modulation of the emitted power of such LEDs:

$$u^2_{mod}(\tau_{ToF}, t) = I(\tau_{ToF})(1 + m_{mod} \cdot \sin(2\pi \nu_m[t - \tau_{ToF}])) \cdot \sin^2(2\pi \nu[t - \tau_{ToF}]) \quad (2.7)$$

$$= \frac{I(\tau_{ToF})}{2} + \frac{m_{mod}I(\tau_{ToF})}{2} \sin(2\pi \nu_m[t - \tau_{ToF}]) + v(\tau_{ToF}, t) \quad (2.8)$$

with a modulation factor m_{mod} and a modulation frequency ν_m and the actual time-of-flight $\tau_{ToF} = 2z/c$, a probing wave packet is created that can resolve a distance $0 \le z \le c/2\nu_m$ before ambiguity occurs[6]. ν is the frequency of the LED corresponding to $c = \lambda \cdot \nu$. The term $v(\tau_{ToF}, t)$ comprises the components of u^2_{mod} of significantly higher frequency compared to ν_m. Additional to the terms given in Equation 2.8 components caused by ambient illumination can be generated, against which detectors

---

[5]Radiation of the sun can be approximated by thermal radiation emitted from a black body, which is modeled by Planck's law. The corresponding temperature at the surface of the sun is $5777\,K$ [PPBS08].

[6]The linear relation of the LEDs intensity with respect to bias variations is caused by the photoelectric effect, that predicts a proportionality of the amount of generated photons with respect to the applied current (electrons per time interval). Since the optical power is proportional to its energy, which is moreover proportional to the amount of photons, that each contribute $\Delta E = h \cdot \nu$, the optical power emitted by a LED is a linear function of the applied bias current. Here h is *Plancks constant*.

Table 2.1 Overview – state of the art time-of-flight range imagers – part one.

Company	Sensor/camera	Technology	Frame rate in fps	Lateral resolution in pixel	Pixel size in μm^2	Operating range in m	Field of view
Panasonic Electric Works Corp.	D-IMager – EKL3104 D-IMager – EKL3105 D-IMager – EKL3106	CCD – CW ToF 3 Modes	15, 20, 25, 30	160 × 120	n.a.	1.2 to 9 1.2–5	60° × 44°
Softkinetic Inc.	DepthSense™ 400 Family DS410	CMOS CW ToF 15 MHz	30	160 × 120	n.a.	1–4	57.3° × 42.0°
Fotonic	T300 D40 D70	Kodak Monochrome CMOS Triangulation CMOS CW ToF 44 MHz	30 ≤50	648 × 488 160 × 120	7.5 × 7.5 50 × 50	0.5–3 0.4–10 0.1–7	131° × 101° 44° × 34° 64° × 48°
PMD Technologies GmbH	PMD[vision]® CamBoard nano PMD[vision]® CamCube3.0	PMD PhotonICs® 19k-S3 CMOS CW ToF up to 80 MHz PMD PhotonICs® 41k-S2 CMOS CW ToF	up to 90 40–80	160 × 120 160 × 120–200 × 200	45 × 45 n.a.	0–2 0.3–7	90° 40° × 40°
odos imaging ltd. MESA Imaging AG	real.IZ 2+3D™ camera SwissRanger™ SR4000	PM ToF CCD/CMOS CW ToF 15 or 30 MHz	up to 50 ≤50	1280 × 1024 176 × 144	n.a. 40 × 40	0.5–10 0.1–5, 0.1–10	50° 43° × 34° or 69° × 56°
IEE	3D MLI Sensor™	CCD/CMOS CW ToF 20 MHz	up to 10	61 × 56	n.a.	non-ambiguity 7.5 @ 20 MHz	130° × 100°
Stanley Electric Ltd.	ToFCam Stanley P-301 ToFCam Stanley P-401 ToFCam Stanley P-411 – preliminary ToFCam Stanley P-421 – preliminary	CW ToF 10 MHz 0.18CIS-PIM6 n.a.	30–180 40 60 20	128 × 128 128 × 126 256 × 240 512 × 480	30 × 30 n.a.	0.5–15 0.5–15, 0.5–5 n.a.	14°, 27° or 52° 27° or 52° 32° × 30° 60° × 57° or 93° × 90°
Advanced Scientific Concepts	DragonEye 3D Flash LIDAR Space Camera TigerEye 3D Flash LIDAR Camera Kit Portable 3D Flash LIDAR Camera Kit	PM ToF PM ToF InGaAs APD	10 n.a. 1–20	128 × 128	n.a.	up to 1500 up to 60/450/1100 up to 70/300/600/100 @ 20% reflectance	45° × 45° 45°, 45° × 22°, 9°, 3° 45° 9° 3° 1°
TriDiCam	Area sensor – preliminary	CMOS PM ToF	14	128 × 96	40 × 40	≈1 without dynamic accumulations	15°

Table 2.2 Overview – state of the art time-of-flight range imagers – part two.

Company	Sensor/camera	Range resolution/accuracy/repeatability	Tolerable ambient illumination	Supply voltage in V	Current consumption in A	Power consumption in W	Light source	Optical filter	Reference
Panasonic Electric Works Corp.	D-IMager – EKL3104	3 cm @ 0lx	20 klx	18–24	0.4	6.8–10	IR LED	n.a.	[Pan]
	D-IMager – EKL3105	14 cm @ 20 klx @ 2 m	100 klx		0.6	10–15			
	D-IMager – EKL3106	90% IR reflektivity			0.35	6–9			
Softkinetic Inc.	DepthSense™ 400 Family DS410	<3 cm @ 3 m	typ. 100 lx	n.a.	n.a.	<15	LED @ 870 nm	n.a.	[Sof]
Fotonic	T300	best ± 0.2 cm @ 30 cm		12	≤2	≤24	658 nm Class 2M	640–670 nm interference	[Fotc]
	D40	<2 cm @ 10 m	n.a.	12–24	n.a.	≤15	845 nm		[Fota]
	D70	<3 cm @ 7 m					8x 0.5W		[Fotb]
PMD Technologies GmbH	PMD[vision]® CamBoard nano	repeatability 5 mm	n.a.	5	≤0.5	≤2.5 (USB powered)	850 nm	n.a.	[PMDa]
	PMD[vision]® CamCube3.0	typ. 03 cm @ 4 m and reflectivity = 75%		12	n.a.	n.a.	870 nm class I		[PMDb]
odos imaging ltd.	real.IZ 2+3D™ camera	<2%	n.a.	n.a.	n.a.	n.a.	class I	n.a.	[odo]
MESA Imaging AG	SwissRanger™ SR4000	<0.7 cm; <0.9 cm	indoor use	12	<1	<12	850 nm class I	n.a.	[MES]
IEE	3D MLI Sensor™	±2 cm @ 15 m	0 to full sunlight	12	n.a.	n.a.	940 nm class I	n.a.	[IEE]
Stanley Electric Ltd.	ToFCam Stanley P-301	1% @ 3 m / 1% @ 3 m	<200 klx	n.a.	n.a.	<15	LED 850 nm / n.a.	n.a.	[Brab]
	ToFCam Stanley P-401								
	ToFCam Stanley P-411 – preliminary	5% @ 3 m	<50 klx			n.a.	1570 nm		[Braa]
	ToFCam Stanley P-421 – preliminary	n.a.							
Advanced Scientific Concepts	DragonEye 3D Flash LIDAR Space Camera	±15 cm 3σ	n.a.	24	n.a.	35	class I 2.5 to 7 mJ/pulse	n.a.	[Adva] [Advc]
	TigerEye 3D Flash LIDAR Camera Kit	n.a.		120 VAC	4.4	240	1570 nm 12 to 20 mJ/pulse		[Advb]
	Portable 3D Flash LIDAR Camera Kit								
TriDiCam	Area sensor – preliminary	±2 cm	1 klx with halogen bulb	12	0.3	3.6	LED 75 W peak 905 nm	820–920 nm	[Tri12]

Table 2.3 Overview – state of the art time-of-flight range imagers – a research perspective.

Company/ Institution	Technology	Frame rate in fps	Image resolution in pixel	Pixel size in μm^2 & Fill factor	Operating range in m	Range resolution/ accuracy/ repeatability	CG, η PDP, $\mathcal{R}$	Operating ambient illumination	Light source	Reference
Shizuoka Univ., Sharp Corp. & Suzuki Motor Corp.	0.35 μm 2P3M CMOS poly on fieldoxide PM ToF	30	336 × 252	15 × 15	1.5–12	1.5 cm @ 3 fps & 1 m			870 nm LED	[IUS05]
Vienna Univ., Institute of Electrodynamics	0.6 μm (Bi) CMOS complex circuitry CW ToF @ 10 MHz	16	16 × 16	125 × 125 66%	0.1–3.2	<5 cm @ 3 m		<150 klx	850 nm LED	[ZDZ10]
EPFL	0.35 μm CMOS SPADs TCSPC @ 40 MHz	20	128 × 128	25 × 25 6%	0.4–3.6	<0.6 cm @ 3.6 m	PDP: 10%	150 lx	637 nm Laser	[NFK08]
Samsung	0.13 μm CMOS concentric gate CW ToF @ 20 MHz	25	192 × 108	28 × 28	1–7	<1% error	7.2 $\mu V/e^-$		850 nm LED	[LLM11]
	0.11 μm 1P4M CMOS PPD CW ToF @ 20 MHz	11	480 × 270	14.6 × 14.6 38.5%	0.75–4.5	<3.8 cm @ 4.5 m		1.5 klx	850 nm LED	[KKK12]
	0.13 μm CMOS PPD CW ToF @ 20 MHz	n.a.	480 × 360	7.75 × 9 14%	1–7	5 mm @ 1 m		1.5 klx	850 nm LED	[KYO12]
Fondazione Bruno Kessler	0.18 μm 1P4M CMOS burried channel CW ToF @ 16.67 MHz	max. 70	320 × 240	14 × 14 48%	0.8–7.5	<16 cm @ 8 fps	7.1 V/spW @ 850 nm	20 klx	850 nm LED	[PMP12]
	0.18 μm 1P4M CMOS CAPD CW ToF @ 20 MHz	7	120 × 160	10 × 10 24%	1.2–3.7	<20 cm @ 3 m	$\eta = 20\%$ @ 600 nm		850 nm LED	[PSM10]
	0.18 μm CMOS in-pixel circuitry PM ToF	30–80 N_{accu} = 128.1	160 × 120	29.1 × 29.1 34%	1–4.5	best 10 cm @ 1 m			905 nm Laser T_p = 50 ns	[PMS11]
Fraunhofer IMS	0.35 m 2P4M CMOS LDPD PM ToF	≤50	128 × 96	40 × 40	<4.5 m	SNR 60 dB voltage domain	$\mathcal{R}$ = 240 $\mu V/W/m^2$		905 nm Laser T_p = 30 ns	[SDS11]

should though ideally be immune. Typical modulation frequencies vary from 10 to 80 MHz which corresponds to measurement ranges up to 1.9 to 15 meter. Assuming a maximal accuracy that is inversely proportional to the measurement range, higher modulation frequencies would become preferable for precise measurements. Since the speed requirements are rather strict, not only for the detectors but also for the driving peripheral circuitry of such high-speed demodulating pixels, large-scale imagers are rarely realized for operation above 20 MHz (cf. Tables 2.1, 2.2 and 2.3). This problem is related to the large capacitive load defined by large pixel matrices.

Compared to pulse modulated ToF principle (cf. Section 2.3.3), an advantage of this method is that neither light sources nor detectors have to be extremely fast, what results in cost-efficient sensor systems [Lan00] (cf. Section 2.3.3). The permanent emission of the sensing light wave, however, limits the maximum signal power for that eye-safety can be guaranteed. In [Ber07] the regulations for eye-safety according to *DIN EN 60825-1 2003* are summarized. The restrictions for mean power emission define the maximum quality of the sensor. Several studies were presented, that demonstrate the physical limitations of range sensors employing the CW method by photon induced shot noise [Sei07; Sei08; FPR09]. This phenomenon is introduced by the quantized nature of light. The quantization is connected to an uncertainty, since the precise number of signal related photons that are collected by a detector is varying arbitrarily with time (cf. Chapter 3). To yield a high-quality picture a sufficient amount of photons that correspond to the actual signal have to be collected. Since background illumination introduces these uncertainties as well and since the maximum signal power is limited by the restrictions to obtain eye-safety, photon induced shot noise defines a fundamental limitation of the accuracy that can be achieved with range sensors – this applies for CW method based sensors especially.

2.3.2.1 Demodulation by mixing

As employed in radio-frequency circuits for communication, homodyne and heterodyne mixing are methods to convert continuously, periodically modulated light into meaningful electrical signals, based on which the time-of-flight can be extracted. In case of homodyne demodulation a sinusoid, that is oscillating at the same frequency at which the intensity of the LED is modulated, is generated, synchronized with the light source and multiplied with a signal that is proportional to the detected radiation at the receiver given by Equation 2.8

$$u_{\text{demodulated}} \propto u_{\text{mod}}^2 \cdot \sin(2\pi\nu_m t) \tag{2.9}$$

$$\propto \frac{m_{\text{mod}} I(\tau_{\text{ToF}})}{4} \cdot \cos(2\pi\nu_m \tau_{\text{ToF}}) + v_{\text{dem}}(\tau_{\text{ToF}}, t) \tag{2.10}$$

results. Here $u_{\text{demodulated}}$ stands for the demodulated signal, not for an intensity measure of waves, while v_{dem} describes the remaining time-dependent terms. If these components are filtered, a non-linear mapping between the objects distance and the output signal is created, that resolves a τ_{ToF} up to $1/2\nu_m$ before ambiguity occurs. This method has the obvious disadvantage of being sensitive to $I(\tau_{\text{ToF}})$. An additive modulation with a cosine reference can extend the ambiguity range by a factor of two and the sensitivity to the amplitude can be eliminated. Considering $u_{\text{dem}_{\text{sin}}}$ as the demodulated signal for mixing with a sine wave and $u_{\text{dem}_{\cos}}$ for cosine wave respectively, the object distance is conducted to[7]

$$z = \frac{c}{4\pi\nu_m} \begin{cases} \arctan\left(\frac{u_{\text{dem}_{\cos}}}{u_{\text{dem}_{\sin}}}\right) & \text{for } u_{\text{dem}_{\sin}} > 0 \\ \arctan\left(\frac{u_{\text{dem}_{\cos}}}{u_{\text{dem}_{\sin}}}\right) + \pi & \text{for } u_{\text{dem}_{\sin}} < 0 \wedge u_{\text{dem}_{\cos}} \geq 0 \\ \arctan\left(\frac{u_{\text{dem}_{\cos}}}{u_{\text{dem}_{\sin}}}\right) - \pi & \text{for } u_{\text{dem}_{\sin}} < 0 \wedge u_{\text{dem}_{\cos}} < 0 \\ \pm\frac{\pi}{2} & \text{for } u_{\text{dem}_{\sin}} = 0 \wedge u_{\text{dem}_{\cos}} \gtrless 0 \end{cases} \tag{2.11}$$

For the transduction of the optical power to an electronic signal, fast photodiodes may be used. The mixing procedures can then be realized using analog circuits as it is well known from RF communication circuits. Although the method of demodulation by mixing with appropriate signals by integrated RF circuitry is very mature, the applicability is rather limited due to the complexity these solutions imply. The integration of analog pixel-wise demodulators and filters would ultimately result in large pixel pitches and the complexity of these solutions would also be connected to a rather poor accuracy. Thus these approaches are not considered as meaningful for further investigation.

In case of heterodyne mixing, the received signal is mixed with a sine wave of slightly different frequency. Solutions were presented that use *image intensifiers* as shutter elements [PCDC06; DCC06; DCC07; CDK09]. These image intensifiers are basically microchannel plates, that allow for a lateral resolving photon multiplication, in combination with an appropriately chosen coating to allow for wavelength conversion, so that e.g. an off the shelf CCD image detector can be used for signal conversion [ST07]. The demodulated signal is

$$u_{\text{demodulated}} \propto u_{\text{mod}}^2 \cdot \sin(2\pi\nu_{\text{dem}}t) \tag{2.12}$$

$$\propto \frac{m_{\text{mod}}I(\tau_{\text{ToF}})}{4} \cdot \cos(2\pi[\nu_m - \nu_{\text{dem}}]t - 2\pi\nu_m\tau_{\text{ToF}}) + \nu_{\text{dem}}(\tau_{\text{ToF}}, t). \tag{2.13}$$

ν_{dem} now describes the terms of higher frequencies than the beat frequency $\nu_m - \nu_{\text{dem}}$, which is typically in the range of 1 to 10 Hz [CDK09]. Conversely to homodyne detection, here the range of the object is within the phase of the demodulated signal, not the amplitude. It is reported that this results in a higher accuracy [PCDC06; DCC06; DCC07; CDK09]. Nevertheless, this is traded for a complex system, which depends on oscillators with sub-ppm frequency precision and image intensifiers operating up to 100 MHz that, however, have to be biased by supply voltages exceeding a few hundred volts.

2.3.2.2 Demodulation by correlation

Correlation is a similar approach to homodyne mixing, but much more general[8]. It relates an appropriate reference signal s_{ref} to the signal detected by the receiver by usage

---

[7]The motivation for this formula was taken from [Lan00].
[8]In some literature the methods that are separately explained here, are gathered under homodyne or correlation detection and are considered as non-different.

of the cross-correlation function for deterministic periodic signals (cf. Chapter 3, esp. Equation 3.40):

$$u_{\text{demodulated}}(\tau) \propto \lim_{T\to\infty} \frac{1}{T} \int_{-\frac{T}{2}}^{\frac{T}{2}} u_{\text{mod}}^2(t) \cdot s_{\text{ref}}(t+\tau)\,\mathrm{d}t. \tag{2.14}$$

Here T defines the correlation time frame. An advantage of this method compared to homodyne detection is that non-ideal pulse shapes can be accounted for. Non-ideality here refers to the parasitic phase error in homdodyne detection with sinusoidal shaped signals that is introduced by harmonics and/or non-linearities. Since cross-correlation describes a similarity measure between two functions, which takes a maximum if both signals are linearly dependent by a positive real number, this method is in principle much more robust. In case of a sine wave and a constant background illumination I_{BG}, the demodulated signals evaluate to

$$u_{\text{demodulated}}(\tau) \propto \lim_{T\to\infty} \frac{1}{T} \int_{-\frac{T}{2}}^{\frac{T}{2}} \left[I_{\text{BG}} + I_{\text{p}}(\tau_{\text{ToF}})\sin(2\pi\nu_{\text{m}}(t-\tau_{\text{ToF}})) \right] \cdot \sin(2\pi\nu_{\text{m}}[t+\tau])\,\mathrm{d}t. \tag{2.15}$$

$$\propto \lim_{T\to\infty} \frac{I_{\text{p}}(\tau_{\text{ToF}})}{2} \cos(2\pi\nu_{\text{m}}[\tau-\tau_{\text{ToF}}]) + \nu_{\text{dem}}(\tau_{\text{ToF}},t) \tag{2.16}$$

and in case of square wave functions with peak value I_{p} and period T_{per} to

$$u_{\text{demodulated}}(\tau) \propto \lim_{T\to\infty} I_{\text{p}}(z)\Lambda(\tau-\tau_{\text{ToF}}) + \nu_{\text{dem}}(\tau_{\text{ToF}},t), \tag{2.17}$$

where

$$\Lambda(t) := \begin{cases} 1 + 4k - \frac{4t}{T_{\text{per}}} & \text{for } kT_{\text{per}} \leq t \leq kT_{\text{per}} + \frac{T_{\text{per}}}{2} \\ 1 - 4k + \frac{4t}{T_{\text{per}}} & \text{for } kT_{\text{per}} - \frac{T_{\text{per}}}{2} \leq t \leq kT_{\text{per}} \end{cases}, k \in \mathbb{Z}. \tag{2.18}$$

In both cases ν_{dem} indicate terms that become negligible with increasing T. A significant attribute of this method is, that it is damping constant ambient illumination. Taking the limits of T to infinity for the given results yields demodulated signals as depicted in Figure 2.5.

For range detection the maximum of these function within the non-ambiguity range has to be found to yield the actual distance of the object. Sweeping τ within the non-ambiguity range would imply a huge calculation effort. A significant reduction of calculation effort can be accomplished by the evaluation of the cross-correlation function at a minimal amount of values. In case of the sine wave modulation, $u_{\text{demodulated}}$ can be sampled at e.g. $\tau \in \{0, 3T_{\text{per}}/4\}$, what yields

$$u_{\text{demodulated}}(0) \propto \cos(2\pi\nu_{\text{m}}\tau_{\text{ToF}}) \tag{2.19}$$

$$u_{\text{demodulated}}(3T_{\text{per}}/4) \propto \sin(2\pi\nu_{\text{m}}\tau_{\text{ToF}}), \tag{2.20}$$

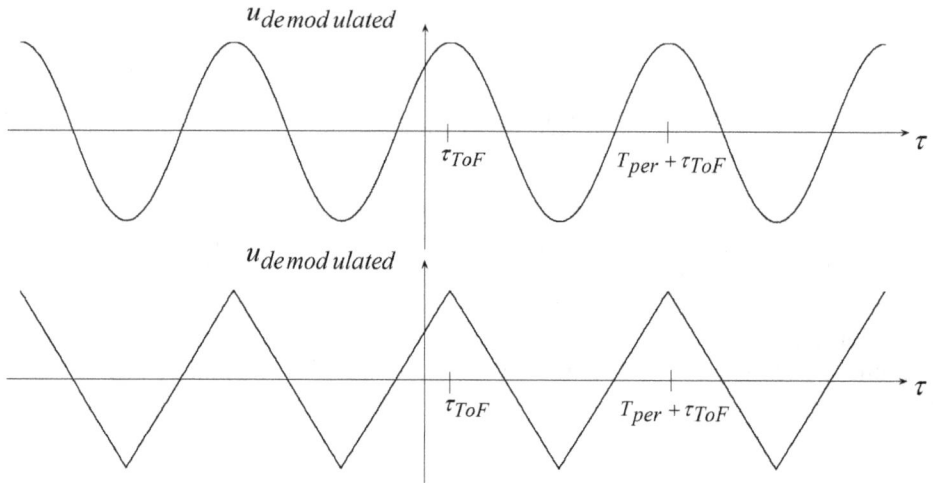

Figure 2.5 Demodulated signals for sine wave (top) and square wave (bottom) modulation.

so that the objects distance can be evaluated according to 2.11. In case of the square wave modulation, evaluation at $\tau \in \{0, T_{per}/4, T_{per}/2, 3T_{per}/4\}$ leads to an object distance of

$$
z = \frac{T_{per}}{8 \cdot c}
\begin{cases}
\dfrac{-u_{dem.}\left(\frac{T_{per}}{2}\right)+2u_{dem.}\left(\frac{T_{per}}{4}\right)-u_{dem.}(0)}{u_{dem.}\left(\frac{3T_{per}}{4}\right)-u_{dem.}\left(\frac{T_{per}}{2}\right)+u_{dem.}\left(\frac{T_{per}}{4}\right)+u_{dem.}(0)} & \text{for} \quad u_{dem.}\left(\frac{T_{per}}{4}\right) > 0 \\
& \qquad \wedge u_{dem.}(0) > 0 \\[2ex]
\dfrac{-2u_{dem.}\left(\frac{3T_{per}}{4}\right)+3u_{dem.}\left(\frac{T_{per}}{2}\right)-u_{dem.}(0)}{-u_{dem.}\left(\frac{3T_{per}}{4}\right)+u_{dem.}\left(\frac{T_{per}}{2}\right)+u_{dem.}\left(\frac{T_{per}}{4}\right)-u_{dem.}(0)} & \text{for} \quad u_{dem.}\left(\frac{T_{per}}{2}\right) > 0 \\
& \qquad \wedge u_{dem.}\left(\frac{T_{per}}{4}\right) > 0 \\[2ex]
\dfrac{4u_{dem.}\left(\frac{3T_{per}}{4}\right)+u_{dem.}\left(\frac{T_{per}}{2}\right)-2u_{dem.}\left(\frac{T_{per}}{4}\right)-3u_{dem.}(0)}{+u_{dem.}\left(\frac{3T_{per}}{4}\right)+u_{dem.}\left(\frac{T_{per}}{2}\right)-u_{dem.}\left(\frac{T_{per}}{4}\right)-u_{dem.}(0)} & \text{for} \quad u_{dem.}\left(\frac{3T_{per}}{4}\right) > 0 \\
& \qquad \wedge u_{dem.}\left(\frac{T_{per}}{2}\right) > 0 \\[2ex]
\dfrac{-2u_{dem.}\left(\frac{3T_{per}}{4}\right)+3u_{dem.}\left(\frac{T_{per}}{2}\right)+4u_{dem.}\left(\frac{T_{per}}{4}\right)-5u_{dem.}(0)}{-u_{dem.}\left(\frac{3T_{per}}{4}\right)+u_{dem.}\left(\frac{T_{per}}{2}\right)+u_{dem.}\left(\frac{T_{per}}{4}\right)-u_{dem.}(0)} & \text{for} \quad u_{dem.}\left(\frac{3T_{per}}{4}\right) > 0 \\
& \qquad \wedge u_{dem.}(0) > 0
\end{cases}
\tag{2.21}
$$

The problem of finding above mentioned Equation can be stated as the search for the intersection point of two straights, that are fitted through the four sampling points.

In case of sine wave modulation, the slope of the cross-correlation function in the vicinity of the maximum is very low. Thus it can be assumed, that square wave modulation yields more accurate results. Nevertheless, the requirements on speed of the illumination source and the detector are more strict due to the necessary harmonics the square wave modulation implies.

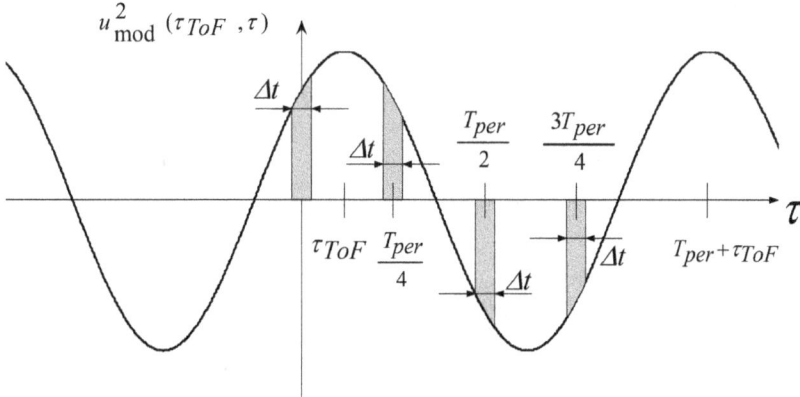

Figure 2.6 Schematic of the sampling method.

Another significant advantage of the correlation method compared to homodyne mixing is that in general several arbitrarily shaped-functions can be used for detection. If they have the attribute of being mutually *orthogonal*, a sensor network can be established in which interference can be limited to an acceptable level. Signals are called orthogonal, if the cross-correlation function vanishes[9]. A class of such mutually orthogonal signals is the set of *Gold codes* [OL07]. Systems based on this method already have been reported [Bü07; Bü08]. Deviating from the ideal robustness against interference though is the increased photon noise level, that each modulated light source implies. This is an even stronger limitation, considering that the over-all illumination has to be kept at a level which guarantees eye safety.

2.3.2.3 Demodulation by sampling

An alternative approach to mixing or autocorrelating detection is the sampling method [Lan00; SSV95]. Here, as depicted in Figure 2.6 the detected sine wave is sampled at the time instances 0, $T_{per}/4$, $T_{per}/2$ and $3T_{per}/4$. Given $u^2_{mod} \propto I_{BG} + I_p(\tau_{ToF}) \sin(2\pi\nu_m[t - \tau_{ToF}])$, the short time integration $u_{dem}(\tau) \propto \int_{\tau-\Delta t/2}^{\tau+\Delta t/2} u_{mod}(\tau_{ToF}, t)^2 dt$ yields e.g. $u_{dem}(0) \propto I_{BG}\Delta t - I_p \sin(\pi\nu_m\Delta t) \sin(2\pi\nu_m\tau_{ToF})/2\pi\nu_m$ for evaluation at $t = 0$. From that it becomes clear that Δt has to be chosen carefully not to attenuate the

---

[9]The term orthogonality is actually defined in linear algebra for the case that the inner product of two vectors vanishes. The inner product itself was invented to measure mutual projections of vectors. However, with the introduction of the Hilbert space, the concept of projection can be transferred to functions. The inner product for functions in the Hilbert space is then defined by the cross-correlation function: $\langle f, g \rangle := \phi_{fg}^E(\tau) := \int_{-\infty}^{\infty} f(t) \cdot g(t + \tau)dt$ [KSW08]. Here, conversely to Equation 3.40, ϕ_{fg}^E is defined for energy signals, what means that f and g are square integrable [OL07].

phase information in the second term. The distance of an object to the sensor can be conducted to [SSV95]

$$
z = \frac{c}{4\pi\nu_{\mathrm{m}}}
\begin{cases}
\arctan\left(\frac{u_{\mathrm{dem}}\left(\frac{T_{\mathrm{per}}}{2}\right)-u_{\mathrm{dem}}(0)}{u_{\mathrm{dem}}\left(\frac{T_{\mathrm{per}}}{4}\right)-u_{\mathrm{dem}}\left(\frac{3\cdot T_{\mathrm{per}}}{4}\right)}\right) & \text{for } u_{\mathrm{dem}}\left(\frac{T_{\mathrm{per}}}{4}\right)-u_{\mathrm{dem}}\left(\frac{4\cdot T_{\mathrm{per}}}{4}\right)>0 \\[2.5em]
\arctan\left(\frac{u_{\mathrm{dem}}\left(\frac{T_{\mathrm{per}}}{2}\right)-u_{\mathrm{dem}}(0)}{u_{\mathrm{dem}}\left(\frac{T_{\mathrm{per}}}{4}\right)-u_{\mathrm{dem}}\left(\frac{3\cdot T_{\mathrm{per}}}{4}\right)}\right)+\pi & \text{for } u_{\mathrm{dem}}\left(\frac{T_{\mathrm{per}}}{4}\right)-u_{\mathrm{dem}}\left(\frac{3\cdot T_{\mathrm{per}}}{4}\right)<0 \\
& \wedge u_{\mathrm{dem}}\left(\frac{T_{\mathrm{per}}}{2}\right)-u_{\mathrm{dem}}(0)\geq 0 \\[2.5em]
\arctan\left(\frac{u_{\mathrm{dem}}\left(\frac{T_{\mathrm{per}}}{2}\right)-u_{\mathrm{dem}}(0)}{u_{\mathrm{dem}}\left(\frac{T_{\mathrm{per}}}{4}\right)-u_{\mathrm{dem}}\left(\frac{3\cdot T_{\mathrm{per}}}{4}\right)}\right)-\pi & \text{for } u_{\mathrm{dem}}\left(\frac{T_{\mathrm{per}}}{4}\right)-u_{\mathrm{dem}}\left(\frac{3\cdot T_{\mathrm{per}}}{4}\right)<0 \\
& \wedge u_{\mathrm{dem}}\left(\frac{T_{\mathrm{per}}}{2}\right)-u_{\mathrm{dem}}(0)<0 \\[2.5em]
\pm\frac{\pi}{2} & \text{for } u_{\mathrm{dem}}\left(\frac{T_{\mathrm{per}}}{4}\right)-u_{\mathrm{dem}}\left(\frac{3\cdot T_{\mathrm{per}}}{4}\right)=0 \\
& \wedge u_{\mathrm{dem}}\left(\frac{T_{\mathrm{per}}}{2}\right)-u_{\mathrm{dem}}(0)\gtrless 0
\end{cases}
\tag{2.22}
$$

The main advantage of this method is its simplicity. Such short time integrators were realized in the charge domain, so that low-noise, high-speed demodulation can be directly achieved in pixels with a pitch of only few tens of micrometers [Lan00; IUS05; LLM11; KKK12; KYO12; PMP12; PSM10]. Published photo detectors for this demodulation scheme initially based on custom CCD/CMOS processes, that enabled devices such as the *Photonic Mixer Device* [SSV95; Lan00]. In these detectors the short time integration is realized by gates that are connected to a photosensitive region and storage nodes, that actually keep the demodulated charge and and transduce it into voltage levels. That can then be non-destructively readout by e.g. a source follower circuit. In those detectors the gates are formed above the photosensitive region. Since these devices are illuminated from the front-side, impinging wave fronts cannot penetrate the devices as deep as in e.g. plain photodiodes or pinned photodiodes. This results in a comparatively low quantum efficiency[10] and thus in less accurate measurements. Photodetectors that define a modulating electric field within the semiconductor region of the detector by forcing a minority current through the device have been developed to enhance the quantum efficiency [NTK05; PSM10]. Furthermore deviations from the sinusoidal shape in the LED or non-linearities in the detector lead to phase-errors. A thorough analysis of these imperfections is given in [Lan00]. If the sensor is well known and properly calibrated, these phase errors can be compensated if instead of the simplified formulas look-up tables are implemented, from which the object distance is estimated. However, this increases the effort and thus costs. Another disadvantage of early CCD/CMOS based demodulating pixels was the integration of ambient light that easily pushed the devices into saturation. Circuits assisting the photodetectors were developed, that suppress the dc ambient-light [Ogg95]. Nevertheless, the photon noise associated with the ambient illumination cannot be compensated, what ultimately limits the resolution.

---

[10]Quantum efficiency defines the ratio of the amount of photon generated electron-hole pairs to the amount of the corresponding amount of impinging photons.

2.3.3 Pulsed wave method

Due to the permanent radiation of the probing wave in case of the CW method, the maximal signal power is very limited for applications that have to guarantee eye-safety. This is caused by the maximal average power, which is far below the peak power, that is tolerable for short time intervals according to [Ber07]. These restrictions to guarantee eye safety led to the development of a modulation scheme, which concentrates the optical power in short time frames, so that the peak signal power can be chosen comparably high with respect to the parasitic ambient illumination. The modulated carrier ideally has the shape of a rectangular pulse[11]:

$$u_{\text{mod}}^2(\tau_{\text{ToF}}, t) \propto \text{rect}[(t - \tau_{\text{ToF}} - T_{\text{p}}/2)/T_{\text{p}}] \sin^2(2\pi\nu[t - \tau_{\text{ToF}}]), \quad (2.23)$$

where T_{p} is the pulse width, which is usually in the range of 30 ns to some microseconds. This method is referred to as *pulsed wave method or pulsed modulated (PM) principle*. Similarly to Equations 2.7 and 2.8 the signal can be approximated by $u_{\text{mod}}^2 \propto \text{rect}[(t - \tau_{\text{ToF}} - T_{\text{p}}/2)/T_{\text{p}}]$, since for integrating detectors the higher frequency components become negligible if wavelengths in the visible or infrared range are used and T_{p} lasts some nanoseconds. To realize these short light pulses lasers or laser diodes are usually used (cf. Tables 2.1, 2.2 and 2.3). Nevertheless, realistic laser pulses suffer from a finite rise-time and a parasitic tail. Typical repetition rates of such lasers are in the range of tens of kilohertz [Lan00]. This limits the maximal frame rate, the resolution that can be achieved by multiple samplings and the non-ambiguity operating range, which equals several kilometres.

2.3.3.1 Multiple short time integration

Multiple short time integration (MSI) is a widely adopted method for range imagers based on PM modulation [ESM04; Koe08; IUS05; PMS11]. The concept is based on a plurality of short time integrators that are synchronized to the emitted light pulse. Assuming an ideally modulated pulse $x_s(\tau_{\text{ToF}}, t) = a \cdot \text{rect}[(t - \tau_{\text{ToF}} - T_{\text{p}}/2)/T_{\text{p}}]$ and short time integrators that implement the operation

$$u_{T_{\text{SW}},\tau}(\tau_{\text{ToF}}) := \int_{\tau}^{\tau+T_{\text{SW}}} x_s(\tau_{\text{ToF}}, t)dt, \quad (2.24)$$

then the response is given as depicted in Figure 2.7. Here, T_{p} is the length of the pulse $x_s(\tau_{\text{ToF}}, t)$ and T_{SW} is the integration period of the short time integrators. Another perspective on this problem can be the utilization of the convolution of the impinging signal with the impulse response function of a short time integrator $h_s = (1/T_{\text{SW}})\text{rect}[(t - \tau - T_{\text{SW}}/2)/T_{\text{SW}}]$. Proper utilization of these signals that are sensitive to τ_{ToF} can resolve the distance of objects under observation. Basically several combinations of τ and T_{SW} exist that allow for range measurements. Examplarily three combinations are stated here. For all these examples the maximal operating range is

---

[11]In this work it is used: $\text{rect}(x) := \begin{cases} 1 & \text{for } |x| \leq 1/2 \\ 0 & \text{elsewhere} \end{cases}$

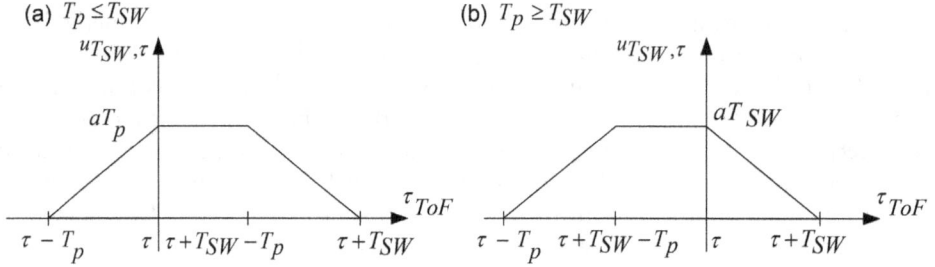

Figure 2.7   Response of the ideal short time integrators to ideal PM pulses.

$0 \leq z \leq cT_p/2$. With variation of T_p the desired operating range can thus be set. This, however, also affects the resolution that can be achieved.

Firstly, it can be shown that the use of $u_{T_p,0}$ and $u_{2 \cdot T_p,0}$ can be utilized to yield

$$ z = \frac{c}{2} T_p \left(1 - \frac{u_{T_p,0}}{u_{2 \cdot T_p,0}} \right) \tag{2.25} $$

This method has been used in [Jer01; ESM04; KHK10]. Second, $u_{T_p,0}$ and u_{T_p,T_p} enables the estimation of an objects distance to the sensor by

$$ z = \frac{c}{2} T_p \frac{u_{T_p,T_p}}{u_{T_p,0} + u_{T_p,T_p}} \tag{2.26} $$

which was demonstrated in e.g. [IUS05]. These two approaches are immune to the actual intensity of the signal. Nevertheless, both are prone to errors introduced by ambient illumination. In [SIK10] it was demonstrated how the impact of this phenomenon was attenuated by appropriate calculation based on two subsequent frames. However, since these frames are affected by uncertainties coming from the entire readout path and are also affected by aliasing, this is not the best solution in terms of achievable accuracy. An advanced solution to circumvent the impact of ambient light is the usage of $u_{T_p,-T_p}$, $u_{T_p,0}$ and u_{T_p,T_p} which result in

$$ z = \frac{c}{2} T_p \frac{u_{T_p,T_p} - u_{T_p,-T_p}}{u_{T_p,0} + u_{T_p,T_p} - 2 \cdot u_{T_p,-T_p}}, \tag{2.27} $$

and can be directly implemented in CMOS range imagers [SIK10]. Here, the advantage is within $u_{T_p,-T_p}$, which does not accumulate the actual signal, but only ambient light. Assuming that this remains constant during $-T_p \leq t \leq 2 \cdot T_p$, the ambient light can thus be subtracted from $u_{T_p,0}$ and u_{T_p,T_p}. Nevertheless this only holds true for deterministic signals. Due to the quantization of photons and the quantization noise associated with this, background illumination will still affect the signal quality negatively as will be pointed out more thoroughly in Chapter 6.

2.4 COMPARISON OF OPTICAL RANGE IMAGING METHODS

As has been pointed out in this work, light is preferred over sound waves as probing element for contactless range measurements, since it is more robust against environmental conditions and yields a higher spatial resolution which is limited by diffraction. Triangulation or interferometry yield precise measures but depend on bulky and large setups and are thus expensive and sensitive to disturbances caused e.g. by vibrations. Triangulation has the disadvantage of being non-coaxial, which results for instance in correspondence problems. The adoption of triangulation or interferometry by scanning detectors, structured light or stereovision causes huge computational effort and thus results in insufficiently low frame rates for real-time imaging. Contrarily time-of-flight based on modulated light waves enables fast depth map acquisition with accuracy of millimetres within operating ranges from decimeters to several tens of meters.

ToF imaging with CW modulation is already relatively mature, but in principle sensitive to ambient illumination since the restrictions on maximal signal power limit the accuracy. In addition, the non-ambiguity range is limited and the presence of high order harmonics due to departure from the ideal sine-wave signals require a considerable calibration effort. PM modulation on the other hand, allows to concentrate signal power in small time frames so that parasitic background light has fewer impact on the measures. The increased calibration effort due to departure from the ideal behavior of the light source and the short time integrators, however, also applies for PM ToF. Tables 2.1 and 2.2 present an overview of the state of the art. Here some major specifications are given to allow for a comparison of the differing approaches. First of all the technology is a key parameter since it defines cost and possible complexity that may allow for a better performance. Panasonic, MESA Imaging AG and IEE base their products on CCD or mixed CCD/CMOS processes that enable detectors based on the CW ToF principle. Softkinetic Inc., Fotonic and Stanley Electric developed range cameras based on CW ToF realized in CMOS. Range cameras based on PM ToF were presented by TriDiCam[12], Advanced Scientific Concepts and odos imaging limited. The specifications listed in the tables can be divided in general targeted restrictions such as *field of view*, *operating range* or *lateral resolution* while other parameters are limited by these such as *pixel pitch*, *frame rate*, *range resolution* or the level of *tolerable ambient light*. Parameters such as *power supply* or *light source* are given for the sake of completeness. These become important for miniaturized or low-cost range cameras. The focus of this work, however, is to increase the performance – especially the range resolution. In general, all the above mentioned parameters can be mutually traded against each other to yield a performance increase in a certain category. Unfortunately, this makes a comparison rather tough. This is worsened by the fact, that manufacturers often do not specify the conditions under which the given performance can be achieved. An extremely high frame rate for instance is directly related to a smaller amount of collected signal related photons and thus results in a worse range accuracy. Similar relations are defined by a large field-of-view and a small pixel

---

[12]TriDiCam is a Fraunhofer IMS spin-off. It develops range cameras in collaboration with the IMS, based on image sensors designed and fabricated at the IMS facilities. The preliminary data about the camera presented here was provided by Stefan Schwope and Matthias Dünninghaus from TriDiCam.

pitch, that both imply a worsened range resolution. PMD Technologies and MESA Imaging developed relatively mature sensors. However, data about tolerable ambient light is not available. Panasonics products contrarily are specified up to 20 klx. For 100 klx though, accuracy is not explained and corresponding acquisition times were not given. Stanley Electric ltd. presented imagers with good lateral resolution that are to work under ambient light levels from 50 to 200 klx. The performance was though not specified for these conditions. PM ToF range imagers by Advanced Scientific Concepts have remarkable operating ranges specified, but the dependence on background light is not explained. Odos imaging limited claim to have developed a range imager with an impressive resolution of 1280×1024, but data is rarely available. TriDiCam so far, develops the only ToF imager intended for outdoor use, that is to be resistant against ambient illumination.

Additional to the overview of the state of the art of available range sensor products, an overview of the latest developments is given from a research perspective in Table 2.3. Except for an approach, where extensive use of in-pixel circuitry ([ZDZ10]) was used for demodulation, a trend towards range imagers with higher resolutions can clearly be extracted from the current publications. The progress regarding SPADs also enabled resolutions of 128×128 and pixel pitches of only a few tens of micrometers for application in ToF imaging [NFK08]. However, the reported tolerable ambient illumination of 150 lx for this sensor does not yet allow for outdoor use. In general the given specifications of tolerable ambient light, range-resolution or frame rate do not allow for proper comparison or further theoretical deduction, since important restrictions like field-of-view, detailed data about the light source like for instance optical power, surface condition of the objects under observation such as reflectance and acquisition time necessary to yield a certain accuracy under the named conditions, are rarely available.

Nevertheless, from the collected data it can obviously be deduced, that real-time range imaging with accuracy from millimeters to few centimeters for operating ranges from decimeters up to tens of meters, under harsh environmental conditions such as ambient illumination in the klx-range is not yet possible with cost-efficient ToF range image sensors. The following parameters are all interrelated. Thus for a certain application the best match has to be chosen:

- field-of-view
- operating range/ambiguity range
- lateral resolution
- pixel size
- depth resolution

- acquisition time/frame rate
- wavelength and optical power of the illumination source
- operation under eye-safe condition
- cost

Physical limitations are defined by the used methodology and the quantized nature of light. For low-light application, however, the sensor is per definition limiting the performance. Thus within this work a detailed analysis of uncertainties arising from CMOS range sensors based on the PM ToF principle is given. Following the mathematical foundation, an explanation of the occurrence and avoidance of uncertainties in CMOS devices, circuits and sensor systems intended for high performance range imaging is presented.

Chapter 3

Temporal noise

In June 1827 Robert Brown (1773–1858) started to study pollen grains and found out that they fluctuate in an arbitrary manner [Lin08]. Actually, he was not the first one to notice this phenomenon of randomly appearing movements of tiny particles, but due to his precise work he was the first one to demonstrate that these movements are neither caused by animalcules, nor by measurement imperfections like vibrations, turbulences or exterior influences from heat or electromagnetism [Lin08]. Brown then studied this phenomenon on inorganic materials like stones, glasses and even a piece of the sphinx and observed these fluctuations in all of those [Lin08]. This ubiquity led to the idea of a rather general underlying process. Albert Einstein (1879–1955) published the first satisfying theory on this topic in 1905, in which he used stochastical approaches to model a frequent bombardment on the pollen grains by molecules [Ein05]. The theory on statistical thermodynamics gathers probabilistic methods to describe the interaction of particles in macroscopic systems. Studies on quantum mechanics proved a very important physical limitation. Not only are stochastic methodologies handy for the description of the sheer incredible amount of particles, a deterministic description of particles is actually not possible. Heisenberg (1901–1976) proved that particles cannot be perfectly characterized. If e.g. the location is measured more precisely, the uncertainty in the impulse is worsening. This is known as the *uncertainty principle*.

Today, stochastic approaches are widely accepted and employed in physics and engineering. These are for example used to model the uncertainties in electronic systems. Unwanted uncertainties are often referred to as noise. Although this could virtually describe any signal imperfection, the term noise will be used in this work to describe non-deterministic temporal noise. Research on the topic of noise modeling and noise reduction in devices, circuits and systems is still ongoing and its importance is growing, since many applications are struggling with the physical limitations that are defined by noise [ITR11a; ITR11b]. However, as has been claimed in [DSV00], many publications present rather unjustified and non-rigorous techniques to estimate these uncertainties. Due to this fact, it is considered necessary to present a deeper insight into the fundamental noise processes, the modeling and characterization of those and the minimization of uncertainty, which will be presented exemplarily for CMOS image detectors in general and CMOS based range imagers based on the pulse-modulated time-of-flight principle in particular.

In imagers intended for acquisition of two-dimensional images some imperfections can be compensated by calibration. For range imagers such imperfections may be non-linearity, delay in the timing or mismatch of the sensing elements. Temporal noise, however, is a non-deterministic random variation and can thus neither be predicted, nor compensated by calibration. It also cannot be fully eliminated, since it is introduced by the devices, that actually serve to acquire the images and to process them. Thus, non-deterministic temporal noise defines a fundamental limitation for sensors in general.

This chapter is presenting insights in several topics regarding the analysis, characterization and prediction of non-deterministic temporal uncertainties. In Section 3.1 the general concepts of mathematical treatment is summarized. Although these concepts are widely employed, this chapter is rather detailed to offer a proper insight into the limitations, which the concepts suffer from. Section 3.2 comprises an overview of methodologies to analyze the propagation of uncertainties in non-linear and time-variant systems. Since general methods for the prediction of diverse processes in e.g. *switched capacitor circuits*, that are used for signal conditioning in e.g. CMOS imagers, may be rather time consuming, a simple method is presented that circumvents such insufficiencies. In the Section 3.3 models for common noise sources are presented. Here, as well, a detailed presentation is preferred to demonstrate the limitations of those models.

Using the words from Alper Demir, it is *borrowed heavily from these references for the discussion in this chapter*.

- Introduction in noise analysis: [Web92; PU02; DR58; OL07; Czy10].
- Noise estimation in non-linear systems: [PU02; DSV00; Gar85].
- Fundamental noise processes: [Ros96; GS01; PU02; DR58; DSV00; Czy10; Buc83].

3.1 INTRODUCTION TO NOISE ANALYSIS

This section is summarizing the fundamentals of noise analysis. Detailed derivations are not given for all relations. The presented material does not include the basic concepts of probability theory like *Borel sets*, *theory of measures*, which provides for instance the widely employed *Lebesgue integral*; or *Kolmogorov axioms*, which are the foundation of the probability theory. The reader is referred to textbooks like [Web92; PU02; Ros96; GS01; DR58; OL07].

3.1.1 Basic probabilistic concepts for the analysis of uncertainties

The estimation of the uncertainty of physical quantities is based on probabilistic measures. The *probability distribution function* or *cumulative distribution function* (CDF) $F_{p-X_r}: \mathbb{R} \to [0,1]$ is a widely employed description of random variables. It is measuring the probability of a random variable $X_r: \Omega \to \mathbb{R}$ being lower than a given boundary $x_r \in \mathbb{R}$. $X_r(\omega)$ is mapping from the sample space Ω to e.g. physical quantities presented by real numbers. Probability distribution functions are then noted as

$$F_{p-X_r}(x_r) := P(X_r \leq x_r) \tag{3.1}$$

or

$$F_{p-X_r}(x_{r1}, x_{r2}, \ldots, x_{rn}) := P(X_{r1} \leq x_{r1}, X_{r2} \leq x_{r2}, \ldots, X_{rn} \leq x_{rn}), \quad n \in \mathbb{N} \tag{3.2}$$

in the multivariate case. Here, $P : \mathscr{A}(\Omega) \to [0, 1]$ denotes the measure of the probability of the sigma-algebra $\mathscr{A}(\Omega) \subseteq \{A | A \subseteq \Omega\}$. If $F_{p-X_r Y_r}(x_r, y_r) = F_{p-X_r}(x_r) \cdot F_{p-Y_r}(y_r)$ holds true for all $X_r, Y_r \in \mathbb{R}$, X_r and Y_r are called stochastically independent. The derivatives of these distribution functions are called *probability density functions* (PDF) $f_{p-X_r} :$ $\mathbb{R} \to \mathbb{R}$:

$$f_{p-X_r}(x_r) := \frac{dF_{p-X_r}(x_r)}{dx_r} \tag{3.3}$$

and

$$f_{p-X_r}(x_{r1}, x_{r2}, \ldots, x_{rn}) := \frac{\partial^n F_{p-X_r}(x_r)}{\partial x_{r1} \partial x_{r2} \ldots \partial x_{rn}} \tag{3.4}$$

respectively. The *conditional probability density function* is defined to

$$f_{p-X_r|Y_r}(x_r | Y_r = y_r) := \frac{f_{p-X_r Y_r}(x_r, y_r)}{f_{p-X_r}(x_r)}. \tag{3.5}$$

Moments m_k are attributes of random variables $X_r(\omega)$:

$$m_k := \mathbb{E}(X_r^k) := \int_{-\infty}^{\infty} x_r^k \cdot f_{p-X_r}(x_r) dx_r, \quad k \in \mathbb{N}, \tag{3.6}$$

where $\mathbb{E}(X_r) := \int_{-\infty}^{\infty} x_r \cdot f_{p-X_r}(x_r) dx_r$ is called the *expectation* of X_r. The first moment m_1 is also called *mean*. Another method for calculating the moments of a random variable X_r is the use of the so-called *characteristic function* M_{X_r}. It is defined as the expectation of $\exp(jv_r x_r)$, where j equals the imaginary unit:

$$M_{X_r}(v_r) := \mathbb{E}(\exp(jv_r x_r)) = \int_{-\infty}^{\infty} f_{p-X_r}(x_r) \cdot e^{jv_r x_r} dx_r. \tag{3.7}$$

This corresponds to the inverse Fourier-transform of $f_{p-X_r}(x_r)$[1]. Expressing the exponential function by its Taylor series yields

$$M_{X_r}(v_r) = \int_{-\infty}^{\infty} f_{p-X_r}(x_r) \cdot \sum_{k=0}^{\infty} \frac{(jv_r x_r)^k}{k!} dx_r = \sum_{k=0}^{\infty} \frac{\mathbb{E}(X_r^k)}{k!} \cdot (jv_r)^k. \tag{3.8}$$

Since the left-hand side of 3.8 corresponds to the inverse Fourier transform of $f_{p-X_r}(x_r)$ and the right-hand side contains all moments of $f_{p-X_r}(x_r)$, it becomes clear

---

[1]In this work the representation of a function $x_s : \mathbb{R} \to \mathbb{R}$, based on an orthogonal basis, that is comprised by exponential-functions rotating in a mathematically negative sense, is referred to as the Fourier transform – so: $(\mathscr{F}x_s)(v) := \int_{-\infty}^{\infty} x_s(t)\exp(-j2\pi t v)dt$. Respectively, the inverse Fourier transform is found to be: $(\mathscr{F}^{-1}X_{\mathscr{F}-s})(t) = \int_{-\infty}^{\infty} X_{\mathscr{F}-s}(v)\exp(j2\pi t v)dv$. In this work it is always assumed that these representations exist, or in other words that $x_s(t)$ and $X_{\mathscr{F}-s}(v)$ are square-integrable: $\int_{-\infty}^{\infty} |x_s(t)|^2 dt \leq \infty$ and $\int_{-\infty}^{\infty} |X_{\mathscr{F}-s}(v)|^2 dv \leq \infty$.

that knowing all moments m_k is equivalent to knowing $f_{p-X_r}(x_r)$. The moments of $f_{p-X_r}(x_r)$ can be readily calculated as

$$\mathbb{E}(X_r^k) := (-j)^n \left. \frac{d^k M_{X_r}(v_r)}{dv_r^k} \right|_{v_r=0} \tag{3.9}$$

The advantage of this method compared to Equation 3.6 is that it is sometimes easier to determine M_{X_r} and carry out the differentiations, when compared to solving the integrals of Equation 3.6. To describe statistical variations from the mean, the *central moments* σ_k are used:

$$\sigma_k := \mathbb{E}([X_r - \mathbb{E}(X_r)]^k) := \int_{-\infty}^{\infty} [x_r - \mathbb{E}(X_r)]^k \cdot f_{p-X_r}(x_r)dx_r, \quad k \in \mathbb{N}. \tag{3.10}$$

The second central moment is also named variance $\text{var}(X_r) := \sigma^2 := \sigma_2$; its square root is called standard deviation $\sigma := \sqrt{\sigma_2}$. These important attributes are widely employed in probabilistic calculus. It is interesting to note, that – using the linearity of the expectation – it can easily be derived, that the variance can be calculated by the first two moments $\text{var}(X_r) = \mathbb{E}(X_r^2) - \mathbb{E}^2(X_r)$. In general, the characterization and modeling of probability distribution functions can become very complex. Nevertheless, the use of the first two moments is often sufficient[2].

Using Equations 3.10 and 3.12, an important relation for the variance of the sum of two random variables can be derived:

$$\text{var}(X_r + Y_r) = \text{var}(X_r) + \text{var}(Y_r) + 2 \cdot \text{cov}(X_r, Y_r). \tag{3.11}$$

Here, $\text{cov}(X_r, Y_r)$ is called the *covariance* of the random variables X_r and Y_r. It is a measure of the affine-linear stochastic relation between two random variables:

$$\text{cov}(X_r, Y_r) := \mathbb{E}([X_r - \mathbb{E}(X_r)] \cdot [Y_r - \mathbb{E}(Y_r)]) = \mathbb{E}(X_r Y_r) - \mathbb{E}(X_r)\mathbb{E}(Y_r). \tag{3.12}$$

X_r and Y_r are called uncorrelated if $\text{cov}(X_r, Y_r) = 0$. To get a better understanding of the covariance, consider two random variables X_r and Y_r, that may be visualized in a scatter diagram like depicted in Figure 3.1. Here, each dot represents a simultaneous measurement of two quantities. Clearly the two processes depicted in Figure 3.1 are somehow affine-linearly related. An increase in X_r is – in average – corresponding to an increase in Y_r.

A measure of the interdependence is the slope of an affine-linear mapping $Y_{fit} = a + bX_r$. The error in the fit can be evaluated by the average quadratic distance χ^2 of the ensembles to the fit:

$$\chi^2 := \mathbb{E}\left[(Y_r - Y_{fit})^2\right] = \mathbb{E}\left[(Y_r - a - bX_r)^2\right]. \tag{3.13}$$

---

[2]This is due to the widespread occurrence of Gaussian and Poisson probability distributions, that are fully characterized by the first two moments in case of the Gaussian and the first moment in case of the Poisson distribution, respectively.

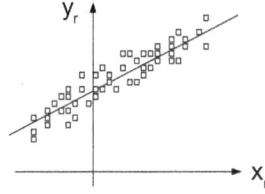

Figure 3.1 Scatter diagram of two random variables.

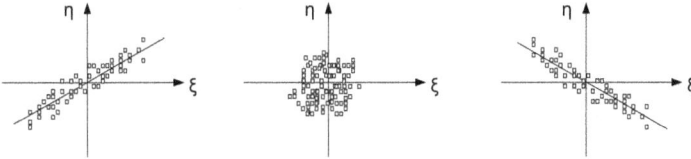

Figure 3.2 Scatter diagrams of two random variables ξ and η with differing correlations.

With the introduction of standardized variables $\xi := (X_r - \mathbb{E}(X_r))/\sqrt{\mathrm{var}(X_r)}$ and $\eta := (Y_r - \mathbb{E}(Y_r))/\sqrt{\mathrm{var}(Y_r)}$, this problem can be translated to finding the slope ρ_* that minimizes the quadratic distance χ^2:

$$\rho_* := \arg \min_{\rho \in \mathbb{R}} \chi^2(\rho) := \arg \min_{\rho \in \mathbb{R}} \mathbb{E}\left[(\eta - \rho \cdot \xi)^2\right]. \tag{3.14}$$

$$= \arg \min_{\rho \in \mathbb{R}} \mathbb{E}(\eta^2) + \rho^2 \cdot \mathbb{E}(\xi^2) - 2 \cdot \rho \cdot \mathbb{E}(\eta \cdot \xi) \tag{3.15}$$

$$= \arg \min_{\rho \in \mathbb{R}} 1 + \rho^2 - 2 \cdot \rho \cdot \mathbb{E}(\eta \cdot \xi) \tag{3.16}$$

$$= \arg \min_{\rho \in \mathbb{R}} \underbrace{(\rho - \mathbb{E}(\eta \cdot \xi))^2}_{\geq 0} - \mathbb{E}(\eta \cdot \xi)^2 + 1 \tag{3.17}$$

$$= \mathbb{E}(\eta \cdot \xi) = \mathrm{cov}(\xi, \eta). \tag{3.18}$$

Apparently the covariance describes a measure for the linear dependence of standardized random variables. According to

$$0 \leq \mathbb{E}\left[(\eta \pm \xi)^2\right] = \mathbb{E}(\eta^2) + \mathbb{E}(\xi^2) \pm 2 \cdot \rho_* \tag{3.19}$$

$$\Leftrightarrow 2 \pm 2 \cdot \rho_* \geq 0 \tag{3.20}$$

$$\Leftrightarrow -1 \leq \rho_* \leq +1 \tag{3.21}$$

it takes values between -1 and 1. In Figure 3.2 scatter diagrams of standardized random variables with differing correlations are given. In the left plot a positive value for ρ_* is given, while the right plot displays a negative one. The plot in the middle visualizes approximately uncorrelated random variables.

Reconsidering the original non-standardized random variables, ρ_* evaluates to

$$\text{cov}(\xi, \eta) = \text{cov}\left(\frac{X_r - \mathbb{E}(X_r)}{\sqrt{\text{var}(X_r)}}, \frac{Y_r - \mathbb{E}(Y_r)}{\sqrt{\text{var}(Y_r)}}\right) = \frac{\mathbb{E}(X_r \cdot Y_r) - \mathbb{E}(X_r)\mathbb{E}(Y_r)}{\sqrt{\text{var}(X_r)} \cdot \sqrt{\text{var}(Y_r)}}. \quad (3.22)$$

Thus, the actual dependence of the random variables is measured by $\mathbb{E}(X_r Y_r)$ – the variances are only normalizing the covariance whereas the mean values introduce a shift. The latter can be eliminated by modeling noise as zero-mean processes.

3.1.2 Stochastic processes

A *stochastic process* is mapping random variables $\omega \in \Omega$ for each e.g. time instance $t \in \mathbb{R}$ on e.g. a physical quantity in $\mathbb{R}$ – so: $X_{r-t} : \Omega \to \mathbb{R}$. An evaluation is then denoted by $X_r(t, \omega)$ or shortly $X_r(t)$, where $\omega \in \Omega$ and $t \in \mathbb{R}$. This is a typical way of describing random processes in physical systems. The general joint probability distribution function $F_{p-X_r} : \mathbb{R}^n \to [0, 1]$ is given by

$$F_{p-X_r}(x_r, t) := P(X_r(t_1) \leq x_{r1}, \ldots, X_r(t_n) \leq x_{rn}), \quad n \in \mathbb{N}. \quad (3.23)$$

Random processes, that are modeled by distribution functions $F_{p-X_r}(x_r, t)$ may be characterized by taking averages that ideally should correspond to the probability model[3]. For example, a good model should yield

$$\mathbb{E}(g(X_r(t))) = \lim_{N \to \infty} \frac{1}{N} \sum_{k=1}^{N} g[^k X_r(t)], \quad (3.24)$$

according to the *law of large numbers*, where $g : \mathbb{R} \to \mathbb{R}$ was introduced as an arbitrary function. Here $^k X_r(t)$ corresponds to an independently drawn sample, measured in one physical system k of N identical physical systems at the time instance t. For a proper statistical evaluation of the model, a sufficient amount N of samples should be taken. However, since the samples have to be measured at the same time instance and they should be exhibited by identical physical systems, this type of evaluation is obviously rather difficult.

Given random processes $X_r(t)$ and $Y_r(t)$ the *cross-correlation function* $R_{X_r Y_r} : \mathbb{R} \times \mathbb{R} \to \mathbb{R}$ is defined by

$$R_{X_r Y_r}(t_1, t_2) := \mathbb{E}(X_r(t_1) Y_r(t_2)) = \text{cov}(X_r(t_1), Y_r(t_2)) + \mathbb{E}(X_r(t_1))\mathbb{E}(Y_r(t_2)). \quad (3.25)$$

According to the explanations for the covariance, this describes a measure for the stochastic affine-linear dependence of $X_r(t_1)$ and $Y_r(t_2)$. The *cross-covariance*

---

[3]At this point, sometimes an insight in the precise formulations of convergence for random processes is given in literature. However, this rather technical detail is omitted here. The interested reader may have a look for e.g. [DSV00; Gar90; PU02; GS01].

$K_{X_r Y_r} : \mathbb{R} \times \mathbb{R} \to \mathbb{R}$ is defined as

$$K_{X_r Y_r}(t_1, t_2) := \mathbb{E}[(X_r(t_1) - \mathbb{E}(X_r(t_1)))(Y_r(t_2) - \mathbb{E}(Y_r(t_2)))] \qquad (3.26)$$

$$= R_{X_r Y_r}(t_1, t_2) - \mathbb{E}(X_r(t_1))\mathbb{E}(Y_r(t_2)). \qquad (3.27)$$

Since noise processes are usually modeled as zero-mean (c.f. Section 3.3) – e.g. $\mathbb{E}(X_r(t)) = 0$, the cross-covariance is a widely employed function. The *autocorrelation* and the *autocovariance* are defined as

$$R_{X_r X_r}(t_1, t_2) := \mathbb{E}(X_r(t_1)X_r(t_2)) \qquad (3.28)$$

and

$$K_{X_r X_r}(t_1, t_2) := \mathbb{E}[(X_r(t_1) - \mathbb{E}(X_r(t_1)))(X_r(t_2) - \mathbb{E}(X_r(t_2)))] \qquad (3.29)$$

$$= R_{X_r X_r}(t_1, t_2) - \mathbb{E}(X_r(t_1))\mathbb{E}(X_r(t_2)) \qquad (3.30)$$

respectively. These tools describe stochastic relations for the actual evolution of random processes and thus play a fundamental role in noise analysis. It is important to note that

$$R_{X_r X_r}(t_1, t_1) = \mathrm{var}(X_r(t_1)) \qquad (3.31)$$

applies for zero-mean processes. If the joint probability distribution functions of a random process $X_r(t)$ are independent of a time shift Δt, the underlying random processes are said to be *stationary*, that is, if

$$P(X_r(t_1) \leq x_{r1}, \ldots, X_r(t_n) \leq x_{rn}) = P(X_r(t_1 + \Delta t) \leq x_{r1}, \ldots, X_r(t_n + \Delta t) \leq x_{rn})$$

$$(3.32)$$

holds true. As mentioned before, often it is not possible to account for the whole distribution function or the probability density function. Often though, it is sufficient to model and characterize only the mean and the autocorrelation function. If these are then independent of the actual time instance, the underlying random processes are said to be *wide-sense stationary* (WSS). In that case $\mathbb{E}(X_r(t_1)) = \mathbb{E}(X_r(t_2))$. Furthermore, the autocorrelation function is fully described by the time difference τ:

$$R_{X_r X_r}(\tau) = \mathbb{E}[X_r(t)X_r(t + \tau)] = \mathbb{E}\left[X_r\left(t - \frac{\tau}{2}\right)X_r\left(t + \frac{\tau}{2}\right)\right], \quad \text{for all } t \in \mathbb{R}. \quad (3.33)$$

Given a period $T_{\mathrm{per}} \in \mathbb{R}$, if $\mathbb{E}(X_r(t)) = \mathbb{E}(X_r(t + T_{\mathrm{per}}))$ and $R_{X_r X_r}(t_1 + T_{\mathrm{per}}, t_2 + T_{\mathrm{per}}) = R_{X_r X_r}(t_1, t_2)$ holds true for all t, t_1 and t_2, $X_r(t)$ is said to be *wide-sense cyclostationary*. Using $t_1 = t + \tau/2$ and $t_2 = t - \tau/2$, the autocorrelation function can also be expressed as

$$R_{X_r X_r}(t, \tau) = \mathbb{E}\left[X_r\left(t - \frac{\tau}{2}\right)X_r\left(t + \frac{\tau}{2}\right)\right]. \qquad (3.34)$$

The defined autocorrelation function is an even function in τ and has a local maximum at $\tau = 0$.

As explained before, the characterization of random processes by measurements of a plurality of physically identical systems is rather difficult, if not impossible. In general, it would be preferable to characterize devices or systems by taking time-averages instead of ensemble-averages. The relation of them is described by the theory of ergodicity. A random process is said to exhibit *mean-square regularity of the mean and the autocorrelation function* if

$$\mathbb{E}\left|\frac{1}{T}\int_{-\frac{T}{2}}^{\frac{T}{2}} {}^{k}X_r(t)dt - \frac{1}{T}\int_{-\frac{T}{2}}^{\frac{T}{2}} X_r(t)dt\right|^2 \to 0 \quad \text{as } T \to \infty \tag{3.35}$$

and

$$\mathbb{E}\left|\frac{1}{T}\int_{-\frac{T}{2}}^{\frac{T}{2}} {}^{k}X_r\left(t - \frac{\tau}{2}\right){}^{k}X_r\left(t + \frac{\tau}{2}\right)dt\right.$$

$$\left. - \frac{1}{T}\int_{-\frac{T}{2}}^{\frac{T}{2}} X_r\left(t - \frac{\tau}{2}\right)X_r\left(t + \frac{\tau}{2}\right)dt\right|^2 \to 0 \text{ as } T \to \infty \tag{3.36}$$

hold true for all $\tau \in \mathbb{R}$ with probability one. Here, k indicates samples, where the terms on the right sides of Equations 3.35 and 3.36 correspond to the limits (c.f. [DSV00]). A random process is then said to have *mean-square ergodicity of the mean and the autocorrelation function* if

$$\mathbb{E}\left|\frac{1}{T}\int_{-\frac{T}{2}}^{\frac{T}{2}} X_r(t)dt - \mathbb{E}\left(\frac{1}{T}\int_{-\frac{T}{2}}^{\frac{T}{2}} X_r(t,\omega)dt\right)\right|^2 \to 0 \quad \text{as } T \to \infty \tag{3.37}$$

and

$$\mathbb{E}\left|\frac{1}{T}\int_{-\frac{T}{2}}^{\frac{T}{2}} X_r\left(t - \frac{\tau}{2}\right)X_r\left(t + \frac{\tau}{2}\right)dt\right.$$

$$\left. - \mathbb{E}\left(\frac{1}{T}\int_{-\frac{T}{2}}^{\frac{T}{2}} X_r\left(t - \frac{\tau}{2},\omega\right)X_r\left(t + \frac{\tau}{2},\omega\right)dt\right)\right|^2 \to 0 \quad \text{as } T \to \infty \tag{3.38}$$

are valid for all $\tau \in \mathbb{R}$. Thus, if a random process is wide-sense stationary and has mean-square ergodicity of the mean and the autocorrelation[4], the following relations

---

[4]In the following text this will be shortly noted as ergodic.

can be used for characterization:

$$\mathbb{E}(X_r) = \lim_{T \to \infty} \frac{1}{T} \int_{-\frac{T}{2}}^{\frac{T}{2}} X_r(t) dt \approx \lim_{N \to \infty} \frac{1}{N} \sum_{k=1}^{N} {}^k X_r(t) \qquad (3.39)$$

and

$$R_{X_r X_r}(\tau) = \lim_{T \to \infty} \frac{1}{T} \int_{-\frac{T}{2}}^{\frac{T}{2}} X_r(t) X_r(t+\tau) dt \approx \lim_{N \to \infty} \frac{1}{N} \sum_{k=1}^{N} {}^k X_r(t) {}^k X_r(t+\tau) \qquad (3.40)$$

respectively. From the last equation it follows that for WSS and ergodic zero-mean random processes the variance equals the power $L_{X_r} \in [0, \infty)$ of the random signal

$$L_{X_r} := \lim_{T \to \infty} \frac{1}{T} \int_{-\frac{T}{2}}^{\frac{T}{2}} X_r^2(t) dt = R_{X_r X_r}(0) = \mathrm{var}(X_r). \qquad (3.41)$$

The term L_{X_r} is also referred to as *noise power*. Its relation to the actual probabilistic quantity *variance* is only valid if the underlying process is WSS and ergodic. For deterministic signals the *Plancherel theorem* relates the time domain definition of the energy to the frequency domain description: $\int_{-\infty}^{\infty} x_s^2(t) dt = \int_{-\infty}^{\infty} |(\mathscr{F}x_s)(\nu)|^2 d\nu$. The Fourier transform of a non-deterministic signal $X_r(t, \omega)$, however, does not necessarily exist. The introduction of a truncated signal

$$X_{r-T}(t, \omega) := \begin{cases} X_r(t, \omega) & \text{for } |t| \leq \frac{T}{2} \\ 0 & \text{for } |t| > \frac{T}{2}, \end{cases} \qquad (3.42)$$

on the other hand, allows for a meaningful relation of the noise power to the frequency-domain. Later on, $T \in \mathbb{R}$ is increased to infinity, which then corresponds to the actual noise process $X_r(t, \omega)$. Since $X_{\mathscr{F}-r-T}(\nu, \omega) := (\mathscr{F}X_{r-T})(\nu, \omega)$ is a random function, meaningful quantities have to be generated by usage of the expectation:

$$\mathbb{E}(X_{\mathscr{F}-r-T}(\nu, \omega) X^*_{\mathscr{F}-r-T}(\nu, \omega)) = \mathbb{E}\left(\int_{-\frac{T}{2}}^{\frac{T}{2}} \int_{-\frac{T}{2}}^{\frac{T}{2}} X_{\mathscr{F}-r-T}(t) X^*_{\mathscr{F}-r-T}(t) \cdot e^{-j2\pi\nu(t-\tau)} dt d\tau \right)$$

$$(3.43)$$

$$= \int_{-\frac{T}{2}}^{\frac{T}{2}} \int_{-\frac{T}{2}}^{\frac{T}{2}} R_{X_{\mathscr{F}-r-T} X_{\mathscr{F}-r-T}}(t-\tau) \cdot e^{-j2\pi\nu(t-\tau)} dt d\tau \qquad (3.44)$$

The substitution $\tau' := t - \tau$ yields

$$\mathbb{E}\left(|X_{\mathscr{F}-r-T}(\nu, \omega)|^2 \right) = \int_{\tau=-\frac{T}{2}}^{\frac{T}{2}} \int_{\tau'=-\frac{T}{2}-\tau}^{\frac{T}{2}-\tau} R_{X_{r-T} X_{r-T}}(\tau') \cdot e^{-j2\pi\nu(\tau')} d\tau' d\tau. \qquad (3.45)$$

To reduce this problem to a single integral, a transformation of the integration space can be done. In the left part of Figure 3.3 the integration space before the substitution is displayed, while the middle part displays the space after substitution.

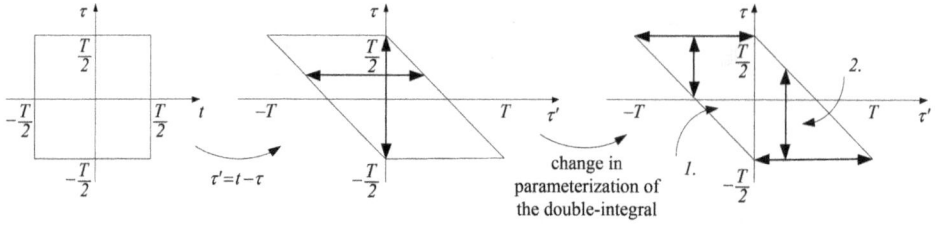

Figure 3.3 Schematic of the transformation of the double integral for the derivation of the Wiener-Khintchine theorem (according to [Czy10]).

According to the boundaries of the integrals in Equation 3.45 the integration with respect to τ is done along the τ-axis, while the integration with respect to τ' is done dependent on the respective τ coordinates. In the right part of Figure 3.3 the space is separated into two parts. Here the integration with respect to τ' is done parallel to the τ' axis, while the integration with respect to τ is now dependent on the respective τ' coordinates. This trick allows to solve the outer integral in Equation 3.45. Applying the mentioned changes in the sense of integration the problem is now expressed as

$$\mathbb{E}\left(|X_{\mathscr{F}-r-T}(\nu,\omega)|^2\right) = \int_{\tau'=-T}^{0}\int_{\tau=-\frac{T}{2}-\tau'}^{\frac{T}{2}} R_{X_{r-T}X_{r-T}}(\tau') \cdot e^{-j2\pi\nu(\tau')}d\tau d\tau' \qquad (3.46)$$

$$+ \int_{\tau'=0}^{T}\int_{\tau=-\frac{T}{2}}^{\frac{T}{2}-\tau'} R_{X_{r-T}X_{r-T}}(\tau') \cdot e^{-j2\pi\nu(\tau')}d\tau d\tau' \qquad (3.47)$$

$$= \int_{-T}^{T} (T - |\tau'|)R_{X_{r-T}X_{r-T}}(\tau') \cdot e^{-j2\pi\nu(\tau')}d\tau'. \qquad (3.48)$$

Now one can define a function in the frequency domain by normalization of Equation 3.46 and increasing T to infinity:

$$S_{X_rX_r}(\nu) := \lim_{T \to \infty} \frac{\mathbb{E}\left(|X_{\mathscr{F}-r-T}(\nu,\omega)|^2\right)}{T} \qquad (3.49)$$

$$= \lim_{T \to \infty} \int_{-T}^{T} \left(1 - \frac{|\tau|}{T}\right)R_{X_{r-T}X_{r-T}}(\tau) \cdot e^{-j2\pi\nu(\tau)}d\tau \qquad (3.50)$$

$$= \int_{-\infty}^{\infty} R_{X_rX_r}(\tau) \cdot e^{-j2\pi\nu(\tau)}d\tau. \qquad (3.51)$$

Taking the inverse Fourier transform and substituting $\tau = 0$ now yields

$$L_{X_r} = R_{X_rX_r}(0) = \int_{-\infty}^{\infty} S_{X_rX_r}(\nu) \cdot e^{j2\pi\nu\cdot 0}d\nu = \int_{-\infty}^{\infty} S_{X_rX_r}(\nu)d\nu. \qquad (3.52)$$

Because $S_{X_r X_r}$ is always positive (c.f. Equation 3.49) and because the integral over $S_{X_r X_r}$ yields the noise power, $S_{X_r X_r}$ can be meaningfully called *noise power spectral density*. Equation 3.51 is well known as the *Wiener Khintchine theorem*.

3.1.3   Propagation of noise in linear time-invariant circuits

Additionally to the major concepts of a probabilistic description of noise processes it is also important to know how noise actually propagates in physical systems like electronic circuits. In the class of *linear time-invariant systems* (LTI systems)[5], signal propagation can be described by the *convolution integral*. In case $X_r : \mathbb{R} \times \Omega \to \mathbb{R}$ is an input referred noise process, which is WSS and ergodic and $h_s : \mathbb{R} \to \mathbb{R}$ describes the impulse response function of an LTI system, the noise will somehow propagate to the output denoted as $Y_r : \mathbb{R} \times \Omega \to \mathbb{R}$. These are then related as

$$R_{Y_r Y_r}(\tau) := \mathbb{E}(Y_r(t, \omega) Y_r(t + \tau, \omega)) \tag{3.53}$$

$$= \mathbb{E}\left[\int_{-\infty}^{\infty} h_s(\tau') X_r(t - \tau', \omega) \mathrm{d}\tau' \int_{-\infty}^{\infty} h_s(\tau'') X_r(t + \tau - \tau'', \omega) \mathrm{d}\tau'' \right]. \tag{3.54}$$

With $R_{X_r X_r}(\tau) := \mathbb{E}(X_r(t, \omega) X_r(t + \tau, \omega))$ and $\mu := \tau' - \tau''$ this evolves to

$$R_{Y_r Y_r}(\tau) = \int_{-\infty}^{\infty} \int_{-\infty}^{\infty} R_{X_r X_r}(\tau - \mu) h_s(\mu + \tau'') h_s(\tau'') \mathrm{d}\mu \mathrm{d}\tau'' \tag{3.55}$$

$$= \int_{-\infty}^{\infty} R_{X_r X_r}(\tau - \mu) \left(\int_{-\infty}^{\infty} h_s(\mu + \tau'') h_s(\tau'') \mathrm{d}\tau'' \right) \mathrm{d}\mu \tag{3.56}$$

$$= R_{X_r X_r}(\tau) * h_s(\tau) * h_s(-\tau), \tag{3.57}$$

where $h_s(t) * x_s(t) := \int_{-\infty}^{\infty} h_s(\tau) x_s(t - \tau) \mathrm{d}\tau$ was used to indicate the convolution. This relation is sometimes also referred to as the *Wiener-Lee relation*. In the frequency domain this corresponds to

$$S_{Y_r Y_r}(f) = S_{X_r X_r}(\nu) \cdot |H_{\mathscr{F}-s}(\nu)|^2 , \tag{3.58}$$

where $H_{\mathscr{F}-s}(\nu) := (\mathscr{F} h_s)(\nu)$ was used. Equations 3.11, 3.57 and 3.58 may be used to model noise propagation in LTI systems, while Equations 3.39, 3.40 and 3.50 can be used for characterization of the underlying WSS and ergodic noise processes. How these processes may look like and how treatment in non-LTI systems may be done will be explained in the following text.

---

[5]Given an input signal $x_s : \mathbb{R} \to \mathbb{R}$ and an output referred signal $y_s : \mathbb{R} \to \mathbb{R}$, which are related by some mathematical operator O, that describes the propagation of x_s to y_s in sense of $y_s(t) = O(x_s(t))$, the system which is defined by O is called linear if $O(\sum_{\eta=1}^{N} k_\eta x_{s-\eta}(t)) = \sum_{\eta}^{N} k_\eta y_{s-\eta}(t)$ and time-invariant if $O(x_s(t - t_\eta)) = y_s(t - t_\eta)$ hold true for all $k_\eta, t, t_\eta \in \mathbb{R}$ [Unb97].

3.2 NOISE ANALYSIS IN NON-LINEAR AND TIME-VARIANT SYSTEMS

Considering Section 3.1.3, linearization at an operating point of e.g. an electronic circuit can yield an LTI system. However, this clearly is limited to applications where *frequency conversion* is negligible[6]. In general, physical systems that are non-linear and/or time-variant can lead to frequency conversion and can thus not be modeled precisely by the methodologies described in Section 3.1.3. Several approaches to accomplish a proper modeling of noise phenomena in non-linear and time-variant systems have been proposed in the past of which some are explained in this section. This allows for an understanding of the limitations each approach involves. Additionally to the overview of these remarkably mature methodologies, a rather simple method is explained that allows for straight-forward analysis of time-sampled systems such as *switched-capacitor filters*, at the cost of simplifying assumptions regarding the underlying systems.

3.2.1 Transformation of probability density functions

Consider a time-invariant, memoryless system that maps an input signal $x_s : \mathbb{R} \to \mathbb{R}$ according to $y_s(t) := g(x_s)(t)$ with $g : \mathbb{R} \to \mathbb{R}$. Defining a set $A(y_s) := \{x_s | g(x_s) \leq y_s\}$ the probability distribution function $F_{p-Y_{r-s}}(y_{r-s})$ can be expressed as [DR58]

$$F_{p-Y_{r-s}}(y_{r-s}) = P(Y_{r-s} \leq y_{r-s}) = P(X_{r-s} \in A(y_{r-s})). \tag{3.59}$$

In case of the existence of a continuous $f_{p-X_{r-s}}$ and a bijective and differentiable function $g(x_s) = y_s$ this simplifies to [DR58]

$$f_{p-Y_{r-s}}(y_{r-s}) = \frac{f_{p-X_{r-s}} \circ g^{-1}(y_{r-s})}{|g' \circ g^{-1}(y_{r-s})|}, \tag{3.60}$$

where $\circ$ indicates the composition of functions and g^{-1} defines the inverse function of g. These equations can readily be generalized to the multivariate case.

The approach of linearization of g around x_{s-0} yields an approximation of the variance according to

$$\text{var}(Y_{r-s}) \approx |g'(x_{s-0})|^2 \text{var}(X_{r-s}), \tag{3.61}$$

or

$$\text{var}(Y_{r-s}) \approx \sum_{k=1}^{N} \sum_{n=1}^{N} \left| \frac{\partial g(x_{s-1}, \ldots, x_{s-N})}{\partial x_{s-k}} \frac{\partial g(x_{s-1}, \ldots, x_{s-N})}{\partial x_{s-n}} \right|^2_{x_{s-1} = x_{s-1-0}, \ldots, x_{s-N} = x_{s-N-0}}$$
$$\cdot \text{cov}(x_{r-s-k}, x_{r-s-n}), \quad N \in \mathbb{N} \tag{3.62}$$

in the multivariate case. This is well known as *Gaussian error propagation*.

---

[6]Frequency conversion is the mixing-process that generates new frequency components due to non-linearity and time-dependence. This can occur in systems either due to time-varying changes in the biasing conditions which directly affects the noise model of the particular devices or due to non-linearity of circuit blocks.

3.2.2 Employing z-transform for noise analysis

Assuming a time-sampled signal $x_{s-sampled}(t) := x_s(t) \sum_{n=-\infty}^{\infty} \delta(t - nT_{per})$, with $n \in \mathbb{N}$ and $T_{per} \in \mathbb{R}$ being the sampling period, the z-transform of the sequence $(x_{s-n})_{n \in \mathbb{N}}$ defined as $x_{s-n} := x_s(nT_{per})$ is motivated from the modification of the Laplace transform[7] $X_{\mathscr{L}-s-sampled}(s) = (\mathscr{L}x_{s-sampled})(s) = \sum_{n=0}^{\infty} x_s(nT_{per})e^{-snT_{per}}$ by the substitution $z_z := e^{sT_{per}}$ if $R_s < 1$:

$$X_{\mathscr{Z}-s}(z_z) := (\mathscr{Z}(x_s - n))(z_z) := \sum_{n=0}^{\infty} x_{s-n}z_z^{-n} \quad \text{for } |z| > R_s \tag{3.63}$$

with the radius of convergence $R_s \in (0, \infty)$. For linear time-invariant systems, the impulse response can be utilized to predict the output signal for any given input signal by the use of convolution, what can be transferred to the z-domain by application of the z-transform. This method allows easy filter design with e.g. switched-capacitor circuits. The transfer function $H_{\mathscr{Z}-s}(z_z) := Y_{\mathscr{Z}-s}(z_z)/X_{\mathscr{Z}-s}(z_z)$ can be transformed by substitution of $z_z = e^{j2\pi\nu T_{per}}$:

$$H_{\mathscr{F}-s}(2\pi\nu) = H_{\mathscr{Z}-s}\left(e^{j2\pi\nu T_{per}}\right). \tag{3.64}$$

For signals that are bandwidth-limited according to the Nyquist-theorem $\nu_{max} < \nu_s/2 = 1/2T_{per}$, this allows for a proper estimation of the filter characteristic. However, if the signal contains frequency components which violate the Nyquist theorem, signals cannot be properly reconstructed. The assumption of a time-invariant system is violated, the transfer function approach is not valid and an estimation of comparably high frequency behavior cannot be made with this method. The utilization of the z-transform for modeling of the propagation of stationary and ergodic noise processes in time-sampled systems can thus only yield meaningful estimates for bandwidth-limited low-frequency behavior. The effect of white noise in switched-capacitor filters, however, cannot be properly modeled by the z-transform.

3.2.3 LPTV methods

For more than two decades methods for simulating the *steady-state* of so-called *linear periodcal time-variant* (LPTV) systems were developed [NTK07; Kun90; Maa03; Kun97; RLF98]. So far, methods like the *harmonic balance* or *shooting methods* are state of the art. The motivation for such tools was originated from *microwave systems*. These systems comprise circuits like e.g. oscillators with high quality factors for which transient simulations based on numerical integration methods are not feasible, due to long settling times and propagation of errors during the simulation [Kun90; Maa03; Kun97; RLF98]. Harmonic balance and the shooting methods contrarily do not evaluate the settling interval but directly compute the steady state of such systems

---

[7]The Laplace transform of a signal $x_s : \mathbb{R} \to \mathbb{R}$ with $x_s(t) = 0$ for $t < 0$ is defined as $X_{\mathscr{L}-s}(s) := (\mathscr{L}x_s)(s) := \int_{-\infty}^{\infty} x_s(t)\exp(-st)dt$, with $s \in \mathbb{C}$ and $\mathfrak{R}(s) > \gamma_s$ where γ_s is the so-called abscissa of convergence. The advantage of the Laplace transform compared to the Fourier transform is that is more widely applicable due to the exponential damping the transform implies.

with periodic excitations. A typical representation of an electronic system that has to be simulated is the set of non-linear *differential-algebraic equations* (DAE)

$$f(v(t), t) = i(v(t)) + \frac{dq(v(t))}{dt} + i_s(t) = 0. \tag{3.65}$$

These equations can be formulated by application of Kirchhoff's current law to each node. Here v are the voltages appearing at each node, $i(v)$ represent the voltage dependent currents, $\dot{q}(v) = dq(v(t))/dt$ represent the dynamic elements and i_s summarizes the current sources. $i(v)$ and $\dot{q}(v)$ may in general define non-linear relations.

3.2.3.1 Harmonic balance

In case of the harmonic balance method, the system of DAEs is divided in linear and non-linear parts and transformed into the frequency domain by Fourier transform:

$$F(V) = YV + j\mathbf{\Omega}(\mathscr{F}q(\mathscr{F}^{-1}V)) + i(\mathscr{F}^{-1}V) + I_s = 0, \tag{3.66}$$

where Y is the matrix representation of the circuits LTI part in the frequency domain, V represent the voltages in the frequency domain and $\mathbf{\Omega}$ is a diagonal matrix that contains the frequency components that are generated by the differentiation with respect to time [Kun90; Maa03]. To actually compute the steady-state, the following procedure is performed by the harmonic balance [Maa03]:

1   guess an initial state V
2   calculate the linear part of the current vector
3   perform inverse Fourier-transform of V to calculate the non-linear part in the time domain
4   transform the non-linear part into the frequency domain and verify if Kirchhoff's current law is violated more than the allowable predefined error constraints allow
5   if the former step shows a non-satisfying violation, vary the initial guess and begin from anew

To provide a fast and proper convergence of this sequence, several methods were developed. Basically, optimization approaches or methodologies to find the roots of Equation 3.66 can be employed, of which the latter is advantageous for numerical reasons [Kun90]. To adopt the harmonic balance method to systems with non-periodic excitation in the sense of the Fourier series, in which all frequency components are harmonics of one single fundamental tone, a generalized Fourier Series was introduced that allowed the application for so-called *almost-periodic signals*[8] [Kun90].

---

[8] Almost periodical is the expression of a signal as the superposition of sinusoids that are not harmonic. This can be done by fitting algorithms. The advantage of these signals compared to classical harmonic signals is numerical efficiency in case a system is stimulated by e.g. non-harmonic sinusoids.

3.2.3.2 *Shooting methods*

For numerical reasons, the harmonic balance approach is predestined for signals that tend to have a shape which is similar to a sinusoid. For systems that have much more abrupt transitions, the so-called shooting methods are advantageous [Kun90]. Shooting methods formulate their simulation in the time domain – conversely to the harmonic balance approach. The periodicity of the signal leads to the boundary condition $v(0) - v(T_{\mathrm{per}}) = 0$. This problem may be readily solved by *finite-differences methods* that simultaneously solve the DAE at each discrete time step. For typical systems, however, this method generates a relatively large amount of equations, what restricts its applicability [Kun90].

For further explanation of the shooting methods a *transition function* $\Phi : \mathbb{R}^{N+2} \to \mathbb{R}$ is introduced, where $N \in \mathbb{N}$ is the number of nodes. The transition function is defined by $\Phi(v(t_0), t_0, t_1)$ and evolves the initial state defined by $v(t_0)$ and t_0 towards $v(t_1)$, so that the boundary condition is now stated as the initial value problem

$$v(0) - \Phi(v(0), 0, T_{\mathrm{per}}) = 0. \tag{3.67}$$

Here, the task is to find the roots $v(0)$ for which Equation 3.67 is valid. For this, an initial guess has to be made for $v(0)$ and a subsequent transient integration has to be performed to evaluate the actual transition Φ. To find the roots of Equation 3.67 methods like Newton-Raphson can be implemented [Kun90]. In principle several numerical approaches can be chosen that all have their advantages and disadvantages as it is thoroughly discussed in [Kun95].

3.2.4 Propagation of noise in non-linear time-variant systems

For applications where the system is stimulated by multiple large-signal excitations that are exposed to comparably small disturbances, a power series of the DAEs given in Equation 3.65 can be developed at the steady-state solution $v_0(t)$, which is limited here to the:

$$df \approx G(v_0(t)) dv(t) + \frac{dC(v_0(t))}{dt} dv(t) + i_{\mathrm{ss}}(t) = 0, \tag{3.68}$$

where G and C are the matrix representations of the linearized i and q. i_{ss} describes the small-signal excitation to which the signal responds with the perturbation $dv(t)$ of the steady-state solution $v_0(t)$. Here, it has to be stressed out that this is a set of linear time-varying differential equations, since Equation 3.68 comprises only linear but time-dependent coefficients.

Such linear but time-dependent differential equations can also be expressed as

$$(\mathscr{L}_t y_{\mathrm{s}})(t) = (\mathscr{K}_t x_{\mathrm{s}})(t), \tag{3.69}$$

where $\mathscr{L}_t$ and $\mathscr{K}_t$ describe time-varying differential operators onto the output signal $y_{\mathrm{s}}(t)$ and input signal $x_{\mathrm{s}}(t)$, respectively. The *Green's function* approach allows to solve this problem. Defining $h_s(t_2, t_1)$ as the systems response to an impulse function applied at t_1:

$$(\mathscr{L}_{t_2} h_s)(t_2, t_1) = (\mathscr{K}_{t_2} \delta)(t_2 - t_1), \tag{3.70}$$

the output response to $x_s(t)$ can be readily calculated as

$$y_s(t_2) = \int_{-\infty}^{\infty} H_s(t_2, t_1) x_s(t_1) dt_1. \tag{3.71}$$

with H_s being the impulse response matrix. This can be easily verified by substituting Equation 3.71 into Equation 3.69, yielding:

$$h_s(t_2, t_1) = \int_{-\infty}^{\infty} H_s(t_2, t_1) \delta(t_2 - t_1) dt_1. \tag{3.72}$$

It has to be pointed out, that Equation 3.71 is an abstraction of the well-known convolution-integral and can also be motivated based on the principle of superposition, which is valid in linear systems.

If the input signal is given in terms of its Fourier transform $x_s(t) = \int_{-\infty}^{\infty} X_{\mathscr{F}-s}(v) e^{j2\pi vt} dv$ and substituted into Equation 3.71, a relation between input and output in the frequency domain can be given [Zad50]:

$$y_s(t_2) = \int_{-\infty}^{\infty} \int_{-\infty}^{\infty} H_s(t_2, t_1) X_{\mathscr{F}-s}(v) e^{j2\pi vt_1} dv dt_1 \tag{3.73}$$

$$= \int_{-\infty}^{\infty} H_{\mathscr{F}-s}(v, t_2) X_{\mathscr{F}-s}(v) dv. \tag{3.74}$$

Here, $H_{\mathscr{F}-s}(v, t_2) = \int_{-\infty}^{\infty} H_s(t_2, t_1) \exp(j2\pi vt_1) dt_1$ was introduced as the generalized version of the transfer function, which is now applicable for linear time-varying systems. This relation enables the prediction of frequency conversion. For temporal noise, however, probabilistic measures have to be used. In general, the output referred correlation matrix $R_{y_s y_s}(t_1, t_2)$ can be expressed as [RLF98]

$$R_{y_s y_s}(t_1, t_2) = \mathbb{E}\left(\int_{-\infty}^{\infty} H_s(t_1, \tau_1) x_s(\tau_1) d\tau_1 \left[\int_{-\infty}^{\infty} H_s(t_2, \tau_2) x_s(\tau_2) d\tau_2 \right]^{\mathscr{T}} \right) \tag{3.75}$$

$$= \int_{-\infty}^{\infty} \int_{-\infty}^{\infty} H_s(t_1, \tau_1) R_{x_s x_s}(\tau_1, \tau_2) H_s^{\mathscr{T}}(t_2, \tau_2) d\tau_1 d\tau_2, \tag{3.76}$$

with $A^{\mathscr{T}}$ being the transpose of $A \in \mathbb{R}^N \times \mathbb{R}^N, N \in \mathbb{N}$. This relation is valid for linearized time-varying systems. It has to be emphasized that no restrictions on the underlying noise processes were made so far. Equation 3.76 can be understood as a generalization of the Wiener-Lee relation given in Equation 3.57. Performing a two-dimensional

Fourier transform of the correlation matrix, this yields

$$S_{y_s y_s}(\nu_1, \nu_2) = \int_{-\infty}^{\infty} \int_{-\infty}^{\infty} \int_{-\infty}^{\infty} \int_{-\infty}^{\infty} H_s(t_1, \tau_1) R_{x_s x_s}(\tau_1, \tau_2) H_s^{\mathscr{T}}(t_2, \tau_2)$$
$$\cdot \, d\tau_1 d\tau_2 e^{-j2\pi\nu_1 t_1} e^{-j2\pi\nu_2 t_2} dt_1 dt_2. \tag{3.77}$$

Making the assumption of having an LPTV system $H_s(t_2 + T_{per}, t_1 + T_{per}) = H_s(t_2, t_1)$, the impulse response function can be represented in terms of a Fourier series: $H_s(t_2, t_1) = \sum_{n=-\infty}^{\infty} H_{s-n}(t_2 - t_1) \exp(j2\pi n t_2 / T_{per})$. The output spectral density for LPTV systems can then be stated as [RLF98]

$$S_{y_s y_s}(\nu_1, \nu_2) = \int_{-\infty}^{\infty} \int_{-\infty}^{\infty} \int_{-\infty}^{\infty} \int_{-\infty}^{\infty} \left(\sum_{k=-\infty}^{\infty} H_{s-k}(t_1 - \tau_1) e^{jk2\pi\nu_0 t_1} \right) R_{x_s x_s}(\tau_1, \tau_2)$$
$$\cdot \left(\sum_{n=-\infty}^{\infty} H_{s-n}^{\mathscr{T}}(t_2 - \tau_2) e^{jn2\pi\nu_0 t_2} \right) e^{-j2\pi\nu_1 t_1} e^{-j2\pi\nu_2 t_2} d\tau_1 d\tau_2 dt_1 dt_2, \tag{3.78}$$

which can be modified to

$$S_{y_s y_s}(\nu_1, \nu_2) = \sum_{k=-\infty}^{\infty} \sum_{n=-\infty}^{\infty} H_{\mathscr{F}-s-k}(\nu_1 - k\nu_0) S_{x_s x_s}(\nu_1 - k\nu_0, \nu_2 - n\nu_0)$$
$$\times H_{\mathscr{F}-s-n}^{\mathscr{T}}(\nu_2 - n\nu_0), \tag{3.79}$$

with $\nu_0 = 1/T_{per}$. Here, so-called *harmonic transfer functions* $H_{\mathscr{F}-s-n}(\nu) = \int_{-\infty}^{\infty} H_{s-n} e^{-j2\pi\nu t} dt$ were introduced. To apply this fundamental equation for numerical simulation purposes, meaningful truncation of the frequencies that have to be accounted for, has to be done.

State of the art noise predictions via periodic small-signal analysis or harmonic balance have their foundation in these relations. For LPTV systems, the steady-state is evaluated and the systems DAE is linearized according to Equation 3.68. The noise propagation in these systems can then be computed with versions of Equation 3.79, that are optimized for numerical purposes. In RF applications, often the knowledge of the time-dependent correlation matrix is not necessary. If the time-varying spectral density is characterized by e.g. a spectrum analyzer, the setup will hardly be able to resolve the time-varying behavior – temporal averaging is done. Since this corresponds to less computational effort, the same is done by e.g. *SpectreRF®*[9] (c.f. [Kun99; CDS11c]). However, for some applications like e.g. the design of switched-capacitor filters, knowledge of the transient evolution of the variance is appreciable, so that time-averaging is not meaningful[10].

---

[9] Virtuoso®SpectreRF® is a circuit simulator developed by Cadence Design Systems, Inc. that implements steady-state solvers based on harmonic balance and shooting methods, small-signal analysis for LPTV systems and transient noise analysis.

[10] SpectreRF® does provide the capability of evaluation of the spectral density at particular time instances (c.f. [CDS11a]).

3.2.5 Noise in the time domain

Noise analysis based on LPTV systems and cyclostationary random processes clearly has its limitations. Free running oscillators for instance do not exhibit a truly periodic behaviour but show *phase noise*, which can be described as a random fluctuation within the phase or the corresponding time period [DSV00; MSV05]. Other systems may not be periodic at all as for instance sigma-delta modulators, memory, asynchronous filters or digital logic circuits (c.f. e.g. [TK99; Sic08]). For these problems several approaches have been proposed in the past.

3.2.5.1 Monte Carlo noise simulation

The method referred to as *transient noise* or *Monte Carlo* method generates sample paths of the noise signals, which are added to the stimulating deterministic signals during time domain simulation. Such sample paths are generated by creation of random numbers in a stochastically meaningful way. For instance, sample paths may be generated as a Fourier series with magnitudes corresponding to the desired spectral density and a random phase for each wave function [CDS06]. However, this approach is not really realistic, since the desired spectral density is only approached in the mean – single sample paths generally deviate from that. It was claimed that the magnitudes are Rayleigh distributed, so that more realistic transient samples can be generated [CDS06]. However, this method still suffers from several disadvantages. Since the noise processes are added to the stimuli, which are often much larger than the noise itself, numerical errors introduced by the simulator have to be excluded by tightening the constraints. To properly account for the noise bandwidth, the chosen time steps of the transient simulation have to be very small. Generated sequences of random numbers often suffer from a finite correlation length so that ideal stochastically independent processes become dependent due to the large number of evaluation time steps. Furthermore, the generation of noise sample paths for all noise sources within an electronic circuit together with the small time steps of the transient simulation define a huge complexity, that limits the complexity of the circuit that can be verified before the computer runs out of memory [DLSV96; DR99].

3.2.5.2 Stochastic differential equations

Consider i_s in Equation 3.65 to include noise sources like e.g. idealized white noise. Such noise processes are not physically realistic but often allow for simplified analysis. If the system that exhibits such noise processes is meaningfully modeled, the system itself will limit the bandwidth, so that realistic predictions can be deduced. A simplified equation can for instance be

$$\frac{dv(t)}{dt} = f(v(t), t, \xi(t)) + \xi(t), \quad v(t_0) = c_i, \tag{3.80}$$

where $\xi(t)$ is an idealized white noise process and c_i is the initial value. A similar equation was first introduced by Paul Langevin (1872–1946) and is thus referred to as *Langevin equation*. Its generic form is written as

$$\frac{dv(t)}{dt} = f(v(t), t) + G(v(t), t)\xi(t), \quad v(t_0) = c_i. \tag{3.81}$$

However, for idealized white noise processes ξ, such DAEs cannot be directly solved in an ordinary sense. Ordinary differential calculus subdivides the domain on which the equation has to be solved in $N \in \mathbb{N}$ parts and takes the limit $N \to \infty$. Nevertheless, since ideal white noise has energy in each frequency component, this limit simply does not exist. The integral $W(t) = \int_0^t \xi(s) ds$ can be identified with the *Wiener process*[11], which is a continuous function in t [Gar85; DSV00]. Symbolically, Equation 3.81 can be written in differential form:

$$dv(t) = f(v(t), t)dt + G(v(t), t)dW(t), \quad v(t_0) = c_i. \tag{3.82}$$

Such equations are referred to as *stochastic differential equations* (SDE) and symbolize the integral equation

$$v(t) = \int_{t_0}^t f(v(s), s)ds + \int_{t_0}^t G(v(s), s)dW(s). \tag{3.83}$$

The first integral on the right-hand side of Equation 3.83 is an ordinary Lebesgue integral. The second integral is a so-called *Itō integral* [Gar85; DSV00]. The calculation of such integrals is not intuitive and far beyond the scope of this work. In some cases SDEs can be directly solved (c.f. e.g. [MSV05; Meh02]), but are, however, considered to be too complex in general for direct solutions [DSV00; DLSV96]. In case of linear SDEs, however, closed general solutions can be found. This was exploited by Demir, who presented an approach where the DAEs were linearized at the deterministic solution according to Equation 3.68, rearranged and exposed to white noise processes. These linear SDEs were then evaluated for stochastic quantities by the expectation, so that systems of ordinary differential equations were yielded for the mean and the covariance matrix, that can be solved by simple calculus [DSV00; DLSV96]. The assumption that was made is that noise processes are small compared to the deterministic excitations, so that the linearization is a meaningful simplification. The advantage of this method compared to the LPTV based methods is that periodicity is not needed, so that more complex systems can be simulated. The employment of white noise processes within this algorithm is not to be considered as a limitation, since basically any spectral density distribution can be obtained by usage of LTI filters as presented in [DSV00].

Alternatively to the description of SDEs that model the transient evolution of sample paths of random processes such as white noise, the random process may directly be described in terms of its probability density function. Differential equations that model the transient evolution of probability density functions of random processes are referred to as *Fokker-Planck equations* (c.f. [Gar85]). These equations can be derived from SDEs, by application of Itō's law [Gar85].

---

[11]The Wiener process is defined such that (i) with probability 1, $W(0) = 0$ and $W(t)$ is continuous in t; (ii) for all $t \geq 0$ and $h \geq 0$: $P[W(t+h) - W(t) \leq w] = (2\pi h)^{-1/2} \int_{-\infty}^w \exp[-u^2/2h]du$ and (iii) if $0 \leq t_1 \leq \cdots \leq t_m$, the increments $W(t_2) - W(t_1) \cdots W(t_m) - W(t_{m-1})$ are independents [Fal03].

3.2.6 A sequential method using a switching time-frequency domain

Conversely to the exact but also complex methodologies, that have been presented in the former text, we developed a rather simple algorithm to predict the propagation of uncertainty in time-varying circuits – preliminary switched-capacitor filters [VSH11]. The presented method can be applied for systems in which the underlying random processes are wide-sense stationary. The system itself does not have to perform a periodical operation but has to be linear. To obtain the variance at particular nodes in the circuit, common frequency-domain calculus is being made use of. If a switching operation is done, variances are distributed in the time domain.

An output signal $y_s(t)$ of a linear time-invariant and stable system, which can be described by its impulse response function $h_s(t)$ is *asymptotically stationary* if the applied input signal $x_s(t)$ vanishes for $t < 0$ and becomes stationary itself for $t > 0$ [Unb97]. This means that in these systems

$$\lim_{t \to \infty} \mathbb{E}[y_s(t)] = \lim_{t \to \infty} \int_0^t h_s(\tau)\mathbb{E}(x_s)d\tau = \lim_{T \to \infty} \frac{1}{T} \int_0^T y_s(t)dt \qquad (3.84)$$

$$\lim_{t \to \infty} \mathbb{E}[y_s^2(t)] = \lim_{t \to \infty} \int_0^t \int_0^t h_s(\tau)h_s(\tau')R_{x_s x_s}(\tau - \tau')d\tau d\tau' = \lim_{T \to \infty} \frac{1}{T} \int_0^T y_s^2(t)dt$$

$$(3.85)$$

holds true. If, for instance, a capacitor C is connected to a white noise $R_{x_s x_s}(\tau) = K \cdot \delta(\tau)$ source at $t = 0$ by a switch exhibiting a resistance R, this yields $\mathbb{E}(y_s^2(t)) = (KRC/2)[1 - \exp(-2t/RC)]$, where $h_s(t) = \exp(-t/RC), t > 0$ is used (c.f. [Unb97]).

If linear switched-capacitor filters are operated in such a way that the assumption of asymptotic stationarity is meaningful, calculus of the propagation of noise in these systems becomes straight-forward. The basic sequence of the proposed algorithm is depicted in Figure 3.4.

Firstly an initial state of the system should be defined at $t = T_0$. This means, that all switches are in a well-defined position, so that the system becomes asymptotically stationary. The application of the Wiener-Lee and Wiener-Khintchine theorems allow for calculation of the variances at particular capacitances $C_m, m = 1 \ldots M, M \in \mathbb{N}$ within the circuit. Therefore, the spectral power densities of the noise processes exhibited by the circuits components are related to all capacitances, which in the charge-domain becomes: Q_{n,C_m,T_0}^2. The calculation of the actual variance by integration of the density function yields values that we refer to as "frozen": $Q_{n,\text{fr},C_m,t_1}^2$. If then a switching operation is initiated, the uncertainty presented in the charge domain remains unchanged in the vicinity of t_1. Once the capacitors are properly reconnected to define the next operation phase of the filter, the accumulated charges are distributed according to the new transfer functions $\psi_{m,T_1} : \mathbb{R}^M \to \mathbb{R}^M$. Due to the linearity, this procedure can be readily evaluated according to Kirchhoff's laws. If the circuit approaches asymptotic stationarity in this phase as well, the distribution of the uncertainty is defined by the ratio of the capacitances only. This can be solved by simple linear algebra. In addition to the distributed variance from the former switching phase, the system exhibits noise

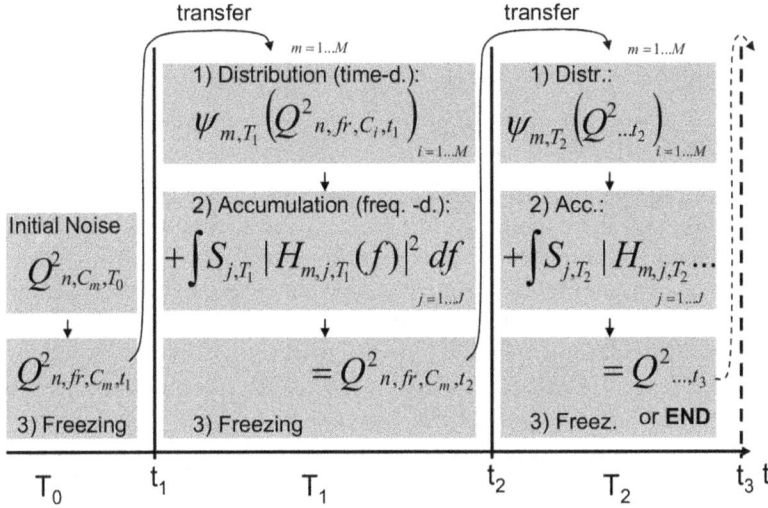

Figure 3.4 Sketch of the proposed algorithm to predict the propagation of noise in switched capacitor filters.

in this phase as well – given by power spectral densities S_{j,T_1}, where j indicates the component that exhibits the corresponding noise. The contribution from the component j to the noise at the capacitance m is defined by the transfer function $H_{m,j,T_1}(f)$. Before the next switching operation is done, the variance at each capacitance is determined by superposition of the uncertainty from the former phase and the actual phase. This sequence is repeated until the circuit reaches its final phase at which the overall noise has to be evaluated. A simple example for this procedure can be found in [VSH11].

The assumption of asymptotic stationarity is of simplifying nature. Nevertheless, it is not mandatory. For systems that do not approach this state formulas may be developed by application of e.g. stochastic differential equations, that predict the evolution of the variance during the transition phases. If a set of such equations for all possible equivalent circuits is obtained, a "look-up table"-approach may be used for eased prediction of noise propagation. In summary, this approach is infantile compared to the direct approach of solving the SDEs or the Fokker-Plank equations, since non-linearity is not accounted for, underlying noise processes have to be wide-sense stationary and formulas have to be generated in case the system is not asymptotically stationary. Nevertheless, the proposed method is superior to the alternative approaches in terms of convenience and thus also in calculation/simulation speed.

3.3 FUNDAMENTAL NOISE PROCESSES IN ELECTRONIC DEVICES

In optoelectronic components due to the quantization of e.g. electronic or light flux, non-deterministic uncertainties in the position and the momentums of these particles arise. These uncertainties of the microscopic particles, which basically form the devices, which then comprise circuits and systems, cause fluctuations of macroscopic quantities like charges, currents or voltages. This section will summarize major concepts, which relate the underlying physical phenomena occurring in physical systems like CMOS devices to analytic descriptions as summarized in Section 3.1.

3.3.1 Thermal noise

Observation of physical quantities as for instance open circuit voltage or short-circuit current of solids at thermodynamic equilibrium expose fluctuations. These fluctuations can be explained in analogy to the Brownian motion. Particles within the solid have kinetic energy and thus move within the material. The particles can collide at the lattice which affects the momentums. Statistical thermodynamics can be used to study the probability distributions of the movement within the solid. It is known that in first order the *Maxwell-Boltzmann distribution function* is a proper approximation [McK93]. During the deduction of the Maxwell-Boltzmann distribution the terms *equipartition theorem* and *temperature* are introduced. Temperature is defined as a linear measure of the mean kinetic energy of particles. It has an absolute zero-point which corresponds to the standstill of the matter. The equipartition theorem describes the relation of the thermal energy and other forms of energy:

$$E_{\text{thermal}} = \frac{n_{\text{dof}} k_B \theta}{2}. \tag{3.86}$$

Here, n_{dof} stands for the degrees of freedom of a system. If one intends to relate e.g. the thermal energy of a solid in which particles can freely move in three dimensions, n_{dof} equals 3. Based on these fundamental physical processes, relations to macroscopic quantities as current or voltage have to be derived to enable further studies. Johnson described in [Joh28] how he related thermal fluctuations to current and voltage. Nyquist was able to demonstrate the first theory on this phenomenon in [Nyq28]. Due to these groundbreaking publications thermal noise is also often referred to as *Nyquist noise, Johnson noise* or *Nyquist-Johnson noise*. Nyquists first theory introduces resistors in thermal equilibrium which exchange energy over a transmission line. Then Fourier analysis is employed to study the power spectral density of the noise processes. Unfortunately, he used several assumptions which are partially not necessary and partially not justified[12]. Alternative derivations are e.g. based on the *Drude model* of electrical conduction and so make according assumptions ([Mü90]). A comparatively easy derivation is presented here, that does not need complex or unjustified assumptions. The goal of the presented derivation is to yield probability distribution functions and power spectral densities of the macroscopic quantities current and voltage to stochastically describe the macroscopic fluctuations that are originated in the microscopic fluctuations within a resistor. For the presented derivation of the model for thermal noise a parallel resonant circuit that employs a noisy resistor is introduced (c.f. Fig. 3.5).

The current source i_n represents the noise exhibited by a resistor, thus the conductivity G is now considered noise-free. The direction of the current is actually

---

[12]For a discussion of this and alternative derivations c.f. [Blu96].

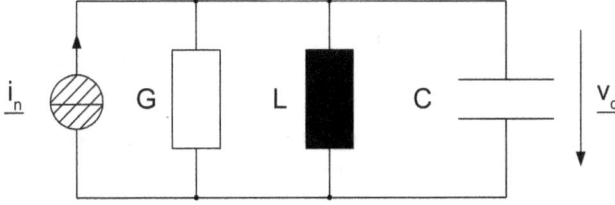

Figure 3.5 Parallel resonant circuit.

undetermined and on average zero – due to the thermal equilibrium. However, a current direction is proposed in Figure 3.5 to allow for complex AC analysis. The resonant circuit is considered to have a high quality factor Q_F so it allows the conversion from the noise current i_n to a voltage v_c at the resonant frequency v_0. Other frequencies do not significantly contribute to the voltage referred noise power $\int_0^\infty |v_c(v)|^2 dv$. This is justified by the high Q_F which can virtually take any value in this thought experiment. This frequency selective behavior allows for a derivation of the power spectral density $S_{v_c v_c}$ as a function of $S_{i_c i_c}$. The use of the equipartition theorem then relates this to the microscopic origin. With the substitutions $v_0 = 1/2\pi LC$ and $Q_F = v_0 C/2\pi G$ the output voltage can be derived:

$$v_c = \frac{i_n}{G(1 + jQ(\frac{v}{v_0} - \frac{v_0}{v}))}. \tag{3.87}$$

Employing the Wiener-Khintchine theorem this results in

$$\int_0^\infty S_{v_c v_c}(v)dv = \frac{1}{G^2} \int_0^\infty \frac{S_{i_n i_n}(v)dv}{1 + Q^2\left(\frac{v}{v_0} - \frac{v_0}{v}\right)^2}. \tag{3.88}$$

Here, a one-sided Fourier transform was used. Since the Fourier transform of a real signal is even, the one-sided Fourier transform relates to the double-sided Fourier transform by division with a factor of 2. For a high Q_F Equation 3.88 is significant only at $v \approx v_0$. Accordingly one can set

$$\int_0^\infty S_{v_c v_c}(v)dv = \frac{S_{i_n i_n}(v_0)}{G^2} \int_0^\infty \frac{dv}{1 + Q^2\left(\frac{v}{v_0} - \frac{v_0}{v}\right)^2}. \tag{3.89}$$

Solving the integral leads to

$$\int_0^\infty S_{v_c v_c}(v)dv = \frac{S_{i_n i_n}(v_0)v_0\pi}{2QG^2} = \frac{S_{i_n i_n}(v_0)}{4GC}. \tag{3.90}$$

Application of the equipartition theorem now yields

$$\frac{1}{2}C\mathbb{E}(v_c^2) = \frac{1}{2}C\int_0^{\infty} S_{v_c v_c}(\nu)d\nu = \frac{1}{2}k_B\theta. \tag{3.91}$$

The stored average electric energy thus corresponds to the thermal noise. Substitution into Equation 3.90 results in

$$S_{i_n i_n}(\nu_0) = 4k_B\theta G. \tag{3.92}$$

Since the derivation did not make any particular assumptions on the resonant frequency the above equation is valid for virtually any frequency, so

$$S_{i_n i_n}(\nu) = 4k_B\theta G. \tag{3.93}$$

Since this relation was first found by Nyquist it is called *Nyquist relation*. Apparently, a noisy resistor can be considered as a noise source in parallel to its conductance G. Alternative one can consider a noise voltage source of

$$S_{v_n v_n}(\nu) = 4k_B\theta R \tag{3.94}$$

in series to the resistance R. As mentioned, the derived noise spectral density is independent of frequency. Noise processes with that characteristic are called *white noise*. Evaluation of the power exhibited from an ideal white noise source will not lead to a definite value but will diverge. This is known as the *ultraviolet catastrophe*. This problem is caused by the equipartition theorem which is not applicable to all frequency components. The first result of the quantum mechanics – the *Planck's law*, however, solves this problem. It accounts for the finite velocities of particles. The thermal noise can now be expressed as

$$S_{v_n v_n}(\nu) = \frac{4Rh\nu}{e^{\frac{h\nu}{k_B\theta}} - 1} \tag{3.95}$$

and

$$S_{i_n i_n}(\nu) = \frac{4Gh\nu}{e^{\frac{h\nu}{k_B\theta}} - 1}. \tag{3.96}$$

For $k_B\theta \gg h\nu$, these relations converge into Equations 3.93 and 3.94. Nyquist already presented this correction in [Nyq28]. Apart from the autocorrelation function or the power spectral density, respectively, the probability density function may be of interest, since it can be used to estimate the probability of a random quantity exceeding a threshold. This can be used to e.g. define a noise margin for digital circuits. Since thermal noise is basically originated by the superposition of a large amount of charge particles which arbitrarily move within a solid, the *central limit theorems* may be applied. The central limit theorems is a set of weak relations about convergence, which claim that an increasing amount of stochastically independent random variables with the same probability distribution function that are additively resulting in another random variable, result in a *normal probability distribution function* which is also referred to as *Gaussian probability distribution function* (c.f. e.g. [Web92; PU02]).

3.3.2 Shot noise and photon noise

Besides thermal noise which is based on the thermal equilibrium, there is another phenomenon called *shot noise* that is based on the exact opposite – the disequilibrium in form of a potential difference. Sometimes mathematical relations are derived coming from an ideal diode tube (c.f. e.g. [DR58]). There, perimeter effects, the interdependency of transported charges and fluctuations of the anode-to-cathode voltage are neglected. The current pulse generating process of electric induction caused by charge packets that transit the potential barrier is the foundation of further derivations. Since this work is about semiconductor devices, diode tubes specific relations are not considered – here the noise process caused by the quantization of current at an "ideal" potential barrier are presented. Here, "ideal" means that the effect of collisions of charge packets or processes as generation or recombination are neglected. Their impact is discussed in Section 3.3.3.

If a potential difference is applied to a medium, a velocity distribution will occur with a mean that differs from zero. The resulting current will consist of the superposition of single impulses which are caused by the electric induction of charge carriers trespassing the potential difference. The integral of the induced current provoked by a single charge-carrier with respect to time will yield the elementary charge. If generation or recombination occurs, it may happen that the result of the integral differs from the elementary charge. These effects are neglected within the presented derivation. Furthermore, there is a probability of emitting charge carriers that carry an insufficient amount of energy to pass the potential barrier. These will transit into the space that is occupied by an electric field and will then travel back to their source. Due to electric induction, this will also result in a fluctuation in the macroscopic quantity current. This can occur for instance within diode tubes, where charges are emitted due to *thermionic emission*. Emitted charge carriers will have initial momentums according to the Maxwell-Boltzmann relation. The current within a diode tube results in a deformation of the potential within the tube, so that a barrier occurs which cannot be surpassed by some particles. These particles will not contribute to the DC current but provoke fluctuations or noise. This is neglected within the presented derivations. The induced current pulses are dependent on the geometry, the material properties and the potential applied at the terminals of the device. In first order it is assumed, that the induced pulses have the same form. If one of the presented assumptions is not justified, the presented derivations cannot provide proper estimation of the phenomena. A series of induced current pulses can appear e.g. as depicted in Figure 3.6.

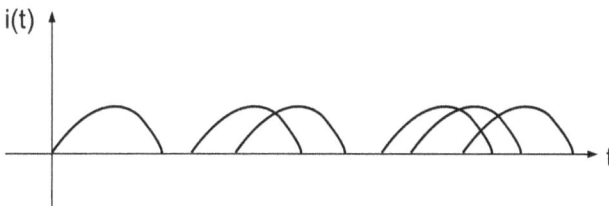

Figure 3.6 Pulse train within a potential barrier.

To analyze this random process, firstly a probability model for the amount of charge-carriers emitted within an interval T has to be developed. This will inherently also yield the probability density function of the current magnitude. These derivations will be followed by the deduction of the autocorrelation function, which finally leads to the spectral noise power density. The presented derivations are adopted from [DR58].

The probability of the emission of one single charge carrier within the interval $\Delta\tau \to 0$ into the space that is occupied by the electric field can be approximated to

$$P(1_{\Delta\tau}) = a\Delta\tau. \tag{3.97}$$

This linear approach is consistent with the power series. The meaning of a will become clear later on. The assumptions made here are, that there is no preferred time instance for the emission of carriers and that the interval $\Delta\tau$ is chosen so small, that the probability of the emission of more than one charge carrier can be neglected:

$$P(0_{\Delta\tau}) + P(1_{\Delta\tau}) = 1. \tag{3.98}$$

With the additional assumption that emitted carriers will not influence the time instances of future emissions the substitution of Equation 3.97 into Equation 3.98 yields

$$P(0_{\tau+\Delta\tau}) = P(0_\tau) \cdot P(0_{\Delta\tau}) = P(0_\tau) \cdot (1 - P(1_{\Delta\tau})) = P(0_\tau) \cdot (1 - a\Delta\tau). \tag{3.99}$$

This can be rearranged to

$$\frac{P(0_{\tau+\Delta\tau}) - P(0_\tau)}{\Delta\tau} = -aP(0_\tau). \tag{3.100}$$

Now taking the limit $\Delta\tau \to 0$ leads to the differential equation

$$\frac{dP(0_\tau)}{d\tau} = -aP(0_\tau) \tag{3.101}$$

which solution is found to be

$$P(0_\tau) = ke^{-a\tau}. \tag{3.102}$$

With the boundary condition $P(0_0) = 1$ the constant can be evaluated to $k = 1$, so

$$P(0_\tau) = e^{-a\tau}. \tag{3.103}$$

The probability of an emission of $K \in \mathbb{N}$ charge carriers within $\tau + \Delta\tau$ equals

$$P(K_{\tau+\Delta\tau}) = P((K-1)_\tau)P(1_{\Delta\tau}) + P(K_\tau)P(0_{\Delta\tau}), \tag{3.104}$$

since either $K - 1$ carriers could be emitted during τ with the last one being emitted in $\Delta\tau$ or all K charge carriers could be emitted during τ. Substitution of Equation 3.97 and 3.98 now yields

$$P(K_{\tau+\Delta\tau}) = P((K-1)_\tau)a\Delta\tau + P(K_\tau)(1 - a\Delta\tau), \tag{3.105}$$

which can be rearranged to

$$\frac{P(K_{\tau+\Delta\tau}) - P(K_\tau)}{\Delta\tau} + aP(K_\tau) = aP((K-1)_\tau). \tag{3.106}$$

The limit $\Delta\tau \to 0$ leads to the differential equation

$$\frac{dP(K_\tau)}{d\tau} + aP(K_\tau) = aP((K-1)_\tau). \tag{3.107}$$

The solution can be found to

$$P(K_\tau) = ae^{-a\tau} \int_0^\tau e^{at} P((K-1)_t)dt + k. \tag{3.108}$$

Since $P(K_0) = 0$ it must follow that $k = 0$. The solution of e.g. $P(89654_\tau)$ has to be evaluated in a recursive manner. Fortunately, it can be shown that the solution can be rearranged to

$$P(K_\tau) = \frac{(a\tau)^K}{K!} e^{-a\tau}. \tag{3.109}$$

This distribution function is referred to as the *Poisson distribution*. Apparently, shot noise does not exhibit a normal distribution. However, according to the central limit theorems, the Poisson distribution will converge into a normal distribution when K is taken to infinity. The moments of the Poisson distribution can be calculated by use of the characteristic function. It follows

$$\mathbb{E}(K_\tau) = a\tau \quad \text{and} \quad \text{var}(K_\tau) = a\tau. \tag{3.110}$$

This is the most important characteristic of the Poisson distribution. Its mean corresponds to its variance. Furthermore, the entire distribution is exactly determined by the first moment. Observing the expectation of the distribution, the meaning of a can be understood – a equals the mean emission rate of carriers.

In the following text the autocorrelation function will be derived. It is assumed that the macroscopic current $i(t)$ can be expressed as the sum of K single independent pulses $i_e(t - t_k)$ at independent time instances t_k:

$$i(t) = \sum_{k=1}^{K} i_e(t - t_k). \tag{3.111}$$

Thus the current is a superposition of $K + 1$ random variables namely the K emission time instances t_k and the actual number K of emissions per time interval. For the deduction of the autocorrelation function, $\mathbb{E}(i(t)i(t + \tau))$ has to be evaluated. However, due to the complexity of the derivation and to the fact that its understanding is not important for the followings text the derivation is given in the Appendix A. There it is shown that the result of the autocorrelation function is

$$R_{ii}(\tau) = \mathbb{E}(i(t))^2 + <n> \int_{-\infty}^{\infty} i_e(t - t_k)i_e(t + \tau - t_k)dt. \tag{3.112}$$

with $<n>$ being the mean rate of emitted charge carriers per time interval. The above given equation consists of the power that is actually associated with the random process under observation and the additive power provoked from the deterministic DC current. Since we are only interested in the former process, the equation is rearranged to

$$R_{i_{AC}i_{AC}}(\tau) = \mathbb{E}((i(t) - \mathbb{E}(i(t)))(i(t + \tau) - \mathbb{E}(i(t + \tau)))). \tag{3.113}$$

From the assumption of having a stationary random process, it follows $\mathbb{E}(i(t)) = \mathbb{E}(i(t + \tau))$ which together with the linearity of the expectation yields

$$\begin{aligned} R_{i_{AC}i_{AC}}(\tau) &= \mathbb{E}((i(t) - \mathbb{E}(i(t)))(i(t + \tau) - \mathbb{E}(i(t)))) \\ &= \mathbb{E}(i(t)i(t + \tau)) - \mathbb{E}(i(t))^2 - \mathbb{E}(i(t + \tau))\mathbb{E}(i(t)) + \mathbb{E}(i(t))^2 \\ &= R_{ii} - \mathbb{E}(i(t))^2. \end{aligned} \tag{3.114}$$

For the AC component, this means

$$R_{i_{AC}i_{AC}}(\tau) = <n> \int_{-\infty}^{\infty} i_e(t)i_e(t + \tau)dt = <n> R_{i_e i_e}. \tag{3.115}$$

This is also known as the second part of the *Campbell theorem*. If the shape of an influenced current pulse is known, the autocorrelation function would be determined. The estimation of such a pulse, however, may be a very complicated task. For the sake of simplicity, it is often assumed that the shape is rectangular (e.g. [Mü90], [BS97]) which can be understood of having charge particles that instantaneously reach some kind of saturation velocity. Alternatively one may even assume pulses of the shape of a Dirac distribution ([Czy10]). Taking the Fourier transform of the general case:

$$S_{i_{AC}i_{AC}}(\nu) = \int_{-\infty}^{\infty} <n> \int_{\infty}^{\infty} i_e(t)i_e(t + \tau)dt e^{-j2\pi\nu\tau}d\tau. \tag{3.116}$$

and substituting $t' = t + \tau$ it follows

$$S_{i_{AC}i_{AC}}(\nu) = <n> \int_{-\infty}^{\infty} \int_{-\infty}^{\infty} i_e(t)i_e(t')dt \, e^{-j2\pi\nu(t'-t)}dt' \tag{3.117}$$

$$= <n> \int_{-\infty}^{\infty} i_e(t) \, e^{j2\pi\nu t}dt \int_{-\infty}^{\infty} i_e(t') \, e^{-j2\pi\nu t'}dt' \tag{3.118}$$

$$= <n> \left| (\mathscr{F}i_e(t)) \right|^2. \tag{3.119}$$

This relation is known as the *Carson theorem*. For low frequencies this can be simplified to

$$S_{i_{AC}i_{AC}}(\nu) = <n> \left| \int_{-\infty}^{\infty} i_e(t) \cdot 1 \cdot dt \right|^2 \tag{3.120}$$

$$= <n>q^2 \tag{3.121}$$

what can be rearranged to

$$S_{i_{AC}i_{AC}}(\nu) = q\mathbb{E}(i(t)) = qI_{DC}. \tag{3.122}$$

where the substitution $\mathbb{E}(i(t)) = <n>q$ is used. This well known relation is named *Schottky relation*. Until which frequency this simplification holds true depends on the details of the structure and its applied bias. In first order it can be assumed that the relation holds for frequencies that are low compared to the reciprocal transit time of charge carriers through the potential barrier. For this range shot noise can be considered as a white noise process. In case the single sided Fourier transform has to be used for calculation, the power spectral density is expressed as

$$S_{i_{AC}i_{AC}}(\nu) = 2qI_{DC}, \quad \nu \in [0|\infty). \tag{3.123}$$

3.3.2.1 Photon noise

There is a strong similarity between the made assumptions used for the derivation of shot noise provoked by the quantization of current flux and the noise that is observed in light sensitive devices due to the quantization of light flux. The elementary particle of the light flux is called *photon*, thus the noise process provoked by impinging photons at a detector is referred to as *photon noise*. This phenomenon is the ultimate limit for radiation detectors and is one of the major contributors to the uncertainty in range image sensors as derived in Chapter 6. Typically the random error due to the quantization of light can be properly modeled by a shot noise or Poisson process. At long wavelengths and high temperatures larger noise values than predicted by Poisson statistics are observed which are associated with *Bose-Einstein statistics* (c.f. [Jan07]). At 300 K, however, this phenomenon is negligible for wavelengths above 1.1 μm. This wavelength corresponds to a photon energy in the vicinity to the bandgap of silicon. Thus it is assumed that this phenomenon is negligible for the design in this work. At higher photon energy a photon has a probability of generating more than one electron, resulting an increased noise level. This phenomenon is referred to as *Fano Noise* (c.f. [Jan07]). Investigations on this topic, however, are beyond the scope of this work.

Interestingly, not only the impinging photon flux is typically modeled by a Poisson process, but the photogenerated charge carrier counts as well. Considering a constant, deterministic photon flux the random process of generating an electron-hole pair can be modeled by a Bernoulli process ([ST07]). Thus there is a probability P_{gen} of generating a charge carrier and a probability $1 - P_{gen}$ of not generating it. Assuming n independent trials for each impinging photon, the probability of generating m charge carriers is known to yield the binomial distribution (c.f. [ST07; Web92; PU02]:

$$P(m) = \binom{n}{m} P_{\text{gen}}^m (1 - P_{\text{gen}})^{n-m}. \tag{3.124}$$

Since the photon flux itself, however, is not deterministic but distributed according to Poisson statistics, the conditional probabilities of generating an electron in case a photon is arrived, has to be evaluated ([ST07]):

$$P(m) = \sum_n P(m|n)P(n) \tag{3.125}$$

$$= \sum_{n=m}^{\infty} \binom{n}{m} P_{\text{gen}}^m (1 - P_{\text{gen}})^{n-m} \frac{\mathbb{E}(n)^n \exp(-\mathbb{E}(n))}{n!} \tag{3.126}$$

$$= \sum_{n=m}^{\infty} \frac{n!}{m!(n-m)!} P_{\text{gen}}^m (1 - P_{\text{gen}})^{n-m} \frac{\mathbb{E}(n)^n \exp(-\mathbb{E}(n))}{n!} \tag{3.127}$$

$$= \left(\frac{P_{\text{gen}}}{1 - P_{\text{gen}}}\right)^m \frac{\exp(-\mathbb{E}(n))}{m!} \sum_{n=m}^{\infty} \frac{\mathbb{E}(n)^n (1 - P_{\text{gen}})^n}{(n-m)!} \tag{3.128}$$

Evaluating the sum yields

$$P(m) = \left(\frac{P_{\text{gen}}}{1 - P_{\text{gen}}}\right)^m \frac{\exp(-\mathbb{E}(n))}{m!} \left(\mathbb{E}(n)^m (1 - P_{\text{gen}})^m + \mathbb{E}(n)^{m+1} (1 - P_{\text{gen}})^{m+1}\right.$$
$$\left. + \frac{\mathbb{E}(n)^{m+2} (1 - P_{\text{gen}})^{m+2}}{2} \cdots\right). \tag{3.129}$$

The sum can be expressed as a factor of $\mathbb{E}(n)^m (1 - P_{\text{gen}})^m$ being multiplied to an exponential function:

$$P(m) = \left(\frac{P_{\text{gen}}}{1 - P_{\text{gen}}}\right)^m \frac{\exp(-\mathbb{E}(n))}{m!} \mathbb{E}(n)^m (1 - P_{\text{gen}})^m \exp(\mathbb{E}(n)(1 - P_{\text{gen}})) \tag{3.130}$$

$$= \frac{\mathbb{E}(m)^m}{m!} \exp(-\mathbb{E}(m)) \tag{3.131}$$

which is, again, a Poisson process. Here $\mathbb{E}(m) = P_{\text{gen}} \mathbb{E}(n)$ was used from what the actual true meaning of P_{gen} becomes clear. It can be identified as the quantum efficiency.

3.3.3 Remarks on thermal noise

The derivation of the Nyquist formula is founded on the equipartition theorem and thus based on thermal equilibrium. Since this assumption is immediately violated once an electric field is applied, it is getting important to get to know when the estimations that can be made by the model are becoming too erroneous. For this thought experiment the velocity distribution has to be studied. In thermal equilibrium

$$\frac{1}{2}m^*\mathbb{E}(\mathbf{v}^2) = \frac{3}{2}k_B\theta, \tag{3.132}$$

applies, where m^* is introduced as the conductivity mass – or the effective mass, which results from the mass tensor. In silicon it can be estimated to $1.08 \cdot m_{e^-}$ which results in 9.84×10^{-31} kg ([Mü95]). The number of degrees of freedom equals to three, since the charge carriers can freely move in the three dimensional space. From this relation the so-called *thermal velocity* can be calculated to

$$v_{therm} = \sqrt{\frac{3k_B\theta}{m^*}}. \tag{3.133}$$

For $\theta = 300$ K this is approximately 112 km s^{-1} for electrons within silicon. The average velocity $\mathbb{E}(\mathbf{v})$, of course, equals zero. In the first order one could make the pragmatic assumption that the Nyquist relation yields meaningful results for the case of having a *drift velocity* which is negligible compared to the diffusion related thermal velocity – thus, when $v_{therm} \gg v_{drift}$ holds true. v_{drift} – in the first order – can be estimated with a linear relation with respect to the applied electrical field:

$$\mathbb{E}(v_{drift}) = \pm\mu_{p/n}E. \tag{3.134}$$

Here, the coefficient $\mu_{p/n}$ is the mobility which typically equals ≈ 1400 cm^2/Vs for electrons in silicon or ≈ 450 cm^2/Vs for holes. However, these are rather rough assumptions. Actually the relation between velocity of carriers in semiconductors and the applied electric field is non-linear which typically is expressed by introduction of a field dependent mobility (c.f. [SN07]). A general relation should converge into the Schottky relation for $v_{drift} >> v_{therm}$ and into the Nyquist relation for $v_{drift} << v_{therm}$, respectively. Often, a very pragmatic way is chosen in literature to take the increase in noise due to the application of an electric field into account – a new definition of a temperature for the charge carriers is introduced which deviates from the room temperature (c.f. [Mü90], [Blu96]). The so defined temperature is no longer a linear measure of the kinetic energy, but a more complex measure for the increase above thermal noise due to the acceleration of charge particles. In literature, here, the term *hot electrons* is introduced. Originated from the electric field, these expose such a high kinetic energy that they are not able to exchange it to the lattice by collisions. A physically motivated approach for the derivation of a formula that relates the new defined temperature T_e to the increase in noise level assumes an approximately isotropic distribution of the velocity and yields a quadratic relation between T_e and the applied electric field ([Mü90], [HP91]). Since this assumption, however, cannot cover the entire range from thermal to shot noise, the presentation of the derivation is omitted. A general derivation could be supported by the Monte-Carlo method and must be based on the velocity distribution in the six-dimensional phase space under consideration of the applied electrical field and the band-structure.

3.3.4   Generation-recombination noise

Until now it was always assumed that generation and recombination processes do not contribute to the noise performance. For the derivation of a model the physics of these processes have to be studied. They can be modeled by the *Master equations* ([Mü90; van70; van86]). Master equations are typically differential equations that model the transients of a system which can adopt one out of a finite amount of states. This approach, however, is beyond the scope of this work. A simpler but less rigorous approach is based on the *Langevin method* (c.f. Equation Section 3.2.5.2, [van70], [van86], [Blu96]). Except for the derivation in [Blu96] which assumes isotropic current density distributions, current flow is not considered in the derivations. In the first order, it can be assumed that those effects lead to a power spectral density component which can simply be added to others. The derivation which employs the Langevin method assumes the simplest model of generation/recombination – the direct recombination process that can be expressed as a differential equation of the relaxation type:

$$\frac{\mathrm{d}n}{\mathrm{d}t} = \frac{-n}{\tau_r} + \xi(t), \tag{3.135}$$

where n corresponds to the electron density, τ_r is the mean recombination rate and ξ models a white noise process. This process is introduced into the differential equation as a noise source that is associated to the generation/recombination fluctuations. This approach is similar to that which was used for the derivation of the thermal noise. In first order the process is assumed to be white, what corresponds to the assumption that the frequency dependency is defined by the actual system – here, the relaxation process described by the differential equation. To develop a power spectral density from the above equation, Fourier transform is applied and the equation is rearranged to[13]

$$\mathcal{F}(n)(v) = \frac{\mathcal{F}(\xi)(v)\tau_r}{1 + j2\pi v\tau_r}. \tag{3.136}$$

The application of the Wiener-Khintchine theorem now yields

$$S_{nn}(v) = \frac{\tau_r S_{\xi\xi}(0)}{1 + (2\pi v\tau_r)^2}, \tag{3.137}$$

where it was used that the ideal white noise process $\xi(t)$ has a flat power spectrum. Noise power spectra with this single-pole characteristic are known as *Laurenzian spectra*. Integration and rearrangement of the above relation yields a value for $S_{\xi\xi}(0)$, so

$$\mathbb{E}(n^2) = \int_0^\infty S_{nn}(v)\mathrm{d}v = \frac{\tau_r S_{\xi\xi}(0)}{4}, \tag{3.138}$$

---

[13] As pointed out in Section 3.2.5 this is not rigorous from a mathematical point of view since the Fourier transform of a white noise process does not actually exist. However, assuming that the process $\xi(t)$ has a bandwidth-limitation that is much higher than the time constants defined by the SDE but still finite, this problem can be circumvented.

so that the resulting single-sided power spectral density for generation/recombination noise becomes

$$S_{nn}(\nu) = \frac{4\tau_r \mathbb{E}(n^2)}{1 + (2\pi\nu)^2 \tau_r^2}. \tag{3.139}$$

The interesting result of the above relation is that the power spectral density became a function of the variance of the random process. This seems to be non-satisfactory, since it cannot readily be expressed as the function of external deterministic relations. However, the underlying system basically hardly describes a real system. Nevertheless, the variance will become a constant value as long as the random process is stationary and ergodic. Then it simply becomes a fitting constant, which can be used for modeling. Stationarity and ergodicity were though already assumed when the Wiener-Khintchine theorem was applied. Furthermore, it is important to note that the above derivation rarely demonstrates any physical characteristics of the process. The frequency dependency was immediately determined by assuming a relaxation type differential equation in combination with a white noise process.

An alternative relation which is neither based on the Langevin method nor on the Master equations assumes ideal rectangular current pulses and Poisson distributed recombination times ([BS97]). Unfortunately, the derivation is erroneous since during the calculation of the expectation, the proposed solution of the integral only follows for $4\tau_r^2 \gg \frac{1}{2(\pi\nu)^2}$. This, however, is not necessary in the other derivations and moreover would not be applicable for further derivations for e.g. the flicker noise (c.f. Section 3.3.6) where especially the low-frequency behavior has to be studied. Without the assumption of having $4\tau_r^2 \gg \frac{1}{2(\pi\nu)^2}$, the derivation from [BS97] would result in an additional term that follows a $\frac{1}{\nu^2}$ characteristic. Fortunately, [BS97] also presents another approach based on the Langevin method that adds another term $\frac{1}{\tau_g}(N_D - n)$ to the SDE 3.135, that corresponds to a generation term with an additional time constant τ_g and uses the donor concentration N_D. Solving that SDE leads to the widely employed solution

$$S_{ii}(\nu) = I_{DC}^2 \frac{N_D - n}{N_D n} \cdot \frac{4\tau}{1 + (2\pi\nu)^2 \tau^2} \tag{3.140}$$

where I_{DC} corresponds to the DC current and τ is an effective time constant.

3.3.5 Random telegraph signal noise – burst noise

Random telegraph signal noise (RTS noise), *random telegraph noise* (RTN) or *Burst noise* are processes which are found in semiconductor devices of all types. This has been observed in simple p/n-junctions, bipolar- and MOS-transistors (c.f. e.g. [MC93]) as well as in photodetectors. The latter will be more thoroughly examined in Section 5.2.4. RTS noise is furthermore becoming a more severe topic as scaling of devices is undertaken.

As for the flicker noise phenomenon described in the next section, there is no satisfactory general understanding of RTS noise. What is often believed, is that the entire conductivity of a device is affected by a random process. For instance, this

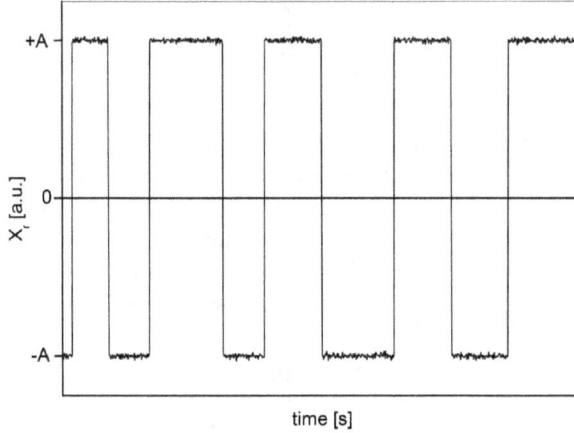

Figure 3.7 Random telegraph signal.

could be caused by a trapping process of a charge particle in a device that is formed
at the intersection of e.g. silicon to silicon-dioxide. This trapping process could then
turn entire surface paths on or off so that conductivity is heavily affected, resulting in a
random modulation of the current in form of a random telegraph signal as it is depicted
in Figure 3.7. There, the case of having two distinct levels is shown. Devices that
exhibited more discrete states were also observed. Assuming there is no preferred time
instance, the probability of a transition into another state follows a Poisson distribution
(c.f. Section 3.3.2). This phenomenon was also verified by several experimental studies
(c.f. [MC93]). More interesting though is the observation, that the average transition
rates that are associated to the time intervals spend in each of the two signal levels may
heavily deviate (c.f. [MC93]). Assuming a stationary, ergodic, symmetrical RTS noise
process $X_r(t)$ which can be described with one mean transition rate a, the derivation of
the power spectral density becomes relatively simple. The product $X_r(t)X_r(t + \tau)$ equals
$+A^2$ for having an even number and $-A^2$ for having an odd number of transitions
in τ. The probability of having K transitions within τ can be calculated by $P(K_\tau) =$
$[(a\tau)^K/K!]\exp(-a\tau)$. The calculation of the autocorrelation function is given by

$$\mathbb{E}(X_r(t)X_r(t + \tau)) = A^2 P(K_\tau = \text{even}) - A^2 P(K_\tau = \text{odd}) \tag{3.141}$$

$$= A^2 \left(\sum_{K=\text{even}} \frac{(a|\tau|)^K}{K!} e^{-a|\tau|} \right) - A^2 \left(\sum_{K=\text{odd}} \frac{(a|\tau|)^K}{K!} e^{-a|\tau|} \right) \tag{3.142}$$

$$= A^2 \sum_{K=0}^{\infty} (-1)^K \frac{(a|\tau|)^K}{K!} e^{-a|\tau|} \tag{3.143}$$

$$= A^2 e^{-a|\tau|} \sum_{K=0}^{\infty} \frac{(-a|\tau|)^K}{K!} \tag{3.144}$$

$$= A^2 \exp(-2a|\tau|), \tag{3.145}$$

where in the last rearrangement the Taylor series of the exponential function was used. Taking the Fourier transform, the power spectrum is yielded to

$$S_{XX}(\nu) = \frac{2A^2/a}{1 + (2\pi\nu)^2/4a^2}.$$

(3.146)

As previously mentioned, semiconductor devices often exhibit largely deviating time constants for the turn-on and turn-off processes. As proven in [Mac54], this, however, leads only to a scaling in the power spectrum – the shape remains the same. With the assumptions, that are used throughout this chapter there is a significant remark to be made – there is no visible difference in the shape of the power spectral density for generation/recombination noise, RTS noise (with the before-mentioned assumptions) or low-pass filtered white noise.

3.3.6   Flicker noise

In nearly every electronic component an additional noise source becomes dominant for low frequencies. Its power spectral density is described by

$$S_{ii}(\nu) \sim \frac{1}{\nu^{\gamma_{\text{flicker}}}}.$$

(3.147)

Since this noise process is ubiquitous, the attempt to explain it by one single phenomenon appears naive. Several different aspects and possible origins are discussed thoroughly in [Buc83]. The presented results are shortly summarized here.

For this random process values from 0.8 to 2 for γ_{flicker} were observed ([Buc83]). Since most often the process exhibits $\gamma_{\text{flicker}} = 1$ – and because frequency is often expressed as f – this phenomenon is referred to as *1/f – noise*. Alternatively, it is also known as *pink noise or flicker noise* for $\gamma_{\text{flicker}} = 2$ it is also referred to as *red noise*[14]. An ideal stationary, ergodic $1/f$ noise process would lead to divergence in the power, since the integral of such a process leads to a relation

$$L_{X_r} \propto \lim_{\nu_{\text{lower}} \to 0} \lim_{\nu_{\text{upper}} \to \infty} \ln\left(\frac{\nu_{\text{upper}}}{\nu_{\text{lower}}}\right).$$

(3.148)

Interestingly – and this is exactly the reason for the importance of the problem – each frequency decade holds the same noise power. For low pass filtered white noise processes this is totally different. There, mainly the higher frequency components contribute to the noise power. In Equation 3.148 both limits will separately lead to divergence. The limit of taking ν_{upper} to infinity is inherently resolved by any physical system, since the bandwidth is always limited. Taking ν_{lower} to zero, on the contrary is not so simply treated. To surpass this problem, one may arbitrarily assume a value for the lower boundary of the integral. Often this picked to 10 Hz. Observations of flicker noise are known, that do not demonstrate any low frequency corner at which the shape

---

[14]The color related naming is of course related to the visible portion of the electromagnetic spectrum. Here, when all frequency components are equally available a spectra is called *white*, whereas it is called *pink* or *red* if more low frequency components are present.

alters from the $1/v$ behavior of measurements as long as half a year [Buc83]. Flinn made the assumption, that an upper corner frequency could be found at 1×10^{23} Hz which corresponds to the time a photon needs to transit the classical radius of an electron (c.f. [Buc83]). A lower limit was chosen in accordance to an estimation of the lifetime of the universe to 10×10^{-17} Hz. In general this seems rather non-satisfactory. Of course one may argue that the time interval in which a measurement is being done is always limited.

The divergence of flicker noise leads to the question if it is actually a stationary process. From the theory on *Chaos and Fractals* it is known, that mathematical non-stationary random processes exist that can exhibit a power spectral density according to Equation 3.147[15]. Here, of course the Wiener Khintchine theorem is violated, so that the interpretation of Equation 3.49 as the noise power fails because it was derived for stationary and ergodic processes only (c.f. Section 3.1.2). Such non-stationary noise processes though could resolve the theoretical divergence but their implementation in compact device models would be very difficult/impractical – if not impossible, since many physically equivalent devices have to be studied simultaneously to yield the time-variant autocorrelation functions. Even worse is that the term time-dependent implies that one cannot in general make sure that any measurements will correspond to any future measurements. The stationary models, however, are widely accepted and embedded in many circuit simulators which leads to the assumption that – even if the underlying processes are non-stationary – their impact on estimation of the probabilistic behavior due to improper treatment is negligible.

Several trials for explanations of the $1/v$ behavior were also undertaken that are founded on stationary processes. As described in Section 3.3.4, it is possible to yield such a characteristic if shot noise pulses of a certain shape in combination with generation/recombination noise are assumed. However, it is difficult to motivate the time constraints as they become necessary to explain the very low frequency range. Another trial for a derivation was presented by Van der Ziel who assumed that the polarization losses of a capacitor would exhibit thermal noise. For low frequencies he assumed that the ratio ϵ''/ϵ' is constant, where $\epsilon = \epsilon' - j\epsilon''$ describes the complex frequency dependence of the permittivity. Van der Ziel yielded the $1/v$ shape with these assumptions but since polarization processes are actually modeling the electric dipoles that can follow an applied electrical field only finitely fast whereas thermal noise is derived as a random process that is originated by the mutual collisions of charge particles, this model does not seem properly physically motivated.

When applying a constant bias current to a specimen, $1/v$ noise can be observed. This can be understood as having fluctuations in the resistance. Possible origins for

---

[15]Benoit Mandelbrot (1924–2010), well known as the founder of the *fractal geometry of nature*, showed that the non-stationary random process known as *Brownian motion/Wiener process* or variants can be used to yield power spectra of the form $1/v^{\gamma_{MB}}$, $\gamma_{MB} \in (0,3)$[Man67]. Loosening the definition of the Brownian motion one can define a so-called *fractal Brownian motion* that is described by a *Hurst exponent* $H \in (0,1)$ which impacts the *roughness* or *fractal dimension* of the random process. It can be shown, that this Hurst exponent also directly determines the power spectra in form of $S(v) \propto 1/v^{1+2H}$, so that actual flicker noise would correspond to $H \to 0$ [HS93; Add97; Fal03].

that can be for instance fluctuations in the charge carrier density or in the mobility. There are theories that are based on either and both of these phenomena.

Hooge empirically found a relation in 1969 which relates the charge carrier density to the power spectral density:

$$\frac{S_r(v)}{R_0^2} = \frac{\alpha_{\text{Hooge}}}{N_{\text{tot}} v}, \tag{3.149}$$

where R_0 is the mean resistance, N_{tot} is the total number of charge carriers and α_{Hooge} was proposed at a material independent constant of approximately 2×10^{-3}. This equation describes the flicker noise as a bulk phenomenon. Later observations on thin film metal resistors demonstrated values for α_{Hooge} that deviated from the original value by more than 1000. Moreover, a strong temperature dependency of α_{Hooge} was observed for metal resistors which was not explained until the publication date of [Buc83].

An alternative model that is founded on surface effects rather than bulk phenomena was developed by McWorther. The presented derivation from [McW55] can also be found in any good noise textbook (e.g. [Buc83; BS97; Blu96]). McWorther assumed that there is a homogeneously distributed impurity concentration at the semiconductor-oxide interface that causes fluctuations in the current flow along the surface. The impurities – also referred to as traps – are basically states within the bandgap of the semiconductor that can cause generation and recombination. As described in Section 3.3.4, the generation/recombination is modeled as

$$S_{\text{GRN}}(v) = \frac{4\tau_x \mathbb{E}(n^2)}{1 + (2\pi v)^2 \tau_x^2}. \tag{3.150}$$

Here, τ_x is assumed to be a random process which corresponds to the homogeneously distributed traps within the silicon-silicon dioxide interface. The tunneling effect is used in the McWorthers model to relate the depth of a trap w_x to the effective recombination time τ_x:

$$\tau_x = \tau_0 e^{\gamma_{\text{tunnel}} w_x} \iff w_x = \frac{1}{\gamma_{\text{tunnel}}} \ln\left(\frac{\tau_x}{\tau_0}\right), \tag{3.151}$$

where τ_0 and γ_{tunnel} are introduced as fitting parameters. According to McWorther, γ_{tunnel} equals approximately 1×10^8 cm^{-1} and τ_0 is not of much importance for the end result. If the traps are evenly distributed within the interval $[w_1, w_2]$ the corresponding time constants are within $[\tau_1, \tau_2]$. The probability distribution function $f_{\text{p}-w_r}(w_x)$ is

$$f_{\text{p}-w_r}(w_x) = \frac{1}{w_2 - w_1} = \frac{\gamma}{\ln\left(\frac{\tau_2}{\tau_1}\right)} \tag{3.152}$$

which results in a probability

$$P(w_r(\omega) \leq w_x) = F_{w_r}(w_x) = \frac{w_x - w_1}{w_2 - w_1} \tag{3.153}$$

of observing a trap in the interval $[w_1, w_x]$. Thus, the probability of observing a recombination time within the interval $[\tau_1, \tau_x]$ can be calculated by

$$P(\tau_r(\omega) \leq \tau_x) = P\left(w_r(\omega) \leq \frac{1}{\gamma_{tunnel}} \ln\left(\frac{\tau_x}{\tau_0}\right)\right) \tag{3.154}$$

$$= F_{w_r}\left(\frac{1}{\gamma_{tunnel}} \ln\left(\frac{\tau_x}{\tau_0}\right)\right). \tag{3.155}$$

Differentiating yields

$$f_{p-\tau_r}(\tau_x) = \frac{1}{\tau_0} \frac{\tau_0}{\gamma_{tunnel}\tau_x} f_{p-w_r}\left(\frac{1}{\gamma_{tunnel}} \ln\left(\frac{\tau_x}{\tau_0}\right)\right) = \frac{\frac{1}{\tau_x}}{\ln\left(\frac{\tau_2}{\tau_1}\right)}. \tag{3.156}$$

The expectation of the power spectral density can thus be evaluated to

$$\mathbb{E}(S_{GRN}(\nu)) = 4\mathbb{E}(n^2) \int_0^\infty \frac{\tau_x f_{p-\tau_r}(\tau_x) d\tau_x}{1+(2\pi\nu)^2\tau_x^2} \tag{3.157}$$

$$= \frac{4\mathbb{E}(n^2)}{\ln\left(\frac{\tau_2}{\tau_1}\right)} \int_{\tau_1}^{\tau_2} \frac{d\tau_x}{1+(2\pi\nu)^2\tau_x^2}. \tag{3.158}$$

Solving the integral yields

$$\mathbb{E}(S_{GRN}(\nu)) = \frac{4\mathbb{E}(n^2)}{\ln\left(\frac{\tau_2}{\tau_1}\right)} \cdot \frac{\tan^{-1}(2\pi\nu\tau_2) - \tan^{-1}(2\pi\nu\tau_1)}{2\pi\nu}. \tag{3.159}$$

For the frequency range of $\frac{1}{2\pi\tau_2} \ll \nu \ll \frac{1}{2\pi\tau_1}$, an approximately $1/\nu$-shaped power spectrum is yielded. This model is widely accepted throughout the literature which is mainly caused by the immense amount of observations of flicker noise from components that are designed at the interface of e.g. silicon-to-silicon dioxide. These exhibit much higher noise than comparable devices that separate the current paths from these intersections. This, for instance, is one major reason why many *buried* devices are widely employed.

Based on the McWorther and the Hooge model, many developments of unified models have been undertaken (c.f. e.g. [Van78; Van88;HKH90;VV00; FXL02]). These try to take into account both phenomena – changes in the number of particles caused by the generation/recombination effects due to traps and changes in the mobility due to scattering. Their presentation, however, is beyond the scope of this work.

Apart from the power spectral density, the probability distribution function of the magnitudes of the flicker noise phenomenon might be of interest. Several experiments have been undertaken – e.g. by Bell and Hooge – which verified an approximate normal distribution within the frequency range of $40\,\text{Hz} < \nu < 100\,\text{kHz}$. But there were also experiments undertaken that observed deviations from that – e.g. by Dissanayake ([Bel85; Buc83]).

Concluding this survey, it is to say $1/f$ – noise is not a well understood phenomenon. But as pointed out in the introductory part this may not be possible at all due to the ubiquity of the flicker noise process.

3.4   NOISE PROCESSES UNDER TIME-VARYING BIAS

The presented noise process models in this chapter all assume stationarity and ergodicity. In many applications, however, the biasing conditions of electronic devices are varied during operation. Thus, time-variant models of the noise processes become necessary to enable optimization of time-varying circuitry. A comprehensive discussion on this topic is presented in [DSV00].

As has been pointed out in Section 3.3.3, the Nyquist relation as a model for thermal noise is derived at thermal equilibrium and for linear resistors. However, often the model is used to derive quantities at non-equilibrium. This is for instance done to model the noise performance of MOSFETs as it is described in Section 4.1. However, one has to be very careful when extrapolating from the range in which a model is verified to the unknown. Studies found on the very general stochastic differential equations demonstrate that e.g. Gaussian thermal noise is not applicable to non-linear resistances because it violates the laws of thermodynamics [WC99]. One simply but actually very non-rigorous assumption is to employ the Nyquist relation if the time-varying bias is slow so that "quasi thermal equilibrium" is achieved and that the random fluctuations are small compared to large-signal excitation of the device [DSV00]. If these assumptions are justified, one can employ

$$S_{v_n v_n}(t, v) = 4k_B \theta R(t) \tag{3.160}$$

as a time-varying power spectral density. Similarly, coming from the Poisson process as the integral of the shot noise process, one can derive a general autocorrelation function which yields

$$S_{i_n i_n}(t, v) = qI(t) \tag{3.161}$$

if the variations in the current are slow compared to the width of an induced current pulse (c.f. e.g. [DSV00]). These assumptions are possible, because thermal and shot noise are usually modeled as white noise processes – which means that there is no correlations between two evaluations of a sample path at arbitrary time instances. For e.g. flicker noise this is not the case. Demir proposed several approaches for noise sources that yield time-variant power spectra which should correspond to actual measurements [DSV00]. During the discussion, he concludes that a noise source that consists of a white noise process $\xi(t)$ which is the input of an LTI system $H(v)$ which is then the input to a memoryless modulator $m(t)$ is capable of modeling frequency conversion as it is e.g. observed in mixers. Here, the LTI system is used to model the frequency dependence of the noise process so that, in general, this approach can be used to model any random process. The modulator $m(t)$ then takes the noise process and generates harmonics of the noise process at higher frequencies. Still, this has to be treated with caution – as the author mentioned it himself, since this concept has not yet been verified with measurements. The possible problem with this approach is that the power spectra that are usually observed in e.g. oscillators or mixers can also be provoked by the non-linearity of the circuit itself. What amount of the total noise is contributed by varying bias and which models have to be used is still a topic to be investigated.

3.5 IMPEDANCE FIELD METHOD

The *Impedance field method* is an approach to study the impact of a multitude of noise sources in complex electronic devices. Therefore, the segmentation of the devices into infinitesimally small portions that exhibit correspondingly infinitesimal small perturbations which are then linearly superimposed to yield the perturbations of the macroscopic quantities such as current or voltages at the device terminals is done. This method was introduced by Walter Shockley in 1966 [SCJ66] and found application in finite-element solvers recently (c.f. e.g. [BG01; Syn09a]). Though the later approaches slightly differ from the original such as they employ the Green's function approach instead of introducing an impedance field (c.f. Section 3.2).

The semiconductor equations are used to describe the physical processes within a device. Here, for instance the *drift-diffusion model* can be used which is composed of the Poisson and the continuity equations:

$$\text{div}[\varepsilon \cdot \text{grad}(\phi)] = q(n - p + N_A - N_D) \quad (3.162)$$

$$\text{div}[q\mu_n(V_T \,\text{grad}(n) - n \cdot \text{grad}(\phi))] - q\frac{\partial n}{\partial t} = qR_{\text{rec}} \quad (3.163)$$

$$-\text{div}[q\mu_p(V_T \,\text{grad}(p) + p \cdot \text{grad}(\phi))] + q\frac{\partial p}{\partial t} = -qR_{\text{rec}}, \quad (3.164)$$

where p is the hole concentration, n the electron concentration, μ_p is the mobility for holes and μ_n for electrons, respectively. $V_T = k_B\theta/q$ is called the thermal voltage, N_A is the acceptor and is N_D the donor concentration. These equations can be rearranged to

$$F(\phi, n, p, \dot{n}, \dot{p}) = 0 \quad (3.165)$$

$$b(\phi, n, p, \dot{n}, \dot{p}, s_e) = 0, \quad (3.166)$$

where s_e were introduced to represent electrical sources and $\dot{n} = \partial n/\partial t$ and $\dot{p} = \partial p/\partial t$ were used. The operator F corresponds to the actual Poisson and continuity equations and b to the boundary conditions. Alternative to the drift-diffusion model, the *hydrodynamic model* or the *Boltzmann transport equation* may be used (c.f. [JM12; HPJ11; Syn09a]). After the large-signal behavior of the above equations is solved the PDEs can be linearized according to the large-signal solution, so that a linear but time-varying and space dependent set of PDEs is yielded. Now the Green's function approach can be applied. Here, similarly as for the LPTV systems, a more-dimensional impulse response function is defined. The difference, though, is that additional components for the space coordinates are introduced. The Green's function $G_{y_s,x_s}(x, x_1, t, t_1)$ is introduced as the response function in the variable y_s to the variable x_s that exhibits $\delta(x_1 - x)\delta(t - t_1)$. Assuming stationary and ergodic random processes, the Fourier transform $H_{\mathscr{F}-y_s x_s}(x, x_1, \nu)$ of that response function can be calculated. Superimposing all infinitesimal noise sources $S_{s_{e-1}s_{e-2}}(x_1, x_2, \nu)$ which are attached to each finite element within the device volume V_{dev}, the total correlation spectral density can be stated as (c.f. [BG01])

$$S_{y_s,x_s}(x,x',v) = \sum_{\chi,\psi=\phi,n,p} \int_{V_{dev}} \int_{V_{dev}} H_{\mathscr{F}-y_s,\chi}(x,x_1,v) S_{s_\chi,s_\psi}(x',x_2,v)$$
$$\cdot H^*_{\mathscr{F}-x_s,\psi}(x',x_2,v) dx_1 dx_2. \tag{3.167}$$

The noise processes that are introduced in each finite element consist for instance of thermal or diffusion noise, generation-recombination noise, flicker noise and trapping or RTS noise [Syn09a]. Therefore, the space dependent variables such as current, number of particles or resistance are translated to the infinitesimal quantities that are attached to each finite element. Though the concept of the impedance field method in general allows for the study of the superposition of multitudes of random processes on a very fundamental level, one problem was not yet solved – the flicker noise phenomenon could not yet be described on a fundamental level. For modeling, still fitting parameters have to be introduced that do not allow for a proper physical interpretation [Bo11].

Comparison of the original impedance field method from Shockley with the Green's function approach shows that they are related to each other. Shockley defined an impedance field $Z_{y_s,x_s}(x,x_1,t,t_1)$ such that it yields the response at the node y_s when a current source injects a unit pulse $\delta x_s \delta(x-x_1)\delta(t-t_1)$. Thus, the impedance field can be expressed as

$$Z_{y_s,x_s}(x,x_1,t,t_1) = \frac{\kappa}{q} G_{y_s,x_s}(x,x_1,t,t_1), \quad x_s = n,p \tag{3.168}$$

where κ is introduced as the conductivity.

Chapter 4

Noise performance of devices available in the 0.35 μm CMOS process[1]

At the Fraunhofer Institute IMS in Duisburg a 2P4M 0.35 μm CMOS process was developed, which enables a large variability for designs e.g. of sensor systems. The process offers for instance several low-voltage (3.3 V) and high-voltage (≥5 V) enhancement and depletion n-type (NMOS) and p-type (PMOS) field-effect transistors (FETs), a vertical pnp and a vertical npn bipolar transistor, several types of diodes, resistors and capacitors, ESD-protection structures and floating-gate transistors. The applicability of the process for the design of opto-electronic devices was investigated in [Du09a] and [Sp10], whereas the latter reference puts more stress on the design possibilities for time-of-flight range imagers. This work intends to characterize the noise performance of available devices and aims to contribute to the foundation for high-performance sensors.

The former chapter gave a detailed overview of the mathematical foundation and basic physical principles, which have to be exploited for proper estimation of the noise performance in e.g. sensor systems. Complementary to these algorithms, proper models have to be developed that enable the prediction of noise exhibition of physical devices using given parameters like geometry, bias and environmental conditions such as temperature. This section thus describes basic models for the components as are available in the above 0.35 μm CMOS process, in which e.g. the time-of-flight sensors that are presented in this work were designed, fabricated and tested. If possible, standard models were preferably used since they are already implemented in standard tools such as SpectreRF® (c.f. [CDS11b])[2]. The results presented in this section were achieved using a low-frequency, low-noise noise measurement setup [Bro10], that adopted the basic principles from [JT07] (c.f. Appendix B). This setup enabled the characterization of the noise performance within a frequency range of approximately 0.1 Hz up to 1–10 kHz depending on the actual noise level, the DC bias and the small-signal output impedance of the device under test (DUT). A characterization of the high-frequency noise performance is beyond the scope of this work. Nevertheless, simply

---

[1]The author wants to acknowledge Xiang Li, Xueyin Chen, Ved Prakash, Andrey Kravchenko, Aliaksandr Andrasiuk and André Schmitz, who contributed to this section with measurements within the framework of theses [Li11] or internships.
[2]Some of the standard models provided by Spectre® are insufficient to properly model effects that were observed in measured noise power spectral densities. However, Spectre® provides the feature of defining a spectral density in form of a text-file. For bias dependent noise sources, models can be implemented by *Verilog-AMS* [Kun04; Sys09].

neglecting this phenomenon can yield a non-satisfying precision in estimates of the noise performance of systems. This especially holds true for systems that are non-linear and/or are driven by time-varying bias since noise folding may occur. Rough estimates of the high-frequency noise performance of certain devices may be made by adoption of "standard" values that have been published in literature. However, this is considered not to be reliable, because in general, several parasitic second-order effects can occur in e.g. MOSFETs that largely depend on the actual CMOS process (c.f. [TM10; EV06]). Since the basic operation of the standard devices are well understood and have been characterized already, the reader is referred to textbooks like [TM10; EV06; SN07].

Resistors are embedded in voltage references, voltage-controlled current sources that are employed in bias networks and are also part of the feedback of operational amplifiers to enable e.g. *Miller compensation*. They are available as metal or polysilicon structures or as diffused layers. These always introduce thermal noise or diffusion noise. Diffused resistors may also exhibit shot noise and generation-recombination noise due to the leakage current of the reverse-biased junctions. In general, resistors can also show flicker noise due to bad contacts and the vicinity to silicon-dioxide layers which corresponds to point defects (c.f. a throughout explanation of point defects occurring at Si-SiO_2-interfaces in [LN05]). The same applies for grain boundaries – especially in case of polysilicon resistors. Diodes are part of ESD-protection structures. In general, they suffer from shot noise or diffusion noise, generation-recombination noise and flicker noise (c.f. [Buc83]). Noise appearing in diodes is usually modeled by a noise current-source in parallel to the device. For pads that are driven by voltages, the current-noise becomes negligible due to the low-impedance of voltage sources and the small leakage current of the diodes that is monotonically linked to their noise spectral power densities. Noise originating from resistors and ESD-diodes is not in the focus of this work. In integrated analog circuitry diodes can be replaced by MOS transistors operating in diode-mode, what is employed within this work; eliminating the need for their characterization. The characterization of noise, that is introduced by transistors or photodiodes though, is crucial, since it defines the performance of the sensor as presented within the investigations of Chapters 1–3. Thus, throughout characterization and analyses are presented in the later text.

4.1 TRANSISTOR NOISE BASICS

In principle the noise that is exhibited by a transistor can be characterized by measurement of the output current fluctuations for different quiescent currents and environmental conditions such as room temperature. A schematic of this setup is depicted in Figure 4.1, in which, exemplary, an enhancement NMOS transistor is illustrated. Throughout the entire characterization the transistors were treated as three-terminal devices. For field-effect transistors the bulk was always short-circuited to the source during measurements. The high-precision current meter can for instance be realized by a low-noise transimpedance amplifier. Its input impedance should be negligible, so the bias voltage of the upper battery depicted in Figure 4.1 equals the drain-source voltage of the DUT. The amplified fluctuations can then be characterized for e.g. amplitude distribution, correlation or power spectral density. This work concentrates on the latter according to Equation 3.50. For low-frequency operation, the transistor can be

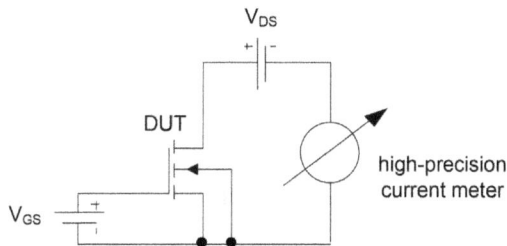

Figure 4.1 Schematic of the measurement principle for noise characterization in transistors.

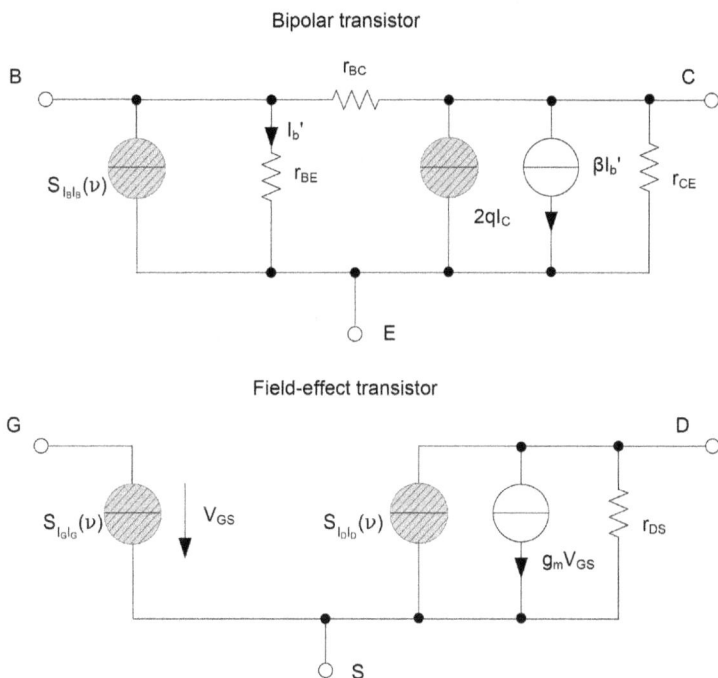

Figure 4.2 Low-frequency small-signal models with embedded noise sources for bipolar and field-effect transistors.

modeled by the standard small-signal models depicted in Figure 4.2. Here, the noise exhibited by series resistances at the drain, gate, source and collector, base, emitter, respectively, is neglected.

4.1.1 Bipolar transistor noise model

Noise in bipolar transistors can be modeled by the addition of a shot-noise source between the collector and emitter electrodes and a noise source $S_{I_B I_B}(\nu)$ between base

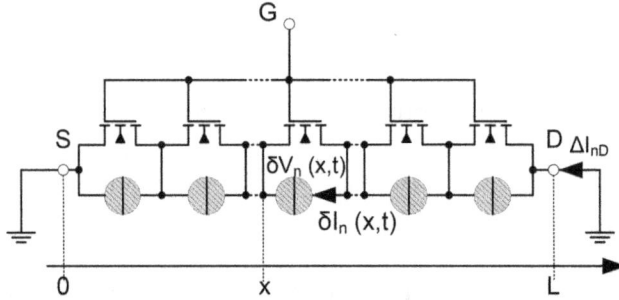

Figure 4.3 Schematic of a segmentation of the channel noise exhibition in field-effect transistors (according to [EV06]).

and emitter, which are assumed to be uncorrelated [CDS11b]. According to [CDS11b], the noise source $S_{I_B I_B}(\nu)$ comprises shot noise, flicker noise and burst noise:

$$S_{I_B I_B}(\nu) = 2qI_B + K_f \frac{I_B^{\alpha_f}}{\nu^{\gamma_{\text{flicker}}}} + \frac{K_b I_B}{1 + (\nu/\nu_c)^2}, \tag{4.1}$$

where I_B is the large-signal current into the base and K_f, K_b, α_f, γ_{flicker} and ν_c are parameters that are fitted to the noise exhibited by the device. In [CDS11b] $\gamma_{\text{flicker}} = 1$, nevertheless, in general this might deviate in real devices as described in Section 3.3.6.

4.1.2 Field-effect transistor noise modeling

The noise contributors in field-effect transistors are often modeled by a current noise source $S_{I_D I_D}$ between drain and source and a current noise source $S_{I_G I_G}$ between the gate and the source electrodes. $S_{I_D I_D}$ is often associated with flicker and thermal noise [CDS11b; EV06; TM10; MF73; MC93] – whereas thermal noise was originally derived for thermal equilibrium (c.f. Sections 3.3.1 and 3.3.3). In Figure 4.3 a segmentation of a transistor is depicted, in which infinitesimal perturbations $\delta I_n(x,t)$ of the current through the channel are modeled by parallel current sources and cause voltage fluctuations $\delta V_n(x,t)$ along the channel.

To derive macroscopic quantities for the channel noise appearing at the drain, the fluctuations are assumed to be small so that it can be assumed that the network is an LTI system. The frequencies for which predictions have to be done, are assumed to be small enough to neglect capacitive coupling through the gate of the MOSFET. Furthermore, the distributed noise sources are assumed to be uncorrelated [EV06]. By associating infinitesimal noise power spectral densities $S_{\delta I_n \delta I_n}(\nu)$ with the distributed perturbations, the total noise appearing at the drain can be evaluated by integration along the channel to

$$S_{I_D I_D}(\nu) = \int_0^L G_{\text{ch}}^2(x) \frac{S_{\delta I_n \delta I_n}(\nu)}{\Delta x} dx, \tag{4.2}$$

where $G_{\text{ch}}(x)$ is the channel conductance.

4.1.2.1 Thermal noise in MOSFET devices

For long-channel devices and the assumption of having $S_{\delta I_n \delta I_n}(\nu) = 4k_B\theta/\Delta R$, this simplifies to

$$S_{I_D I_B-\text{thermal}}(\nu) = 4k_B\theta G_{nD}, \tag{4.3}$$

with ΔR being the channel resistance per unit length at point x that exhibits thermal noise, the *thermal noise conductance at the drain* $G_{nD} := \mu \frac{W}{L^2} \int_0^L [-Q_i(x)]dx$ with μ being the mobility of the carriers, W and L being width and length of the device, respectively, and $Q_i(x)$ is the inversion charge of the transistor [EV06]. G_{nD} can also be related to the transconductance of the transistor $g_m := \partial I_D/\partial V_{DS}$ by

$$G_{nD} = g_m \cdot \frac{2}{3} \frac{\frac{3}{4}\left(1 + \frac{\sqrt{2qN_b\epsilon_{Si}}}{2C_{ox}\sqrt{\Psi_S}}\right) - \frac{Q_{is}}{2C_{ox}V_T}}{1 - \frac{Q_{is}}{2C_{ox}V_T\left[1 + \frac{\sqrt{2qN_b\epsilon_{Si}}}{2C_{ox}\sqrt{\Psi_S}}\right]}} \frac{Q_{is}}{Q_{is} - Q_{id}} = g_m\gamma_{nD}, \tag{4.4}$$

with Q_{is} and Q_{id} being the inversion charge at the source and drain electrodes, the gate oxide capacitance per unit area C_{ox}, the threshold voltage of the transistor V_T, the doping concentration of the substrate N_b, the electrical permittivity of silicon ϵ_{SI} and the surface potential Ψ_S [EV06]. The parameter γ_{nD} is referred to as the *thermal noise excess factor* related to the drain.

For short-channel devices second-order effects arise, so that one-dimensional modeling of the transistor is not accurate anymore. *Channel-length modulation, velocity saturation, mobility reduction* and *hot-carriers* are phenomena that have to be accounted for. This can be done by a more general definition of the excess factor – now simply referred to as γ_e [EV06; TM10], so that Equation 4.3 becomes

$$S_{I_D I_D-\text{thermal}}(\nu) = 4k_B\theta g_m\gamma_e. \tag{4.5}$$

For long channel devices γ_e converges to γ_{nD}. The standard Berkeley BSIM3v3 model assumes $\gamma_e = 2/3$ and thus does not take into account the complex relation between the channel noise and the applied voltages, the devices' geometry, doping concentrations etc. [CDS11b].

Many experiments have been undertaken in various processes, under varying geometry and biasing conditions. An overview over many of these observations is given in [DJ06]. Mostly, $\gamma_e \approx 0.67$–3.5 were observed. However, there is also one measurement stated that shows $\gamma_e = 8$. Spectre® provides more complex models for thermal noise. Nevertheless, these should be verified by characterization to enable reliable predictions. This especially holds true, since for the derivation of those formulas several assumptions and approximations are done and, in general, the theory about the thermal noise exhibited by short-channel devices is not yet fully resolved [TM10; ITR11b].

4.1.2.2 *Flicker noise in MOSFET devices*

Flicker noise, or 1/f-noise, is the dominant noise source in the low-frequency domain in MOSFETs [CDS11b; EV06; TM10; MF73; MC93; Buc83]. As described in Section 3.3.6, two major theories were developed to explain this phenomenon – the *McWorther model* and the *Hooge model*. The first one associates 1/f-noise to carrier number fluctuations caused by trapping centers in the vicinity of the Si-SiO2-interface of e.g. MOSFETs, while the second model relates the noise to mobility variations. These two models have been combined in *unifying flicker noise theories*, which for instance simply superimpose the two phenomena [EV06] or truly combine them by relating the number fluctuation of the McWorther model to *Coulomb scattering* which is known to cause mobility fluctuations [TM10]. Several derivations have been presented in the past, that aim to relate the geometry, doping profiles and applied bias to the exhibited noise power spectra. Nevertheless, largely differing results led to the necessity of the introduction of proportionality factors, that are known to be process-dependent (c.f. e.g. [CDS11b; TM10]) or sometimes also bias dependent (c.f. e.g. [EV06]). The *Level2 SPICE model* implements for instance [CDS11b; EEC; CH99]

$$S_{I_D I_D - \text{flicker}}(\nu) = \frac{K_f I_D^{\alpha_f}}{C_{ox} L_{eff}^2 \nu^{\gamma_{\text{flicker}}}} \qquad (4.6)$$

as the output related current noise spectral density, where K_f, α_f and γ_{flicker} are process dependent variables which are fitted to measurement results and L_{eff} is the effective channel length, which differs from the drawn length L due to process related reasons (c.f. [CH99]). Considering the transconductance of a transistor modeled by the *Level1 SPICE* model, represented by $g_m = \sqrt{2\mu C_{ox} I_D W/L}$ the flicker noise can be expressed as a gate-source referred voltage noise spectral density:

$$S_{V_{GS} V_{GS} - \text{flicker}}(\nu) = \frac{K_f I_D^{\alpha_f - 1}}{2\mu C_{ox}^2 WL \nu^{\gamma_{\text{flicker}}}}. \qquad (4.7)$$

Since the input referred noise density is inversely proportional to the gate area $W \cdot L$, it can clearly be seen that flicker noise becomes a major issue for small-size devices. Deviations from this have been reported for devices for which $S_{V_{GS} V_{GS} - \text{flicker}}(\nu) \propto L^{\beta}$ with $-1 \leq \beta \leq -1/2$ was observed [MC93]. C_{ox} scales inversely proportional with the gate oxide thickness, defining a design variable for future enhancement. Since the widely adopted McWorther model associates flicker-noise with multiple trapping processes which in average lead to the well-known 1/f-shape (c.f. Section 3.3.6), the actual flicker noise behavior gets lost if the device is significantly scaled down. The averaging process is often described by *Multi-Laurenzian spectra*, of which one describes a generation-recombination process in the frequency domain which can be expressed as $S(\nu) \propto 1/(1 + (\nu/\nu_c)^2)$. For submicron MOSFETs, however, there may only few trapping-centres be located at the gate, making the device demonstrate single bursts in the current flow which is referred to as *Burst noise, random-telegraph signal* (RTS) noise or simply *generation-recombination noise* (c.f. e.g. [Mil11]).

Contrary to Equation 4.6 several flicker noise models were employed in standard MOSFET models for circuit simulation based on SPICE or Spectre®. Some popular

Table 4.1 Overview – popular flicker noise models (extracted from [CDS11b]).

Type-1
$$S_{I_D I_D\text{-flicker}}(\nu) = \frac{K_f I_D^{\alpha_f}}{C_{ox} W_{eff} L_{eff} \nu^{\gamma_{flicker}}} \tag{4.8}$$

Type-2
$$S_{I_D I_D\text{-flicker}}(\nu) = \frac{K_f I_D^{\alpha_f}}{C_{ox} L_{eff}^2 \nu^{\gamma_{flicker}}} \tag{4.9}$$

Type-3
$$S_{I_D I_D\text{-flicker}}(\nu) = g\,(W, L, V_{DS}, V_{GS}) \cdot \frac{I_D}{C_{ox} L_{eff}^2 \nu^{\gamma_{flicker}}}$$
$$+ h(W, L, V_{DS}, V_{GS}) \cdot \frac{I_D}{W_{eff} L_{eff}^2 \nu^{\gamma_{flicker}}} \tag{4.10}$$

Type-4
$$S_{I_D I_D\text{-flicker}}(\nu) = \frac{K_f g_m^2}{C_{ox} W_{eff} L_{eff} \nu^{\gamma_{flicker}}} \tag{4.11}$$

Type-5
$$S_{I_D I_D\text{-flicker}}(\nu) \propto \frac{K_f g_m^{\alpha_f}}{C_{ox} W_{eff} L_{eff} \nu^{\gamma_{flicker}}} \tag{4.12}$$

models are listed in Table 4.1. The flicker noise model of Type-1 for instance is used in the *Level-1* to *Level-10* models and the EKV model[3] (the level-10 model corresponds to the BSIM3v2 model); the Type-2 flicker noise model is embedded in e.g. BSIM3v3, BSIM4 and the EKV model [CDS11b]. The Type-3 model is a more advanced model, which is provided by BSIM3v3 and BSIM4 – which however have slightly different functions g and h. These are bias and geometry dependent and comprise three and four fitting parameters, respectively (c.f. [CDS11b]). Type-4 is employed by the EKV model and differs from Type-1 and Type-2 in the current dependence [CDS11b]. For instance in saturation: $S_{I_D I_D} \propto I_D$ holds true for long-channel devices, since $g_m \propto \sqrt{I_D}$, whereas Type-1 and 2 allow for a more flexible modeling due to γ. Type-5 accounts for this, since the exponent of g_m is now a variable [CDS11b]. As pointed out in Section 3.3.6 flicker noise is not yet fully understood.

Another noise source in MOSFETs is shot noise, which is introduced by parasitic leakage currents through the gate oxide, modeled as $S_{I_G I_G}$. From the previously mentioned models only the EKV3 model actually accounts for this (c.f. [CDS11b]). Shot noise can be characterized either by DC-measurements of the leakage-currents or by direct noise measurements [MF73; MC93]. In this work, the first approach is preferred although the verification is not as proper as real noise-measurements. The direct method has the difficulty of being highly sensitive to the thermal noise exhibited by the source resistance used for biasing V_{GS}. Assuming $S_{I_G I_G} = 2q I_{G-leak}$ and a source resistance R_{source} of the battery which exhibits thermal noise, it can readily be derived that $R_{source} > 4 k_B \theta / 2q I_{G-leak}$ has to be valid to yield the dominance of the leakage current's shot-noise over the thermal noise of the source resistance [MC93]. Considering a leakage current of 1 pA, this corresponds to R_{source} being higher than 52 GΩ. This

---

[3]The acronym EKV stands for Enz-Krummenacher-Vittoz, who developed the EKV model, which aims at accurate modeling of the weak-inversion and sub-threshold region operation of the MOS transistor, conversely to BSIM which is more precise in the strong inversion region [EV06; TM10].

discussion also demonstrates, that a very high source impedance is necessary for the shot noise of the leakage current to become non-negligible, so that its influence on applications is limited.

4.2 NOISE PERFORMANCE OF STANDARD MOS FIELD-EFFECT TRANSISTORS

Within the framework of this work the low-frequency noise performance of several devices available in the $0.35\,\mu m$ CMOS process was characterized. The MOS field-effect transistors under investigation were:

- enhancement, n-type transistor with thin gate oxide (NEDIG)
- natural depletion, n-type transistor with thin gate oxide (NNDIG)
- well-in-well enhancement, n-type transistor with thin gate oxide (well-in-well NMOS)
- enhancement, n-type transistor with thick gate oxide (MOSNE)
- natural enhancement, n-type transistor with thick gate oxide (MOSNN)
- enhancement, p-type transistor with thin gate oxide (PEDIG)

In this chapter the principle noise performance is presented. To define a meaningful foundation for the discussion of the performance comparison, first details about the realization of the devices under test have to be presented. This data was obtained by process and device simulation using a technology CAD program (TCAD from Synopsis) that models the entire process flow and the device physics based on finite-element simulation. With that, discussion based on the standard models, as they are presented in the former section becomes possible. The vertical doping profiles of the n-type transistors that are realized with a thin gate oxide of about 9.4 nm are presented in Figure 4.4. To enable a proper comparison only data in the vicinity of the surface is presented, since this mostly defines the devices performance. It can be observed, that the doping profiles of the enhancement-type transistor, which is directly embedded in the p^--type epitaxial layer (NEDIG), is similar to the enhancement-type transistor, which is embedded in its own p-well (well-in-well NMOS). The doping concentration of the natural depletion-type transistor (NNDIG), however, is two orders of magnitudes below them and is of n-type. The donator concentration in the vicinity of the Si-SiO$_2$-interface is result of a phosphorus implant at the end of the process, partially implanted through the thin gate oxide. This process is enhanced by the relatively low saturation level of phosphorus in SiO$_2$ compared to phosphorus in silicon (*pile-up*). In Figure 4.5 the electrostatic potential distribution is displayed. The noise models are all expressed as functions of the drain-current or transconductance. The boundary conditions in the TCAD simulations, contrarily, were defined as so-called *Dirichlet-boundaries* that define potentials, which in combination with the *Poisson equation* and doping profiles form the relations to calculate the electrostatic potential. The measurements, actually define electric field intensities by biasing with a current (which is proportional to the derivative of the electric potential). These are referred to as *Neumann boundaries*. For an according simulation of the electrostatic potential distribution corresponding Dirichlet boundaries were derived. It was assumed that the devices operate in the saturation regime

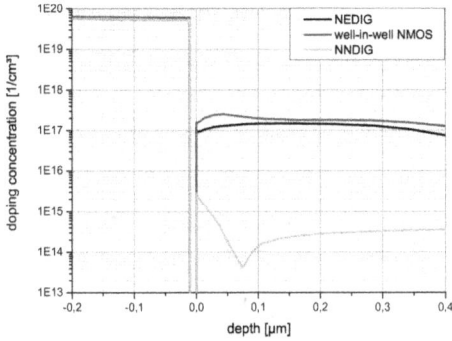

Figure 4.4 Vertical total doping profiles of standard NMOS-FETs with thin gate oxide.

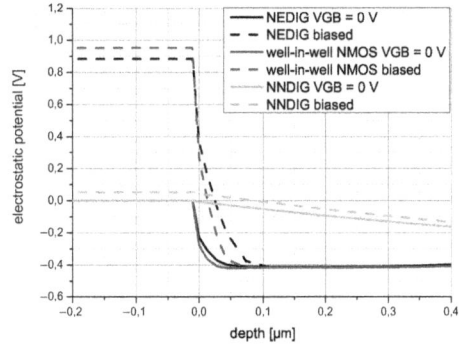

Figure 4.5 Vertical electrostatic potential profiles of standard NMOS-FETs with thin gate oxide.

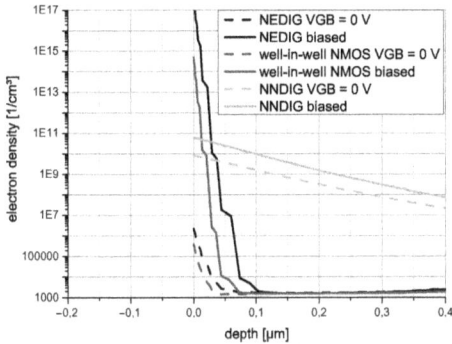

Figure 4.6 Vertical electron density profiles of standard NMOS-FETs with thin gate oxide.

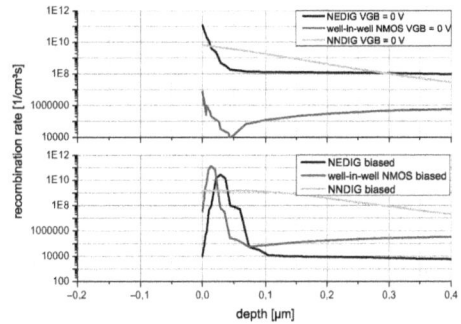

Figure 4.7 Vertical recombination rate profiles of standard NMOS-FETs with thin gate oxide.

so that long-channel device Level1 model could be used to derive a proper gate-bulk bias for the simulation, using given bias currents. For the comparison of the different NMOS-FETs two-dimensional TCAD simulations were done for all transistor-types with $10\,\mu m$ gate-length and a gate-source/bulk bias which causes a drain-current of $5\,\mu A$. However, to omit the arbitrariness in the choice of the coordinates at which vertical cuts have to be extracted, no drain-source bias is applied and profiles are extracted perpendicular to the surface. It can be observed that the potential maxima within the silicon bulk are located at the $Si-SiO_2$ interface, what corresponds to the electron-density distribution depicted in Figure 4.6. This indicates that the inversion layer is directly formed at the surface, what explains the shift of the maxima of the recombination-rate per unit-volume below the surface when the gate-bulk voltage is more positive as it is illustrated in Figure 4.7.

The recombination-rate of the well-in-well transistor is slightly higher compared to the NEDIG. The recombination-rates of the depletion-type transistor, however, are

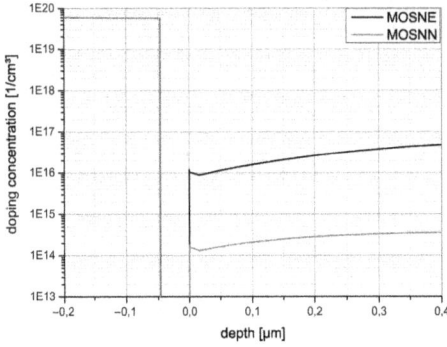

Figure 4.8 Vertical total doping profiles of standard NMOS-FETs with thick gate oxide.

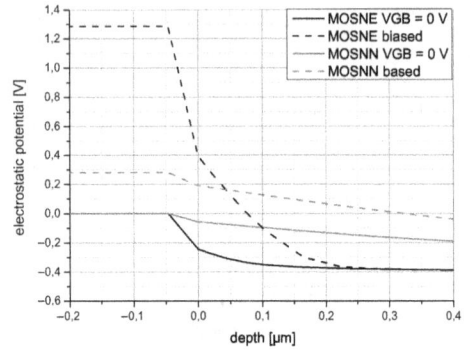

Figure 4.9 Vertical electrostatic potential profiles of standard NMOS-FETs with thick gate oxide.

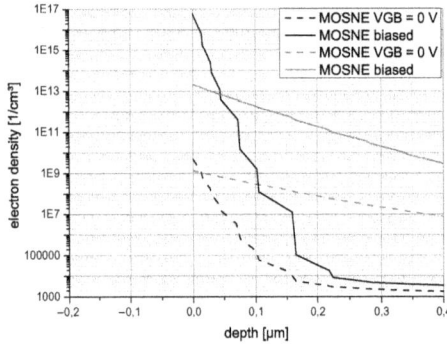

Figure 4.10 Vertical electron density profiles of standard NMOS-FETs with thick gate oxide.

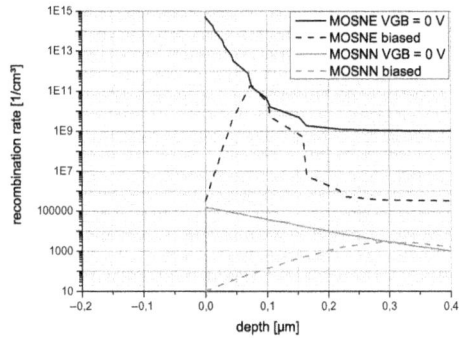

Figure 4.11 Vertical recombination rate profiles of standard NMOS-FETs with thick gate oxide.

far below the enhancement-types'. In Figure 4.8 the doping profiles of the FETs with the thick gate oxide of about 45.5 nm are depicted. The doping concentration of the natural enhancement-type transistor (MOSNN) is approximately two orders of magnitude below the doping concentration of the MOSNE, for which a threshold adjustment implantation was carried out.

In Figure 4.9 it can be observed, that the potential maxima are again at the surface of the device, which corresponds to the maxima of the electron-densities given in Figure 4.10. The recombination rates of the natural MOSNN are below $1 \times 10^6 \, \text{cm}^{-3} \, \text{s}^{-1}$ and thus far below the MOSNE's with approximately $1 \times 10^6 - 1 \times 10^{11} \, \text{cm}^{-3} \, \text{s}^{-1}$ when biased. Figure 4.12 depicts exemplary noise measurements. For all the noise measurements that are presented in the following text, the devices were operated in saturation. Additionally, for the devices larger than $10 \times 10 \, \mu\text{m}^2$ it was verified that the noise exhibition of each device was independent of varying drain-source voltage bias.

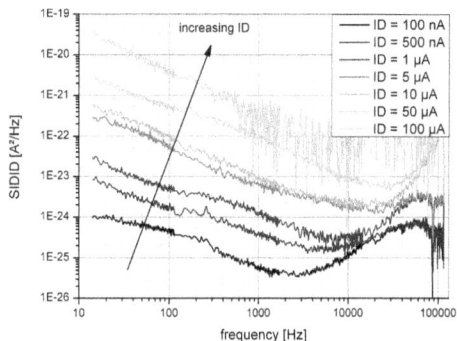

Figure 4.12 Output referred noise current spectral density for $10 \times 10 \, \mu m^2$ enhancement-type NMOS-FET with thin gate oxide (NEDIG).

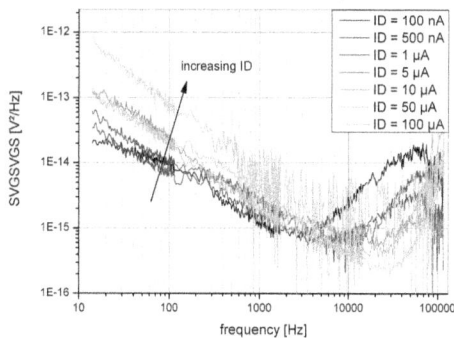

Figure 4.13 Input referred noise voltage spectral density for $10 \times 10 \, \mu m^2$ enhancement-type NMOS-FET with thin gate oxide (NEDIG).

In the following text the results of the noise characterization of the available MOS-FETs is presented. The data is referred to the input of the amplifier, which was used for noise characterization. This corresponds to the drain current referred power spectral noise density $S_{I_D I_D}$. Additionally, the gate-source referred power spectral noise densities are given – $S_{V_{GS} V_{GS}} = S_{I_D I_D}/g_m^2$. Since the presented devices are all rather large with a size of $10 \times 10 \, \mu m^2$, they are all operated in saturation and strong inversion and are biased with the same constant current, the transconductance can be given as $g_m = \sqrt{2 \mu \epsilon_{SiO_2} W I_D / t_{ox} L}$. The mobilities applied to this formula were extracted by TCAD simulations.

The pole-frequencies of the "current meter" were modeled only with finite accuracy and because the accuracy of the actual noise measurements significantly drop beyond the cut-off frequency, the measurements partially show an increase at higher frequencies – conversely to the ideal $1/v$-shape which should transit into white noise. The measurements may thus only be trusted up to frequencies of some kilohertz, dependent on biasing, noise level, selected amplification of the amplifier and the output impedance of the actual DUT.

The demonstrated noise measurements of Figure 4.12 are referred to the drain current. It can clearly be seen, that a higher bias current results in an increase of the flicker noise level. If Equation 4.6 or 4.8 is used to fit the current dependency, a factor α_f which is close to 1 can be observed. If the noise is referred to the gate-source nodes – as it is depicted in Figure 4.13 – it demonstrates only a weak bias dependency as it is given in Equation 4.7. This has much impact on the actual circuit in which the device is embedded. If the transistor is used as an amplifier, a higher bias current results in an improved noise performance if $\alpha_f < 1$. However, if $\alpha_f > 1$, the input referred noise level is rising with increased bias. If the transistor is diode-connected, the noise exhibition of the device is directly modeled by $S_{I_D I_D}$ in parallel to the small-signal equivalent impedance of $1/g_m$ (for long-channel devices), so that an increase in the noise level has to be expected. According to the Type-1 noise model, the flicker noise component is expected to be inversely proportional to the width of the device, while

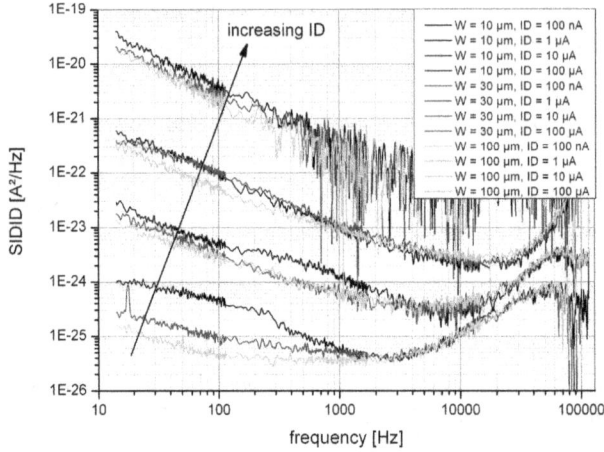

Figure 4.14 Output referred noise current spectral density for enhancement-type NMOS-FET with thin gate oxide (NEDIG) for various geometry – part 1.

Type-5 predicts a dependency $S_{I_D I_D} \propto W_{\text{eff}}^{\alpha_f/2-1}$ for large transistors in saturation, since $g_m \propto \sqrt{W}$. Type-2 and Type-4 – conversely to that – prognosticate independence with respect to the width while Type-3 models a rather complex dependency. The channel length dependence of the noise spectral density is also known to vary for different types of devices. In [MC93] it was reported, that $S_{V_{GS}V_{GS}-\text{flicker}} \propto L^{\beta}$ with $-1 \leq \beta \leq -1/2$. In the flicker noise model of Type-1 the relation $S_{I_D I_D} \propto L^{-1}$ is proposed, in Type-2 and Type-4 it is modeled as $S_{I_D I_D} \propto L^{-2}$ where in the model Type-5 $S_{I_D I_D} \propto L^{-2-\alpha_f/2}$ is used for the prediction of the length dependence. Again, the flicker noise model of Type-3 describes a rather complex relation with altogether four fitting parameters, so that it allows for a flexible fit.

However, for a proper fit also a sufficient amount of different samples have to be characterized to yield precise estimates of the model parameters. In Figure 4.14 it can be observed, that the $S_{I_D I_D}$ is slightly dependent on the width. If $S_{I_D I_D} \propto W^{-\beta}$ is proposed as a model to predict this dependency, β can be estimated to approximately 0.3 what does not match to the models of Type-1, Type-2 or Type-4. If the Type-5 model is used instead, α_f can be evaluated to approximately 1.4. The verification of the feasibility of the Type-3 model needs much more effort. Here, a global fit has to be done for a meaningful set of measurement samples in order to simultaneously fit all model parameters. However, as it will be shown in later text, this fitting procedure is not meaningful so far due to largely varying measurement results and the presence of an insufficient amount of samples to properly apply statistical calculus. Figure 4.15 demonstrates the geometry dependence of $S_{I_D I_D}$ of small-scale devices that demonstrate second-order effects as they are described in the former section. To allow for a proper comparison, these devices were picked from the same chip. Conversely to the presented data in Figure 4.14, $S_{I_D I_D}$ is not a monotonous function of the width anymore. This, however, may be a statistical phenomenon. The length dependency can be modeled by $S_{I_D I_D} \propto L^{-\beta}$ with $\beta \approx 1.3$. This neither corresponds to the models of Type-1, Type-2

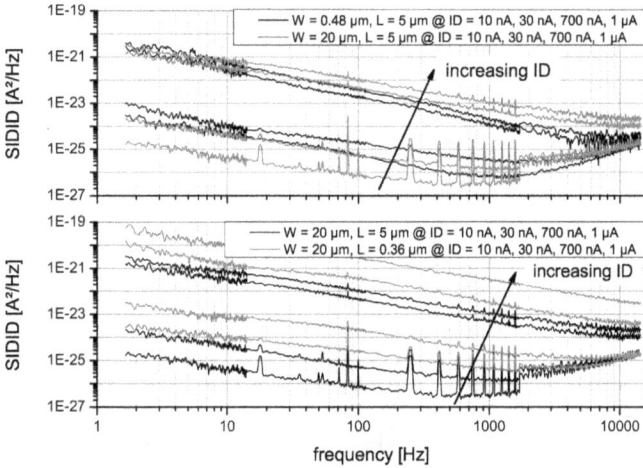

Figure 4.15 Output referred noise current spectral density for enhancement-type NMOS-FET with thin gate oxide (NEDIG) for various geometry – part 2.

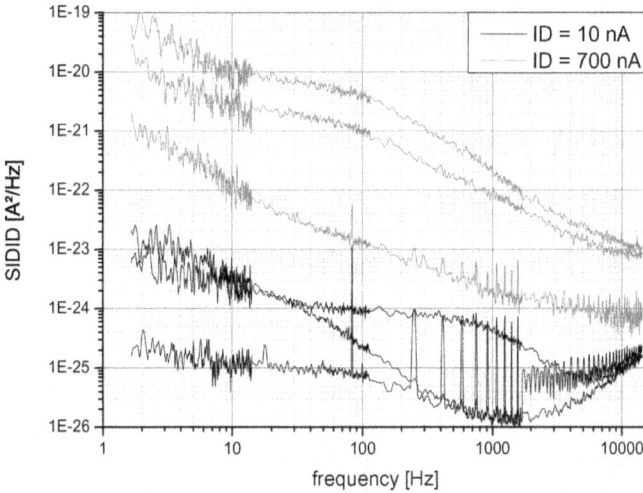

Figure 4.16 Output referred noise current spectral density plots of 20 × 0.36 μm^2 enhancement-type NMOS-FET from different samples.

nor to Type-4. If model Type-5 is used, α_f approximately equals 0.6, which does not correspond to the width dependency that was extracted before, indicating that this model is not sufficient for proper estimates of flicker noise in these devices. For proper evaluation a large amount of devices should be characterized. In [Mil11] the noise performance of 10 samples from one single wafer was characterized. The DUTs had a size of 0.36 × 0.36 μm^2 and varied over approximately three orders of magnitudes. Similar results were obtained in this process as well.

Figure 4.16 presents results from 3 samples of size 20 × 0.36 μm^2 which showed variation of 2-3 orders of magnitude. This is proving the need for a large amount of

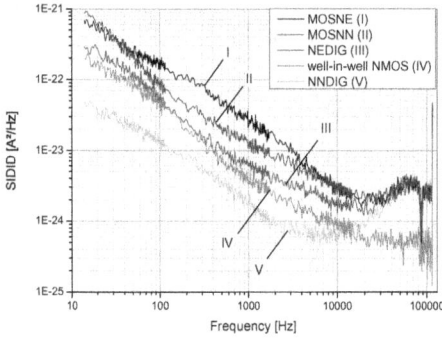

Figure 4.17 Output referred noise current spectral density for $10 \times 10\,\mu m^2$ NMOS-FETs of different types at $I_D = 5\,\mu A$.

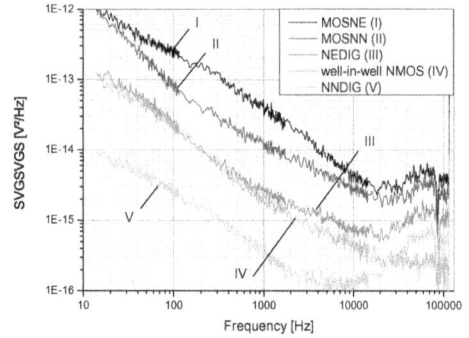

Figure 4.18 Input referred noise current spectral density for $10 \times 10\,\mu m^2$ NMOS-FETs of different types at $I_D = 5\,\mu A$.

sampled devices, which can then be analyzed for minimal, maximal and average noise exhibition, so that corner parameters can be extracted to allow for a proper noise estimation in sensor systems.

Figure 4.17 demonstrates first results of the noise performance for varying types of n-type MOSFETs. Clearly the transistors with the thick gate oxide demonstrate the highest noise, which is in accordance to all typical models. The natural transistor MOSNN shows a slightly better performance than the MOSNE. This is consistent with the theory claimed in [Mil11], which proposes a smaller spatial variation of the dopant distribution for smaller total concentrations. It is argued that this causes smaller threshold variations of the transistor and thus results in a better noise performance. The transistors with the thinner gate oxide (approximately a factor of 4.5) all show superior performance compared to the MOSNE/MOSNN transistors. Again, the natural device demonstrates the lowest noise exhibition. The NEDIG and well-in-well transistor present a similar noise level what might be explained by the similar doping-concentrations and recombination-rates depicted in Figure 4.4 and 4.7, respectively.

As it was explained in the former part of this section, TCAD simulations can be employed to estimate and compare the noise performance of different devices. For the sake of simplicity, up to now only one-dimensional profiles were presented. There $V_{DS} = 0\,V$ was chosen to allow for proper comparison so that the arbitrary choice of the point to do vertical cuts have no impact. However, as soon as those devices are conducting a current the two-dimensional profiles may be studied to yield a more precise model of the performance. A physically meaningful tool for such estimations is the impedance field method. Unfortunately though, as it is described in Section 3.5 the flicker noise model has to be carefully calibrated. This, however, is beyond the scope of this work.

In Figure 4.19 the total doping profile of an enhancement p-type FET (PEDIG) with a thin gate oxide is compared to one of an NEDIG. It can be extracted, that the PEDIGs level in the vicinity of the Si-SiO_2-interface is approximately one order of magnitude lower. Conversely to PMOS transistors of former times, its potential maximum is not

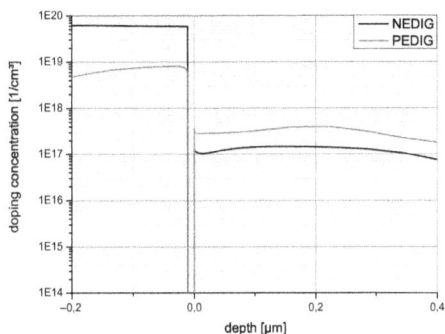

Figure 4.19 Vertical total doping profiles of standard enhancement-type NMOS- and PMOS-FETs with thin gate oxide.

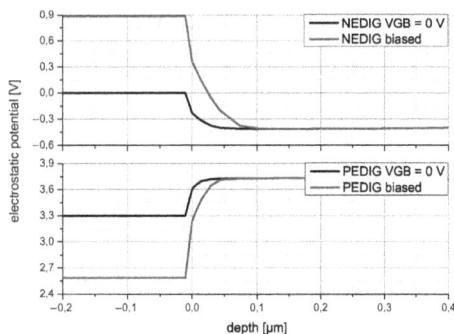

Figure 4.20 Vertical electrostatic potential profiles of standard enhancement-type NMOS- and PMOS-FETs with thin gate oxide.

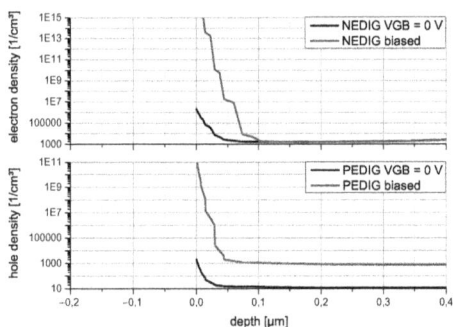

Figure 4.21 Vertical electron density profiles of standard enhancement-type NMOS- and PMOS-FETs with thin gate oxide.

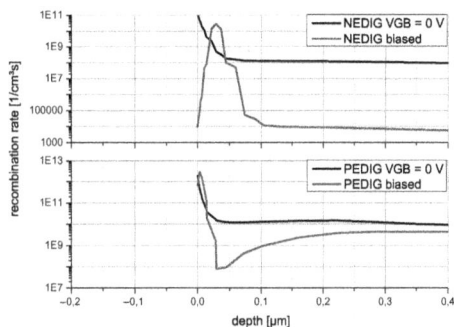

Figure 4.22 Vertical recombination rate profiles of standard enhancement-type NMOS- and PMOS-FETs with thin gate oxide.

buried below but is directly formed at the surface as presented in Figure 4.20 (c.f. discussion in [TM10]). This fact causes the maximum of the hole-concentration to be located at the surface as well – what is depicted in Figure 4.21. In Figure 4.22 the recombination rates of the PEDIG are compared to the NEDIGs. Though the recombination rates that were yielded by simulation are far worse for the PEDIG, the PMOS has shown a better noise performance $S_{I_D I_D}$. In former times, PMOS devices were considered to exhibit less flicker noise than their NMOS counterparts due to the channel formation below the surface [MC93; TM10]. Although this does not apply for the presented PEDIG, it exhibits less noise as it is exemplarily depicted in Figure 4.23. Considering the lower doping concentration of the PMOS and the theorem of [Mil11] which claims a correspondingly more homogeneous spatial dopant concentration, this may explain the superior performance of the device. Referring the spectral density to the gate-source nodes the advanced noise performance of the PEDIG over the NEDIG

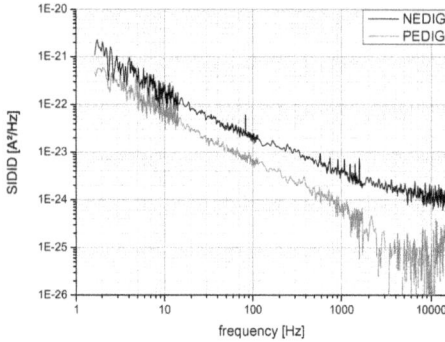

Figure 4.23 Output referred noise current spectral density for $20 \times 5\,\mu m^2$ enhancement-type NMOS- and PMOS-FETs at $I_D = 1\,\mu A$.

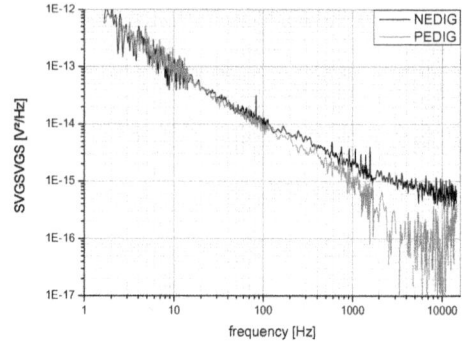

Figure 4.24 Input referred noise voltage spectral density for $20 \times 5\,\mu m^2$ enhancement-type NMOS- and PMOS-FETs at $I_D = 1\,\mu A$.

vanishes, because of the smaller transconductance. This is caused by difference in the mobility which is worse for holes in silicon compared to electrons.

4.3   NOISE PERFORMANCE OF AVAILABLE BIPOLAR DEVICES

Bipolar devices are typically advantageous over field-effect transistors with respect to the transconductance, which is usually higher and directly proportional to the bias current compared to FETs where $g_m \propto \sqrt{I_D}$ applies in strong inversion and saturation. Moreover, bipolar devices are said to exhibit lower low-frequency noise than MOS-FETs. This can be explained by the fact, that MOSFETs by definition form their current leading channels close to the surface, while bipolar transistors conversely allow for vertical structures, thus allowing the channel to be displaced from the impurities that are associated to the Si-SiO$_2$-interfaces. However, this does not apply for junction-FETs (JFETs) which can also be formed in a vertical manner and do allow for relatively large distances to such interfaces – even if they are implemented horizontally. Nevertheless, these advantages are exchanged for an increase in area and power consumption and parasitic effects like current paths through the substrate.

In Figure 4.25 cross-sectional views of available bipolar transistors in the process are depicted. The pnp-bipolar transistor is widely employed in e.g. voltage references [Raz03; Hol94; RM01; Bak10]. It basically uses a p$^+$-type drain/source diffusion of e.g. standard MOSFETs as an emitter-region, a standard n-well to define the base-region and the p$^-$-substrate as the collector. All necessary components are available in standard CMOS processes based on a p-substrate.

This device, though, is usually considered to be a parasitic device of e.g. standard PMOS transistors, that may even cause *latch-up* in combination with an npn-device, comprised of an n-well, the p-substrate and an n$^+$-type diffusion of e.g. a MOSFET structure. In typical processes the depth of the well is some micrometers deep and has a low doping concentration so that the current amplification of this device is rather

vertical npn – bipolar transistor

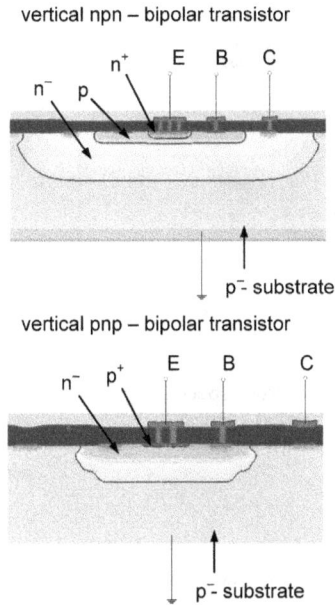

vertical pnp – bipolar transistor

Figure 4.25 Cross-sectional view of vertical bipolar transistors available in the process.

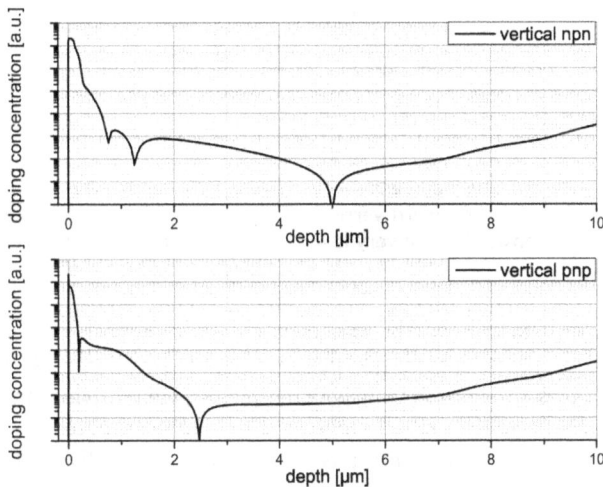

Figure 4.26 Vertical total doping profiles of vertical bipolar transistors available in the process.

poor. In the process that has been employed in this work, for instance, the current amplification factor amounts to only 10–15. The presented npn-structure is composed of an n^+-type diffusion as emitter-region, a p-type well as base-region and an n-well to embed these and act as the collector-node. Since the base region is much more narrow,

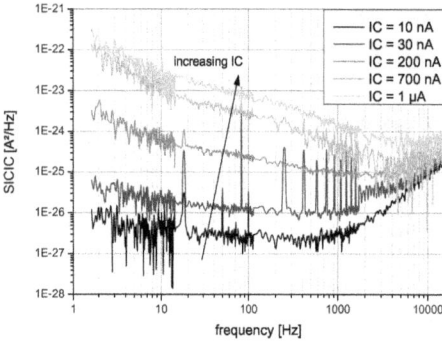

Figure 4.27 Output referred noise current spectral density for the standard vertical npn bipolar transistors with very low source resistance.

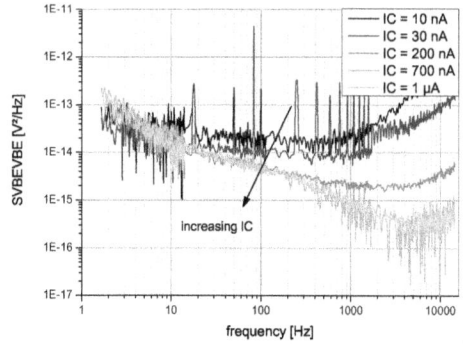

Figure 4.28 Input referred noise voltage spectral density for the standard vertical npn bipolar transistors with very low source resistance.

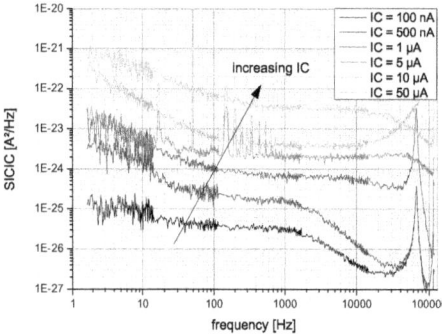

Figure 4.29 Output referred noise current spectral density for the standard vertical pnp bipolar transistors with very low source resistance.

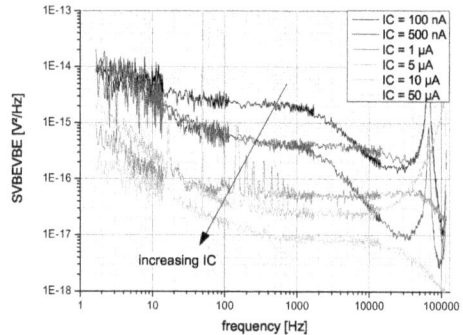

Figure 4.30 Input referred noise voltage spectral density for the standard vertical pnp bipolar transistors with very low source resistance.

current amplification factors in the range of 290–470 have been measured for this device. The corresponding doping profiles are depicted in Figure 4.26. Since current amplifications factors of bipolar transistors typically vary in the range of 100–1000 and collector currents rarely exceed some tens of micro-amperes, for sensor systems with meaningful power consumption base currents are typically in the order of some hundreds of pico amperes up to some hundreds of nano amperes. This input bias current by far exceeds the parasitic leakage current of MOSFETs which is in the order of femto to some pico amperes. Since the input current of bipolar devices is associated with noise as it is discussed in the Section 4.1, the necessity of its characterization arises. If the device is input biased by an battery, the source resistance is vanishing so that the measured noise spectral density is dominated by the collector shot noise which can be observed in Figures 4.27–4.30. $S_{I_B I_B} = 2qI_B + K_f I^{\alpha_f}/v^{\gamma_f} + K_b/(1 + (v/v_c)^2)$ is always higher than the well known shot noise component of the base current $2qI_B$.

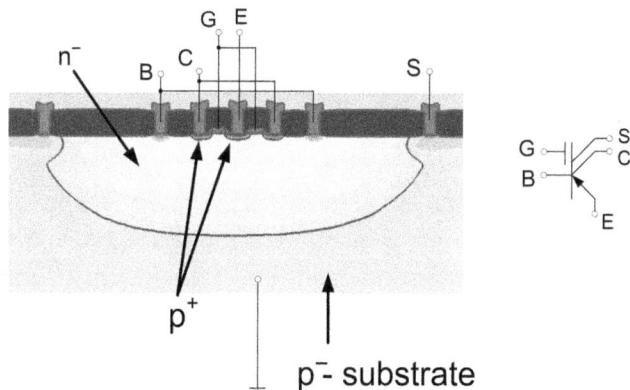

Figure 4.31   Cross-sectional view of a gated bipolar transistor.

To make sure the noise level $S_{I_B I_B}$ dominates over the current noise of the source resistance $4k_B\theta/R_{\text{source}}$, the source impedance has to be chosen as $R_{\text{source}} > 2qI_B/4k_B\theta$. Nevertheless, this corresponds to a high source impedance of several giga ohms. This, however, is beyond the scope of this work. Clearly the bipolar devices exhibit less low frequency noise then the presented MOSFET devices if the source resistance is very low. For increasing source resistances, however, the MOSFET devices become superior since the input current noise is negligible compared to the bipolar input current noise.

Alternatively to the presented devices, bipolar transistors may also be realized as lateral devices relying on horizontal charge transport. The diffusions of the device can partly be realized self-aligned, so that the base length is defined by the accuracy of lithography rather than thermal diffusion. This can become advantageous in terms of reproducibility of e.g. the current amplification factor. In terms of noise, though, this may worsen the performance due to the vicinity of the device to the Si-SiO$_2$-interface. A compromise was presented in [Vit83]. Here a spherical MOSFET was proposed to operate in bipolar-mode by forward biasing of the bulk-source diffusion – which then corresponds to the base-emitter junction. Therefore, of course, the device should be embedded within a well, to avoid DC-current through the substrate. The actual compromise that is offered here, is that the gate, which is used for the base length definition by self-alignment, can also be biased.

If it is biased in such a manner that the charge carriers that define the channel below the surface are pushed deep into the silicon substrate, the noise performance can be improved at the cost of decreasing the current amplification factor or the transconductance, respectively, what is caused by a longer effective base length. Initially, these principles were directly employed within actual circuits such as high accuracy current-mirrors, low-noise amplifiers or voltage-references. The verification of the improved noise performance on a device level was published more recently [DM02]. Because the standard vertical pnp transistor that is available in the process suffers from a low current amplification and flexibility (because the collector is defined by the substrate), the proposed hybrid transistor was realized based on the PEDIG which was laid out spherically. A cross-sectional view is depicted in Figure 4.31. The layout can be done

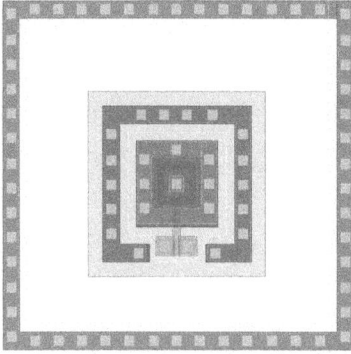

Figure 4.32 Rectangular layout of the gated bipolar transistor.

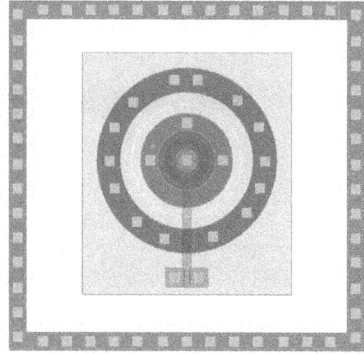

Figure 4.33 Spherical layout of the gated bipolar transistor.

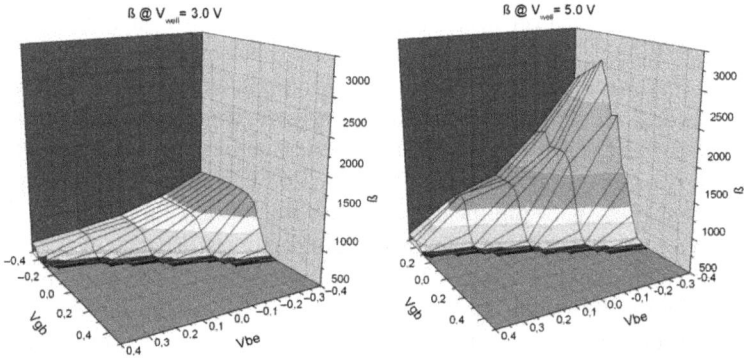

Figure 4.34 Current amplification of the gated rectangular bipolar transistor.

for instance in a rectangular or a spherical manner as demonstrated in Figures 4.32 and 4.33. In the framework of this work it was observed, that the rectangular variant was sensitive to break-through. A possible explanation of this phenomenon is an increased electrical field intensity at the corners of the polysilicon gate above the thin gate oxide. This is avoided by the spherical layout. Moreover, the collector-to-emitter ratio is increased resulting in a higher amplification. For the rectangular variant high current amplifications β were already observed as indicated in Figure 4.34. The potential-difference of the collector and emitter was kept at 3.3 V to ensure the operation in the saturation region. An interesting phenomenon that was observed, is the increase in the current amplification by augmentation of the reverse-bias of the embedding n-well. This may be explained by the expansion of the depletion zone, increasing the probability, that emitted charge-carriers get drained by the collector-base junction – or conversely – that the parasitic path from the emitter towards the base electrode gets partially pinched off.

Chapter 5

Noise in active pixel sensors

This chapter is dedicated to the noise performance of optoelectronic photodetectors. Since the basic operation principles of these devices – including the generation and propagation of electromagnetic radiation, its conversion to electronic charge by the photoelectric effect and the fundamentals of carrier transport in semiconductors – are well-known, the reader is referred to textbooks like [SN07; PN86; ST07; The95]. This chapter intends to explain the origin of noise sources and reduction methods in up-to-date CMOS active pixel sensors (APS). Additionally a novel readout structure is presented that is superior to the widely employed source follower readout implemented by enhancement MOSFETs. It yields a high output voltage swing and low noise, while requiring no additional processing steps. The readout structure consists of a low-noise JFET whose gates are formed by a floating diffusion, thus preserving in-pixel accumulation capability – which additionally improves noise performance. This structure outperforms a simple in-pixel implementation of a JFET and a photodetector in terms of the necessary area consumption, thus improving fill factor. For pixels with a pitch of several microns this readout structure is a good trade-of between area, output voltage swing and, most important, noise performance. Furthermore, since only one ground connection is needed for application, fill-factor and power-grid disturbances like DC-voltage drop can be additionally improved.

5.1 PHOTODETECTOR PRINCIPLE

Compared to *photogates* or standard pn-junction photodiodes, pinned photodiodes are used very often due to their low noise behavior. Nevertheless, in terms of speed all these approaches may become insufficient for applications based on large-area detectors. Such detectors become necessary under low light conditions where long integration times are rather unacceptable. This finds application e.g. in x-ray panels, imagers for spectroscopy or range imaging. A sensor called *lateral drift-field photodetector* (LDPD) has thus been developed in which a lateral electric field is introduced to speed up the transfer process of photon generated charges into the storage and readout node – e.g. a floating diffusion[Du10][1]. For creation of this lateral drift-field,

---

[1] [Kos96; Kos97] was probably the first one to introduce potential gradients into CCD elements by multiples of implantations. This idea was adopted for CMOS photodiodes and realized with

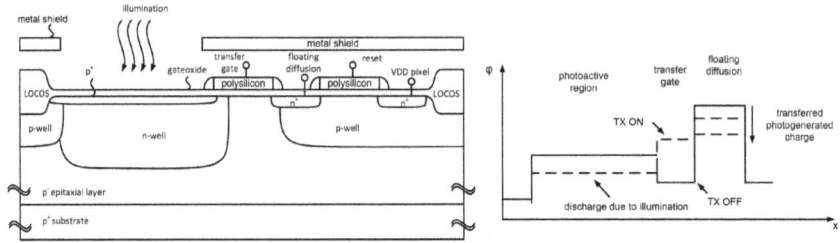

Figure 5.1   Cross-sectional view of a PPD (left) and its potential distribution (right).

which is formed by a doping gradient, successively increasing (towards the storage node) windows are formed in the mask used for structuring the implant. By thermal annealing the implanted dopants diffuse, creating a smooth gradient [MFH10; Du10]. With this approach of creating drift-fields into photo detectors transfer-times of some nanoseconds were achieved for pixels with a size of $40\,\mu m \times 40\,\mu m$ [SDS11; Du11].

Figure 5.1 displays a cross-sectional view of a pinned photodiode (PPD) and its potential distribution. The photosensitive region is formed by an at least partially depleted n-well implant operating as a photo diode. This well is protected against generation-recombination centers located at the Si-SiO$_2$ interface by the p$^+$-pinning layer. The well is separated from the readout node by a transfer gate (TX). The separation of the photosensitive well and the floating diffusion serving as a storage node enables the user to define the integration for the photogenerated charge in the pixel itself by shutting off the transfer gate. Furthermore, reset noise is reduced since the storage node capacitance is often very low compared to the common pn-junction based photo diode counterpart, where the capacitance is formed by the whole device [TKI82]. In PPDs a plurality of storage nodes may be applied to enable the demodulation of optical signals. For this the different storage nodes may be connected to the photosensitive region directly via transfer gates. Alternatively a connection node formed by an additional gate (CX) may be applied which connects the different storage nodes by transfer gates in CCD manner.

In case of the LDPD – depicted in Figure 5.2 – photon generated electron-hole pairs are separated by a drift-field enabling a high transfer speed compared to diffusion based transfer. Electrons are accelerated towards a collection gate which is connected to a constant potential. By enabling one of the transfer gates, charge is then transferred into the corresponding storage node. To avoid blooming at least one draining gate may be applied to the collection gate. In general, more gates may be used enabling the demodulation of illumination by the photo detector itself. For proper operation of the device the n-well may be connected to the storage nodes so that it can be properly

---

a plurality of gates to enable demodulation as it is necessary for e.g. time-of-flight imaging. Alternative approaches employ for instance current flow through a polysilicon layer that is structured above the photodetector to introduce a potential gradient into the device [Loh] which though is very power consuming; multiples of polysilicon fingers on top of the detector which are biased with constant potentials [Bü06] or varying lateral expansion of the well which also results in a lateral potential gradient [TSI10].

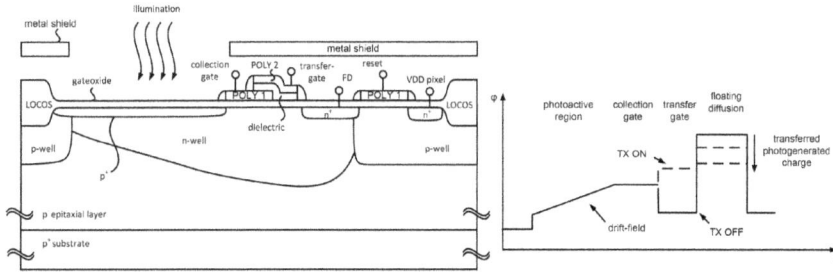

Figure 5.2 Cross-sectional view of a LDPD (left) and its potential distribution (right).

depleted what is necessary for a smooth and sufficiently high potential gradient. Thus the well for the device has to be designed in such a way that storage nodes may be disconnected from the detector by switching OFF the respective transfer gates and that the gradient for the connected diffusion is sufficiently high to quickly transfer charges and to fully deplete the photoactive region in reset operation.

A standard APS readout circuit for a PPD or a LDPD is the 5 transistor readout architecture. This is illustrated in Figure 5.3 where M1-M4 and TX form the 5T cell. In that M1 is the PMOS reset transistor which defines the potential at the storage node before charge transfer. Using a PMOS as a reset transistor is called *hard reset* operation; in this mode the diode is connected to the positive supply voltage rail by a low impedance, since the PMOS is operating in triode region when the reset potential is reached. Alternatively an NMOS transistor can be chosen for reset functionality. Choosing the positive supply power rail lower than $\phi_{g-reset} - V_T$ will keep the device in triode region for the entire reset process. On the contrary – if $\phi_{g-reset} - V_T < v_{vdd-pixel}$ the device will transit from saturation regime to weak-inversion; this operation scheme is referred to as *soft reset*. It is reported that this results in less uncertainty regarding the reset potential (c.f. Section 5.2). However, due to the high impedance connection to the power supply voltage, the storage node reset potentials are rather sensitive to disturbances during the reset phase that can be introduced by capacitive crosstalk, parasitic illumination or the collection of charges by diffusion. If during charge transfer too many charge carriers are accumulated in the floating diffusion – working as a storage node – M2 preserves the pixel from blooming into other diffusions. In case of an NMOS reset transistor this device becomes dispensable. M3 is forming the actual driver for the readout. It is operating as a source follower. For a good noise performance its current conducting channel should best be formed deep below the surface to avoid interaction with generation recombination centres [WSR08]. If the channel is formed in such a way that the transistor is working in depletion mode the output swing is further increased leading to even more dynamic range [CWM09]. Alternative readout schemes will be examined in the following text. M4 simply works as a switch and is thus separating M3 from the output drivers of other rows as indicated in Figure 5.3. However, for achieving a higher output swing a transmission-gate instead of a simple transistor switch may be used [CWM09]. The output of the presented $M \times N$ pixel matrix is defined column-wise in terms of voltage levels across the nodes of the current sinks, which are depicted in the lower section of Figure 5.3. These may be further

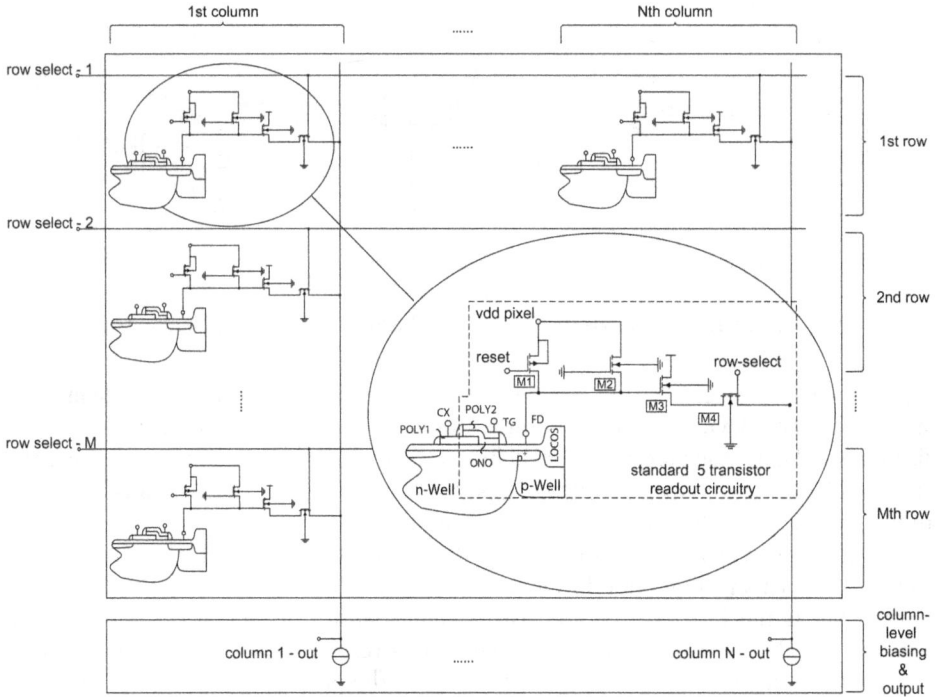

Figure 5.3 Schematic of a 5T APS readout circuit.

connected to signal conditioning circuitry like *correlated double sampling stages* or other switched-capacitor filters.

5.2 PHOTODETECTOR NOISE AND REDUCTION TECHNIQUES

5.2.1 Dark noise

Dark noise is defined as the uncertainty exhibited by a photodetector at zero illumination. Photodiodes in imagers often employ reverse-biased pn-junctions to achieve a large depletion zone for a fast separation of photogenerated electron-hole pairs and the avoidance of significant bias currents which may cause a high power dissipation. Nevertheless, the reverse-bias of the junction is causing a parasitic leakage current which is discharging the storage nodes and exhibiting shot noise. The leakage currents of diodes ideally originate only from the diffusion of minority carriers. For higher impurity concentrations in the vicinity of the depletion zones generation-recombination processes cause additional charge carrier flow causing increased dark current. Especially at Si-SiO$_2$ interfaces the impurity concentrations are considered to be relatively high, so that buried devices such as the pinned photodiodes become preferable in terms of noise performance [Lou03; EKC92; SA90; PCH05; ITY03]. The photodiodes itself can be separated in bias dependent area and perimeter components [Lou03]. It was demonstrated, that due to different generation-recombination mechanisms the leakage current is a strongly non-linear but monotonic function of the reverse bias voltage [Lou03]. This means that high electric fields, as they may occur for instance at the edges of photodiodes, have to be avoided. In actual CMOS APS additional phenomena can occur. For PPDs or LDPDs, for instance, in addition to the leakage current of the photoactive area, leakage current flow is introduced by the floating diffusion as well. If an NMOS transistor is introduced for reset or antiblooming functionality a gate-leakage current can also be introduced. Moreover, if NMOS transistors are connected to the floating diffusion and are in the OFF-state, *gate-induced source leakage* is introduced which is a phenomenon causing an increased leakage current due to band-to-band tunnelling processes [PCH05; CWW01].

Another effect is arising during the transfer process of charges from the photoactive region towards the floating diffusion. When the transfer-gate of a pinned photodiode or an LDPD is turned ON, an inversion layer is formed below the Si-SiO$_2$ interface. Due to the p$^+$-pinning layer of the photoactive region, though, this inversion layer is not formed at the side of the transfer-gate at which the photoactive region is connected. This may cause high electric fields introducing hot-carriers that can harm the Si-SiO$_2$ interface and thus cause an increase in the *trap-induced dark current* [WRT06]. This may be suppressed using an extension of the pinning layer by additional implants or gradient in the gate oxide thickness [BMY] but may also be limited by cycling the transfer-gate from accumulation to inversion [MST08].

In addition to shot noise associated with dark-current, separation noise can be introduced during the transition from the ON-state of the transfer transistor to the OFF-state. During this phase the charge forming the inversion layer is reduced and partially spills into the photoactive region or the floating diffusion. This transition is also associated with an uncertainty. A method to reduce this effect by a gradient of the threshold voltage along the transfer gate was reported by [PBK10]. This method intends to increase the probability of the transition of the inversion-layer forming charges into the floating diffusion by the introduced threshold gradient. Alternatively, the slope of the transition may be altered as it is known from switched-capacitor filters (c.f. e.g. [AH02]).

Dark current can also stem from the source follower. It is known that short-channel MOSFETs operated in saturation can cause hot carrier generation in the vicinity of the drain region, where the electric field is at maximum. Such hot electrons can cause additional minority carriers in the substrate due to *impact ionization* or *electroluminescence* that may spill into the photosensing area or the floating diffusion [WS01; MML97]. Electroluminescence corresponds to the spontaneous emission of photons from electrons – such as hot electrons – which loose energy. This phenomenon is largely bias dependent. The induced leakage current is increasing with ongoing discharge of the photodiode or the floating diffusion, respectively. Its impact can be decreased by limiting the time frame of the bias-current, which may be easily realized since it is only needed during the readout-phase.

5.2.2 Photon noise

Photon noise is basically the shot noise induced by the actual illumination. Because this is the actual signal which has to be detected, it describes a fundamental limit for the detection accuracy. Since shot noise is a Poisson process, a mean number $N \in \mathbb{N}$ of photon-generated electron-hole pairs corresponds to a standard deviation and signal-to-noise ratio of $\sqrt{N}$. Though the shot noise of the illumination that has to be detected cannot be reduced, additional shot noise introduced by parasitic illumination should be avoided. For modulated light as it is used e.g. for time-of-flight applications, it is thus meaningful to concentrate the light power of the modulated signal within a limited time window and to avoid integration of background light outside the window.

In the case of the PPD or LDPD this can be done by draining the photogenerated charge into a dump diffusion. For this a plurality of transfer gates can be connected to the actual photosensitive region, from which at least one is connected to a region which is used for draining. The storage node which is intended to keep the photogenerated charge should ideally not be light sensitive. For front-side illuminated image sensors this may be realized by proper metal shielding. Nevertheless, electron-hole pairs can be generated deep below the depletion zone of the actual photodiode, so that diffusion becomes the major transport mechanism. To avoid parasitic charge transfer into the floating diffusion, this may be shielded by a buried p^+-implant which defines a potential barrier for the photongenerated electrons. Alternatively a deep n^+-implant may be realized to attract parasitic charges. This has to be connected to a n^+-plug as it is common for BiCMOS processes to define a low-impedance path which is necessary to remove all charges (c.f. [Has05]). The creation of an additional attractor for photon-generated charges, though, has to be realized carefully not to remove any signal charges. For back-side illuminated imagers, background light can basically be drained by embedding the structures that are not to be photosensitive in wells with low-impedance connections to drain parasitic charges.

5.2.3   Reset noise

This noise is associated with the resetting operation described above. A capacitor which is charged to a certain voltage level by an arbitrary source resistance R_{source} can be evaluated with respect to the voltage uncertainty by application of the Wiener-Lee relation if the steady-state is reached, so that asymptomatic stationarity can be assumed. The source resistance will then be in thermal equilibrium so that it exhibits a power spectral density $S_{v_{R_{source}} v_{R_{source}}} = 4k_B\theta R_{source}$. The voltage uncertainty across the nodes of the capacitor can be further evaluated to

$$\text{var}(v_C) - \int_0^\infty \frac{4k_B\theta R_{source} \cdot dv}{1 + (2\pi v R_{source} C)^2} = \frac{k_B\theta}{C}. \tag{5.1}$$

In the charge-domain this can be expressed as $\text{var}(Q_C) = k_B\theta C$. This used to be a major noise source in CMOS active pixel sensors and is widely referred to as *reset noise*. Early detectors suffered from high reset value uncertainty introduced by the high storage capacitance. With the introduction of pinned photodiodes the equivalent capacitance of the storage node was separated from the photoactive region and thus reduced dramatically [INY99; ITY03]. Nevertheless, reset noise still remains the dominant noise source in CMOS APS.

With adoption of *correlated double sampling (CDS)* it can be reduced furthermore [PCH05; Whi74]. CDS is a filtering process that subtracts two samples; the signal after reset from the uncertain reset. This is assuming that the same reset value uncertainty is present in both values or, in other words, that they are correlated. However, CDS is usually implemented by peripheral circuitry and can thus only reduce noise after the signal has already been affected by it. Furthermore, for imagers in global shutter mode used for applications, where smear effects have to be avoided, CDS circuitry becomes quite complex, since storage nodes have to be implemented for each pixel to save the reset value of those.

Alternatively to reset noise reduction via CDS, circuitry can be realized that allows for less reset noise by actively disturbing the steady-state within the actual reset phase. A possible implementation can for instance measure the actual reset value which can then be adjusted by current sources or current sinks. Three different topologies have been analytically studied by application of Itō's law [FGM06]. A similar but simpler approach is the so-called *soft-reset* which also avoids the steady-state by application of feedback during the reset phase. For soft-reset the gate of an NMOS reset transistor is set to HIGH, so that a load capacitance is charged. The transistors overdrive voltage is lowering for increasing voltage level across the load - thus the transistor will finally operate in weak-inversion. It can be derived that the uncertainty in the voltage level of the storage capacitor will be lower than $k_B\theta/2C$ [Tia00; TFG01]. Nevertheless, due to the high-impedance connection of the storage capacitor to the power supply the node is sensitive to disturbances that may couple e.g. capacitively.

5.2.4 Thermal, flicker and RTS noise

Analyses showed that for APS with reset noise reduction by CDS, the source follower transistor (SF) often becomes the dominant noise source [KHL05]. The noise of such SF transistors can be modeled as a combination of thermal noise, flicker noise and RTS noise sources. Flicker noise and RTS noise dominate at low frequencies and can be thus at least partially filtered by the CDS but for the same reasons as already described this is not necessarily the best approach. Even worse, it has been deducted, that RTS cannot be fully eliminated with CDS at all [WRM06].

Thermal noise, on the other hand, is basically a frequency independent noise process. If undersampled- and that is basically the case in analog sampled-data circuits, such as CMOS imagers – the thermal noise power spectral density can be substantially increased in the baseband. Hence, the bandwidth control is essential in applications requiring low noise, since this limits the amount of noise aliased into the baseband. Naturally, the reset noise discussed in the former section is nothing else but a bandlimited undersampled thermal noise.

To reduce RTS and flicker noise it is important to understand how they originate. Those noise sources are often associated with defects (traps) appearing in the substrate or interfaces like the $Si\text{-}SiO_2$-interface (c.f. Sections 3.3.5 and 3.3.6). Such traps may capture or release charges that take part in the current flow, what causes RTS noise. If the region, in which current is flowing, suffers from many of those traps – according to the McWorther model – flicker noise is formed [McW55], while low trap density causes RTS noise (c.f. Sections 3.3.5 and 3.3.6 and Chapter 4).

On circuit level such noise processes can be reduced by switching off the bias current before readout. This switched biasing method is based on the assumption that traps are empty if no current is present what still holds true for a certain time after

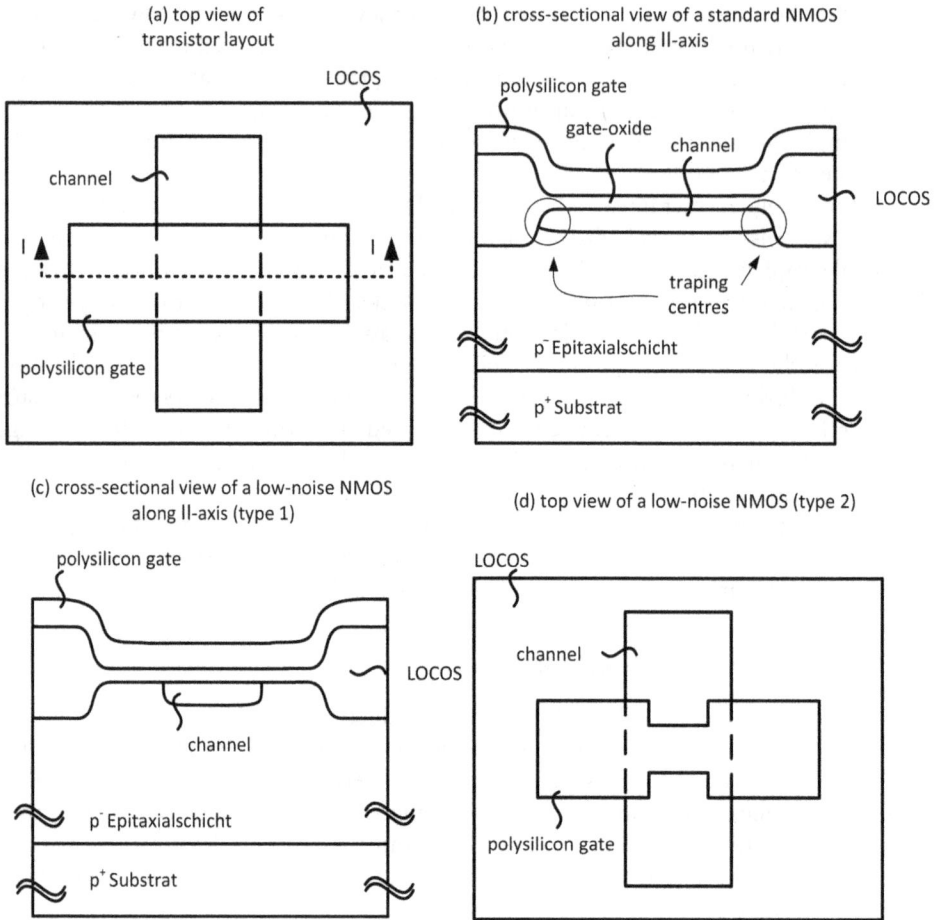

(a) top view of
transistor layout

(b) cross-sectional view of a standard NMOS
along II-axis

(c) cross-sectional view of a low-noise NMOS
along II-axis (type 1)

(d) top view of a low-noise NMOS (type 2)

Figure 5.4 (a) top view of a standard NMOS FET layout, (b) cross-sectional view of a standard NMOS
transistor layout, (c) cross-sectional view of a low-noise NMOS transistor (type 1) and
(d) top view of a low-noise NMOS transistor (type 2) according to [MHM10].

switching on the bias [MM11]. Nevertheless, this approach does not reduce the amount
of traps themselves but the impact those may cause by a trade-off against deterministic
disturbances.

For reduction of traps the source follower itself has to be designed carefully. Its
reduction by variation of the geometry has been studied in detail [LVFS07; MM09].
Furthermore, it has been shown that RTS noise can be reduced by clever layout without
changing the size of the transistor. To achieve this, the channel of the transistor is
formed dislocated to LOCOS which usually suffers from higher defect densities causing
RTS and flicker noise [MHM10; SA90; EKC92] as can be observed in Figure 5.4. Here,
in the two leftmost figures a standard NMOS transistor is depicted in two perspectives
while figure (c) illustrates how trapping centres at the channel can be avoided by
displacing the channel from the LOCOS isolation. Another method to avoid trapping

(a)

(b)

drain   gate   source
    gate oxide

transfer   transfer
   in   out

G   D

electron
flow

S

n-well   p⁺   p⁻
       p⁺ ⊕ →

charge
packet

modulated surface
hole current

blocking
electrode

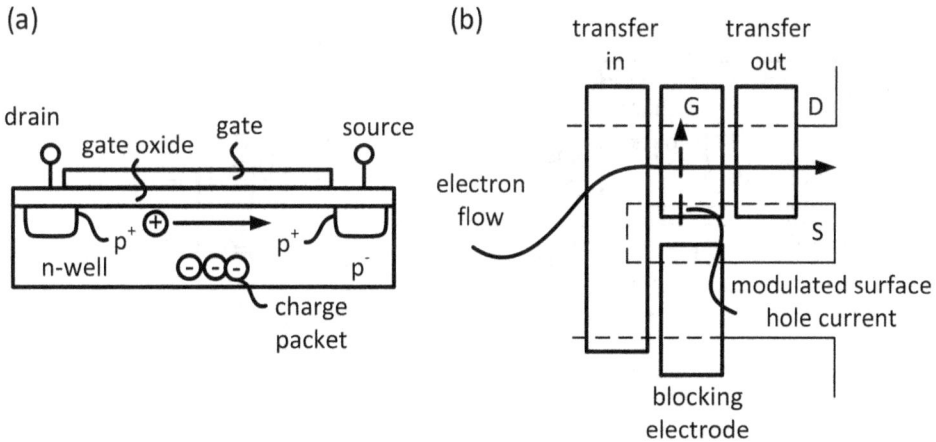

Figure 5.5 (a) cross-sectional view and (b) top view of the Brewer readout structure (according to [Bre78]).

centres is to shrink the gate length at the middle of the transistor to concentrate the current in the centre of the device. The reported improvements are significantly while no process modifications were necessary.

Compared to these approaches, that are CMOS compatible but thus also limited, more creative approaches may be implemented. Several rather complex approaches have been undertaken on to minimize noise – many of them for CCD detectors [Fos93]. All of them aim for less contact with generation-recombination centers that are caused by defects and are – according to the McWorther model – mainly located at the Si-SiO₂-interfaces. This may explain why bipolar transistors and JFET transistors are often reported to exhibit less low frequency noise than standard MOSFETs [SN07; MC93].

Standard bipolar transistors require non-negligible base currents and may thus not be applicable for simple readout structures. Also, base current noise affects significantly the transistor noise performance when the input source exhibits a high source impedance. It has been reported, how MOS transistors may operate in bipolar mode without pulling significant input current [Vit83]. For example, if a PMOS transistor is formed by two p^+-diffusions in a n-well, a lateral pnp-transistor is present as well, which is usually considered as parasitic. This can be enabled by forward bias of the source-well diode. The current can then be pushed below the gate oxide by applying a bias voltage to the gate, thus contact to traps located at the interface may be avoided. Nevertheless, the device suffers from an inherent parasitic additional pnp-transistor formed by p^+-implant, n-well and p-substrate, which pushes current through the substrate that can then increase the dark current and thus degrade the noise performance. This effect can be reduced by proper biasing of the well. However, it requires an additional non-standard power supply level. Also, the resulting transconductance may be too low. For a high sensitivity the collector to emitter ratio has to be maximized, what usually results in round and thus large transistors. A bipolar floating base detector has been reported [RCS92]. As amplifying stage it uses a vertical bipolar transistor that yields the input current from a MOSFET, which is modulated from the back-gate by

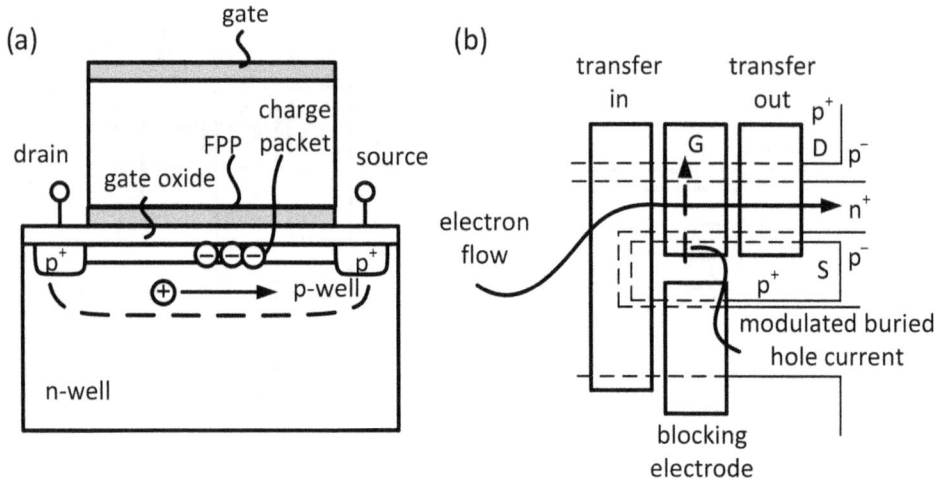

Figure 5.6   (a) cross-sectional view and (b) schematic of the Matsunaga readout structure (according to [MOI87]).

the accumulated charge in a JFET manner. The base current is in close proximity of the gate oxide affecting the noise performance. MOSFETs modulated from backside were used in large variety for CCD imagers [Bre78; Bre80; RCB95; RCB96; YMIMH88]. The Brewer readout structure, as it is depicted in Figure 5.5, was proposed in 1978. Here, the channel of the readout structure is modulated by a "back-gate" which stores the photongenerated charges. This structure was intended for CCDs and was reported to exhibit an input referred equivalent noise level of approximately 16 electrons rms. In all the reported approaches lowly doped sections taking part in either readout current or charge storage, are in proximity or direct contact to Si-SiO_2-interfaces. Some of them also need additional poly-silicon gates for proper biasing of the storage node, so that charges can be further transferred in CCD manner.

In Figure 5.6 an alternative readout structure is depicted that was presented by Matsunaga in 1987 [MOI87]. In this structure the readout channel is displaced from the Si-SiO_2 interface, so that contact with trapping centres is avoided. The channel is modulated from above by the n-type region which stores the photon-generated charges. Since this region has to be lowly doped to ensure sensitivity with respect to the controlling gates, it can be fully depleted by the reset transistor. Thus not only flicker noise and RTS noise is omitted but reset noise is eliminated as well. For this device an equivalent input referred noise level of only 1 electron rms is reported. Nevertheless, the complexity of this device hinders its use for low-cost applications. All the referred structures intended for CCD readout have to be designed carefully to yield a smooth gradient in transfer mode, while using meaningful potentials. This, however, may have impact on metal-wires and inter-metal dielectrics.

Compared to above explained rather complex approaches, though, the noise performance was much improved by only minor process modifications. Creation of a buried MOSFET as the source follower device yielded a satisfactory noise performance, while using only one additional implantation and one mask [WSR08; CWM09].

5.3   CORRELATED DOUBLE SAMPLING

As described in the former section *correlated double sampling (CDS)* is a readout scheme which can e.g. be implemented by switched capacitor filters. Firstly, it was introduced by [Whi74]. The functionality of CDS can be interpreted either in the time or in the frequency domain, whereas one has to be sensitive to what frequency domain means since this is a time-discrete filter. In the time domain CDS can be understood as the subtraction of two consecutive samples of an analog signal. Analyzing a CDS filter with the z-transform method a low-pass filter characteristic is yielded for signals below half of the sampling frequency [Whi74]. However, as noted in Section 3.2.2 this method is not feasible to study the impact of white noise.

Operation principle

Figure 5.7 displays a typical implementation of a CDS filter as it is typically imple-mented at column-level. The necessary timing to operate this circuit is given in the right part of the same figure. With beginning of t_0 the *operational transconductance amplifier (OTA)* is operated in unity-gain mode by closing the switch ϕ_1, pre-charging the sample-capacitor C_S to $V_{in}(t) - V_{ref_{OTA}}(t)$. Simultaneously the switch ϕ_3 is closed and the switch ϕ_2 remains opened. This pre-charges the feedback capacitor C_F to $V_{ref_{CDS}}(t) - V_{ref_{OTA}}(t)$. The signals are sampled on C_S and C_F at t_1 by opening the switches ϕ_1 and ϕ_3. Assuming a complete charge transfer, the sum of the stored charges equals:

$$Q(t_1) = (V_{in}(t_1) - V_{ref_{OTA}}(t_1))C_S + (V_{ref_{CDS}}(t_1) - V_{ref_{OTA}}(t_1))C_F. \qquad (5.2)$$

After closing ϕ_2 at t_2 the capacitor C_F is introduced into the feedback path of the OTA. Assuming $V_{ref_{CDS}} \neq V_{ref_{OTA}}$ – now the load capacitance C_L which was charged to $V_{ref_{OTA}}(t_1)$ and left floating within the interval $t_1 - t_2$ is charged by C_F through the on-resistance of switch ϕ_2. This displacement current causes the potential at the high-resistance node at the negative OTA input, between C_S and C_F to drop so that the OTA turns on – caused by the potential difference between the inverting and non-inverting input. This affects the output voltage, so that the potential difference at the input of the OTA is compensated. Thus, assuming the interval $t_3 - t_2$ is sufficient for the output voltage to settle and additional parasitic effects can be neglected, the sum of the stored charges amounts

$$Q(t_3) = (V_{in}(t_3) - V_{ref_{OTA}}(t_3))C_S + (V_{out}(t_3) - V_{ref_{OTA}}(t_3))C_F. \qquad (5.3)$$

Of course, the amount of charges is kept constant, since the capacitors cannot be discharged through the high-resistance node. With $Q(t_3) = Q(t_1)$ the output voltage at t_3 becomes

$$V_{out}(t_3) = V_{ref_{CDS}}(t_1) + \left(1 + \frac{C_S}{C_F}\right)\left(V_{ref_{OTA}}(t_3) - V_{ref_{OTA}}(t_1)\right) + \frac{C_S}{C_F}\left(V_{in}(t_1) - V_{in}(t_3)\right),$$

$$(5.4)$$

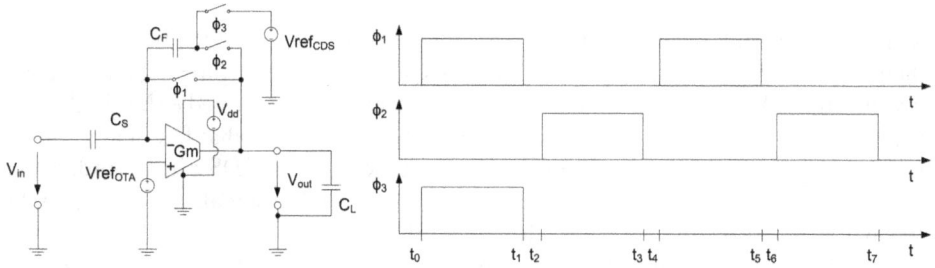

Figure 5.7 Schematic of an analog CDS filter (left) and timing (right).

which amounts to[2]

$$V_{out}(t_3) = V_{ref_{CDS}}(t_1) + (C_S/C_F)(V_{in}(t_1) - V_{in}(t_3)). \qquad (5.5)$$

for $V_{ref_{OTA}}(t_1) = V_{ref_{OTA}}(t_3)$. Optionally, the circuit can be used to accumulate signal at the feedback capacitor by executing the sampling phase without resetting the feedback capacitor during $t_4 - t_5$. Afterwards the charge has to be transferred into the feedback capacitor C_F by closing ϕ_2 again ($t_6 - t_7$). This sequence can be repeated until the signal is intolerably altered by non-linearity as it can be caused by the OTA when the output approaches v_{vdd} or *gnd*. In the following text, however, it is assumed that the standard CDS operation without multiple accumulations is performed.

A DC offset of the amplifier is apparently canceled as well as low-frequency noise components. However, the accuracy at the output is ultimately limited by the remaining input referred noise level of the OTA, the noise exhibited by the reference $V_{ref_{CDS}}$ and the remaining input noise from V_{in} which is transferred to the output. Additionally, noise may arise from the positive power supply rail and *gnd*. The impact from v_{vdd} is damped by the power supply ripple rejection ratio of the OTA. Typically disturbances mostly affect the signal chain through the source follower transistor, where they can capacitively couple onto the floating diffusion node. Disturbances on *gnd* are also damped by the OTA but directly affect the output, because that is ultimately referred to *gnd*.

A typical simplification that originates from Equation 5.4 is that flicker noise components arising from V_{in} or $V_{ref_{OTA}}$ are negligible because of the subtraction of correlated noise processes. Since the white noise, however, is not correlated subtraction of those components is assumed to double the white noise components. These simplifications, of course, imply that the sampling frequency of the CDS filter has to be sufficiently high to justify the elimination of the flicker noise. Since the thermal noise,

---

[2]Applying $T_s = t_3 - t_1$ Equation 5.5 can be expressed in z-domain. Substitution of $z_z = \exp(j2\pi\nu T_s)$ will then allow for interpretation of the filter in the frequency domain. However, this method relies on time-invariance and linearity of the system. Unfortunately, this is not valid for frequency components that are not significantly lower than $1/2\pi T_s$. Thus, this method should not be employed for the study of noise propagation within switched-capacitor filters (c.f. Section 3.2).

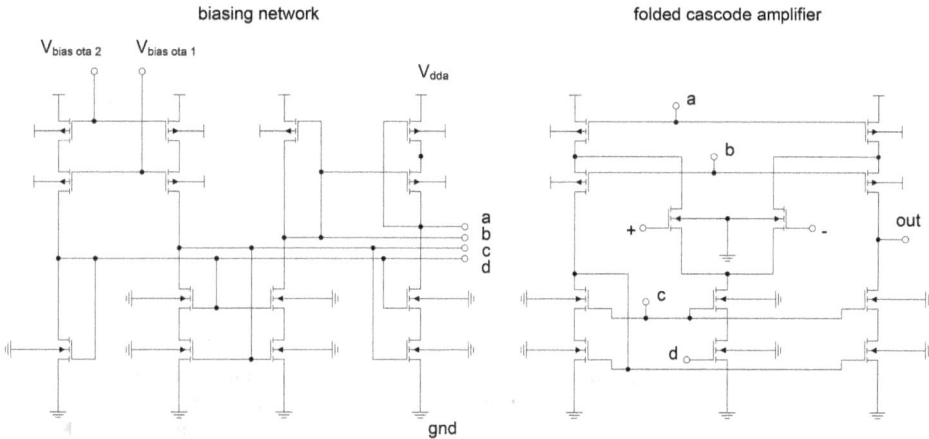

Figure 5.8 Schematic the folded cascode amplifier (right) and its biasing network (left).

however, is always undersampled in switched-capacitor filters such a justification is not simply made for the propagation of the white noise components.

Case study to demonstrate noise filtering performance

Within the framework of this work a CDS circuit as it is depicted in Figure 5.7 was fabricated and characterized. The purpose of these measurements was the demonstration of the filtering performance to various noise processes. Therefore, firstly, the measurement setup was characterized, followed by the characterization of the OTA and finally noise sources were applied to the input of the filter to study the filtering performance.

To attenuate parasitic impact arising from the power supplies and voltage references, these were realized with low-noise batteries which yield stable output voltages through the use of low-dropout regulators and by-pass and buffer capacitors. The entire measurement setup was carefully shielded to avoid pick-up of any interference. Furthermore, a low noise voltage-to-voltage amplifier – *HVA-200M-40-F* – with an input impedance of 13 pF in parallel to 1 MΩ, a bandwidth of 200 MHz and an output impedance of 50 Ω was used to enhance the noise of the DUT over the noise floor of the spectrum analyzer (*Agilent 35670A*) or oscilloscope (*LeCroy LC684DL*) by amplification of 20 dB. The oscilloscope was used to measure the time-varying variance of the CDS filter, whereas the spectrum analyzer was employed to study the performance of the *low noise amplifier (LNA)*, the OTA, the power supplies, references and the noise sources. The 1/f noise sources were generated with the spectrum analyzer and the white noise sources with an *Agilent 33110A waveform generator*.

The presented CDS circuit employs a folded cascode OTA, which typically offers a very high output impedance, a wide voltage swing and a high transconductance (c.f. e.g. [AH02]). The biasing network basically comprises large voltage-swing current mirrors that propagate meaningful branch current to the OTA (c.f. e.g. [AH02]). The capacitive load defined by the interconnects at the output of the OTA was measured to 54 pF (a *Keithley 590 CV analyzer* was used). With the OTA being operated in unity-gain

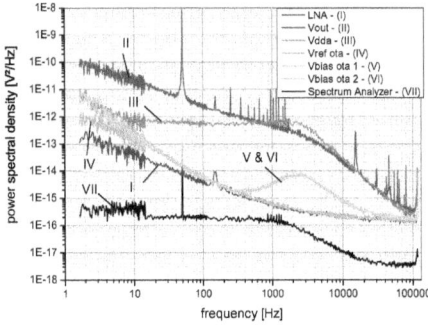

Figure 5.9   Power densities of the power supplies, the references, the LNA, the spectrum analyzer and the OTA.

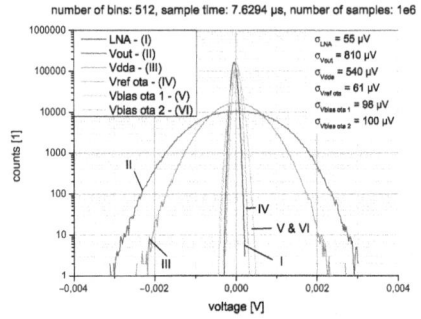

Figure 5.10   Histograms of the power supplies, the references, the LNA, the spectrum analyzer and the OTA.

mode (ϕ_1, ϕ_3 = ON, ϕ_2 = OFF) and connected to the LNA the bandwidth was evaluated to 23 kHz using the scope and the waveform generator. Using $v_{3dB} = G_m/C_{load}$, this corresponds to a transconductance of approximately 1.5 µA V^{-1}.

The noise performance of the measurement setup and the OTA is displayed in Figures 5.9 and 5.10. The displayed data is referred to the output of the CDS filter which corresponds to the input of the LNA. From noise floor of the measurement setup, the power supplies and voltage references were designed to lie below the noise exhibited by the OTA, making that the dominant contributor (c.f. Figures 5.9 and 5.10). The noise performance of the OTA approximately follows a 1/f shape for frequencies below 10 kHz. From then on the noise spectral density is mostly arising from the $V_{ref_{OTA}}$ which is propagated to the output of the amplifier in unity-gain mode. This is also covered by the bandwidth limitation which was evaluated to ≈23 kHz.

Interestingly, the histograms of all noise processes seem to be in good agreement with a Gaussian approximation.

The waveforms that are applied to the CDS circuit during the actual CDS operation are displayed in the upper part of Figure 5.11. Within this case-study the ϕ_3-signal follows ϕ_1 so that no multitudes of accumulations are performed but the standard CDS-operation is yielded. The stimuli were generated by an *Agilent 16902 Logic Analysis System*. Because the switched capacitor filter is by definition time-variant, measurements were done in time domain using an oscilloscope. By picking a *time/dev* of 10 ns and zooming in the vicinity of t_3, transients are gated out so that by employing the histogram-option of the scope statistics of the relevant time instance is yielded. That means that a certain amount of data points for the histogram are sampled within the displayed time frame of e.g. 8 dev of 10 ns/dev while the next sequence of data points is sampled after the period $t_3 - t_0$. This, however, implies that while keeping the total number of samples and the *time/dev* constant, increase of the period will cause that more low-frequency components are taken into account while high-frequency components are filtered. Thus, comparison along the $t_3 - t_1$-axis has to be treated carefully.

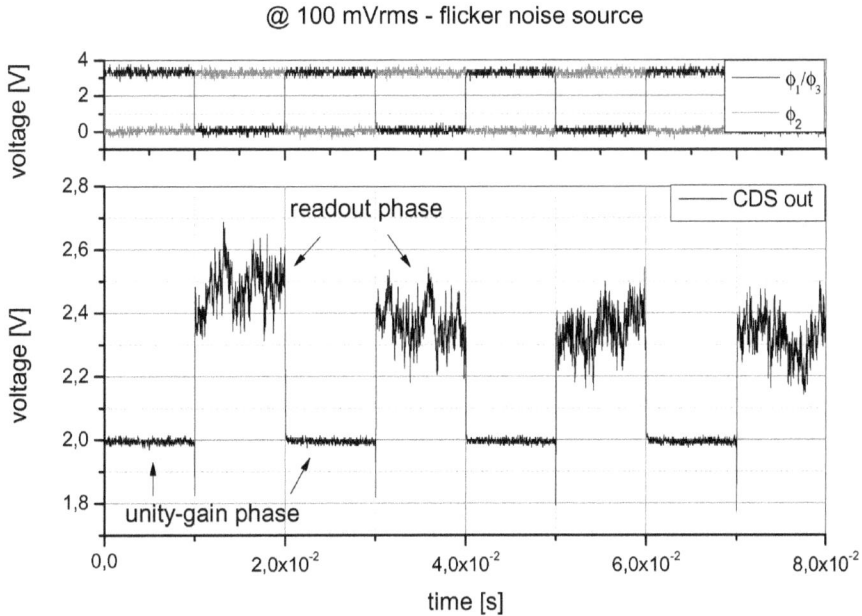

Figure 5.11 Applied waveforms and measured response of a CDS circuit.

To study the noise shaping performance of the CDS stage noise sources are applied to the input. The most perfect white noise source, of course, is an electrical resistor in thermal equilibrium. However, connecting a resistor to the input of a CDS stage also affects the deterministic behavior of the circuit because the time constant to charge C_S varies. In Figure 5.12 one can observe that the noise exhibition is rarely varying for different resistors though these were varied within 5 orders of magnitude.

Thus, a white noise generator with a low output impedance was connected to the CDS-stage. To study the impact of a white noise process at the input of the DUT, three different noise levels were produced. Their spectral densities and histograms are given in Figures 5.13 and 5.14, respectively.

Within the observed frequency-range the generator almost perfectly produces frequency-independent white noise processes. These were also observed to be in good agreement with a Gaussian process.

The response of the CDS-stage to varying white noise sources was studied by sampling the signal at the end of the feedback phase (t_3) multiple times and performing statistical analysis. In addition to the variation of the intensity of the white-noise process, the correlation-times $t_3 - t_1$ were varied throughout the measurements. Figure 5.15 displays the obtained results.

The measurements with $t_3 - t_1 = 10\,\mu s$ have to be treated carefully, because of the limited bandwidth. This may result in a lower noise level due to the fact, that high-frequency noise components cannot any longer assumed to be uncorrelated. However, to yield accurate estimates, proper calculus should be employed. The remaining results can be considered as independent of the correlation times. For the white noise sources

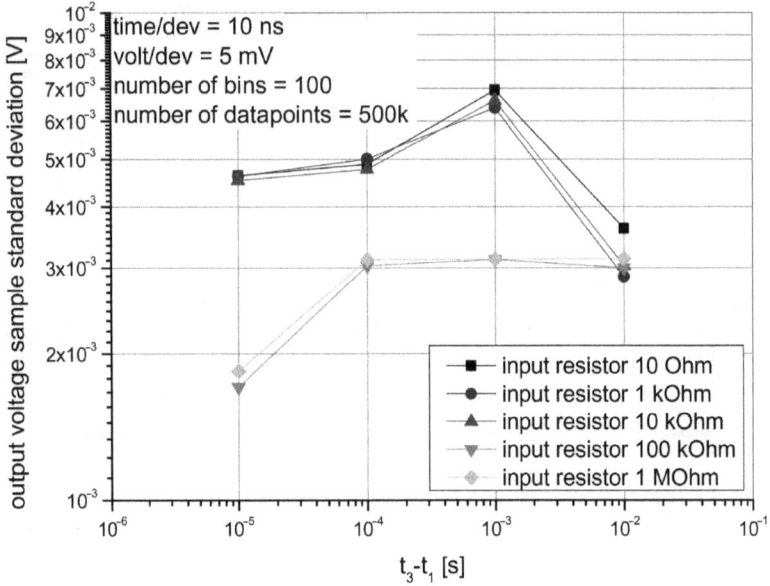

Figure 5.12 Noise shaping performance of a CDS circuit with its input connected to gnd via defined resistances.

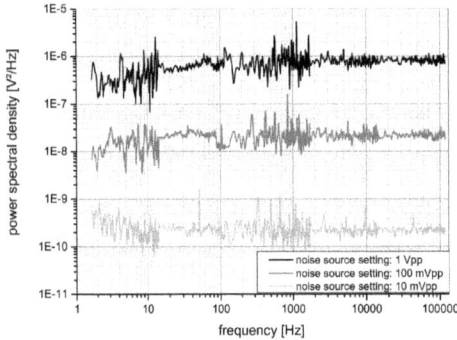

Figure 5.13 Power spectral densities of the white noise sources.

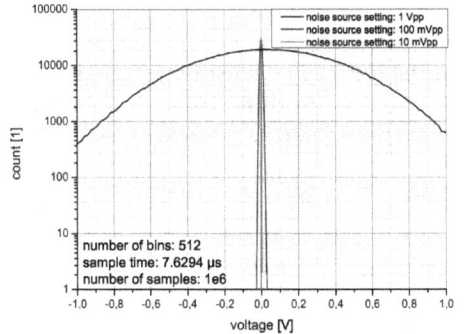

Figure 5.14 Histograms of the white noise sources.

of 10 mVpp and 100 mVpp, the output of the CDS has slightly lower variance than the input process, which does not correspond to the typical assumption that white noise is simply doubled. This may be caused by high-pass filter characteristic in combination with the bandwidth limitation. For 1 Vpp, the output of the CDS filter has slightly higher variance than the noise source. Here, the approximation of doubling the white noise component is justified. Unfortunately though, the transition from 100 mVpp to 1 Vpp cannot be explained without further investigations. This, however, may be resolved by proper use of models and calculus as explained throughout the Chapters 3 and 4.

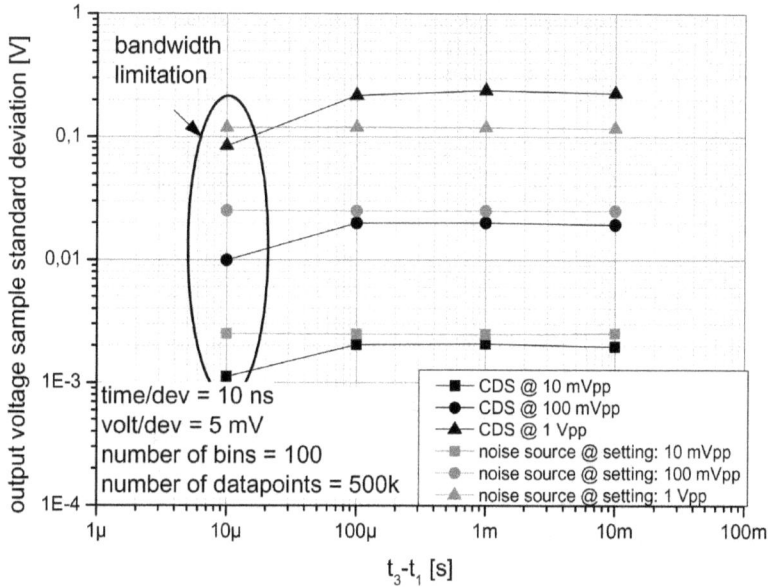

Figure 5.15 Noise shaping performance of a CDS circuit with its input connected to white noise sources.

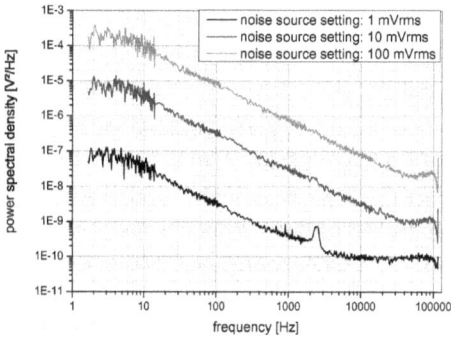

Figure 5.16 Power spectral densities of the flicker noise sources.

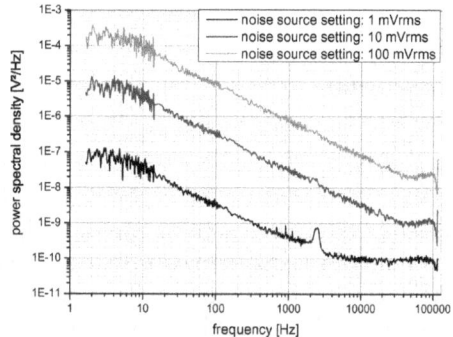

Figure 5.17 Histograms of the flicker noise sources.

In Figures 5.16 and 5.17, the flicker noise processes which were applied to the CDS are depicted. Within the bandwidth limitation of the CDS filter these exhibit a proper 1/f shape. Interestingly, it is also to be noted, that these random processes approximately follow a Gaussian distribution.

The noise shaping performance of the CDS circuit to flicker noise processes is given in Figure 5.18. Here again, $t_3 - t_1 = 10\,\mu s$ is associated with the bandwidth limitation. One has to be careful with deductions along the $t_3 - t_1$-axis, because the sampling period is varied simultaneously resulting in a filtering process which is taking more

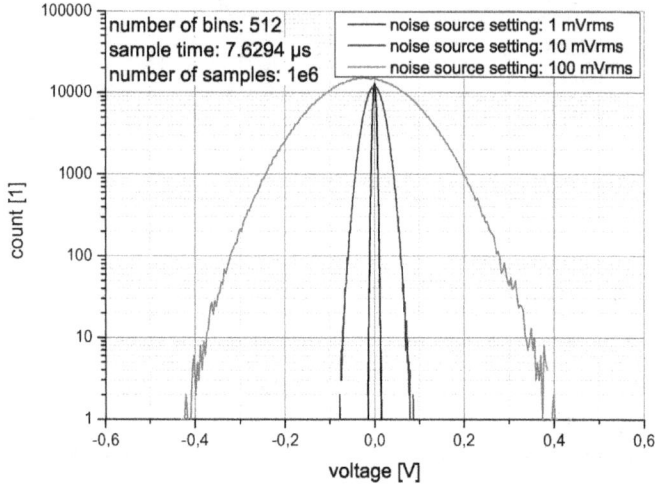

Figure 5.18 Noise shaping performance of a CDS with its input connected to flicker noise sources.

low-frequency components into account for the histograms when $t_3 - t_1$ is increased. What can be concluded from Figure 5.18 is that with increasing $t_3 - t_1$, the CDS noise performance is worsening compared to the input noise process. What cannot be simply explained is that for 10 mVrms the output of the CDS exhibits less noise than the input process for all $t_3 - t_1$, whereas for 1 mVrms and 100 mVrms a transition can be observed.

Conclusion

Surely, the widely employed CDS filter has its use for e.g. image sensors. During the design, however, one has to be careful about the noise arising within the OTA and about the mixing of undersampled random processes such as white noise. Also the speed of the CDS has to be carefully chosen, since the impact of flicker noise component on the one hand may be reduced by operation at higher sampling frequencies but the necessary bandwidth will simultaneously increase the impact of the white noise. Obviously, there must be an optimum which should be aimed at. Proper calculus has to be employed to ensure the performance of the CDS filter. Otherwise, it might happen that the performance deviates from the expectation and may even exceed the noise level of the input signal.

5.4   NOVEL JFET READOUT STRUCTURE FOR CMOS APS[3]

Based on the achievements of the past research concerning low-noise readout structures for photodetectors that are feasible for integration in CCD or CMOS processes, a novel

---

[3]The author wants to acknowledge Andrey Kravchenko who supported this section with measurements and TCAD simulations within the framework of a student assistant position.

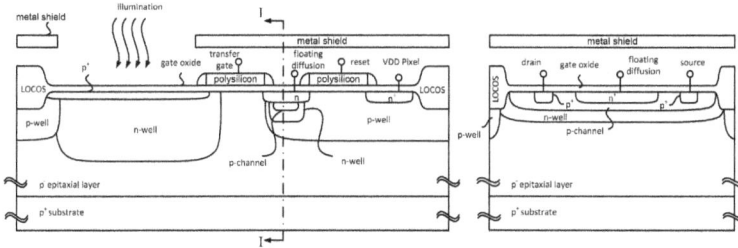

Figure 5.19 Cross-sectional views of a JFET structure implemented in a PPD – (left) along charge transfer direction, (right) orthogonal to cut II.

readout structure was proposed in [Sü13a; Sü12] and developed within the framework of this work. Although this work intends to complement the high precision sensor development in CMOS processes, the proposed readout structure is predestined but not limited to such processes and typical detectors associated with them such as PPDs and LDPDs. The proposed low-noise readout structure is based on the idea of the integration of JFETs within CMOS APS. Since the charge carrier transport within the proposed JFET readout structure takes place deep in silicon, low frequency noise will be suppressed, as discussed in the previous section. Depending on e.g. available threshold adjustment implantations and voltage levels, the implementation may be undertaken without process variations. If the proposed structure has to be used within a PPD, in worst case two additional implants become necessary. For application in combination with the LDPD only up to one additional implant is needed.

Figures 5.19 and 5.20 show how a p-channel JFET readout structure can be implemented in PPDs or LDPDs. Instead of using a standard floating diffusion in combination with MOSFETs for readout, the floating diffusion directly controls the conducting p-channel located deep in silicon by varying the expansion of the depletion region. In case of the PPD the channel is placed inside of a n-well serving as a back-gate, while in case of the LDPD the channel may be directly surrounded by the already existent well. The current flows e.g. perpendicular to the transfer direction of the photon generated electrons. The respective cross-sectional view is displayed in the right part of Figure 5.19. Drain and source are defined by p^+ diffusions, that have to be separated from the floating diffusion if the doping concentration of the latter is too high. This way Zener or avalanche effects are avoided. The separation should though not be too large, so that a contact of the current forming channel and the oxide in the lightly doped region is avoided. The alternative approach of separation using LOCOS or STI is not considered as useful, since the defect density of the gate oxide compared to LOCOS or STI is usually low [SA90; EKC92]. To avoid the traps located in proximity of the region between drain/source and the floating diffusion, additional implants may be introduced, for which it has to be made sure that no Zener or avalanche effects can occur. In standard APS the floating diffusion usually has to be highly doped in order to avoid a Schottky diode when contacted to the source follower or a PMOS reset. If, however, a NMOS reset is used, whose source is defined by the floating diffusion itself in combination with the JFET-readout structure, there is no longer need for the

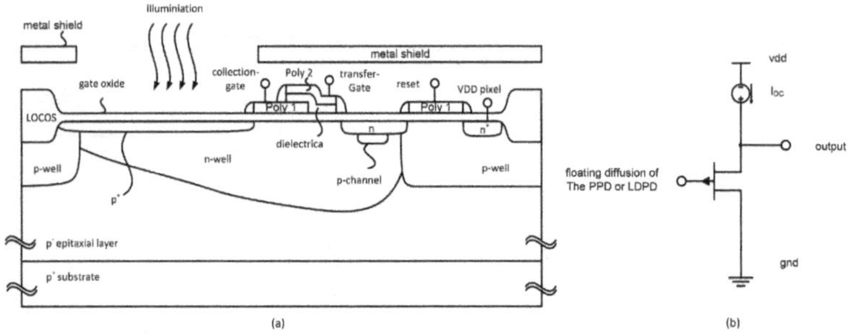

Figure 5.20   (a) cross-sectional view of a p-channel JFET embedded in a LDPD and (b) source follower implementation.

high doping concentration. Thus Zener and avalanche effects can be suppressed by using a lower doping concentration. In such a device there would be no contact of a lightly doped region to the surface at all. Moreover, if the reset transistor is designed accordingly and full-well capacity is not an issue in certain applications (since this will decrease with lower doping of the storage node), reset noise can be eliminated similar to the Matsunaga readout, which was explained in Chapter 5.2.

For an easy implementation in a CMOS process, however, the doping concentration of the channel should be higher than for the well, but lower than for the floating diffusion. For this purpose e.g. p-threshold adjustment implants for NMOS transistors may be used. The additional n-well for the PPD may result from the same implantation, from which the photodiode is formed or from a n-well used for embedding PMOS transistors. The channel-depth and its doping concentration should though directly be adjusted to the circuit including the JFET structure. Thus, in worst case the improvement in noise performance is traded against two additional implantation steps and necessary costs for the masks in case of the PPD, while only up to one additional mask and implantation step may become necessary in case of the LDPD.

In general, several readout circuits using the JFET structure are possible. For instance, the transistor may be used for realization of common-source, common-gate or differential amplifiers, if a high voltage gain is needed. However, since junctions may not be well controlled as e.g. gate oxides, the transconductance may vary heavily for those circuits. A source follower implementation suffers less from device parameter variation since – provided a sufficiently high transconductance and output resistance – its closed-loop voltage gain exhibits low sensitivity to the loop gain variations. As depicted in the right part of Figure 5.20, the drain current through the channel is set to a constant value by a bias current source. The circuit works properly as long as the JFET is in saturation and the current is kept constant. Figure 5.21 shows the operation of the device. In the plot of the transfer characteristic it is shown that for a constant bias current I_{DC}, the gate-source voltage does not vary significantly due to the feedback action. Its value will approximately meet the pinch-off voltage V_P for low drain currents and zero for I_{IS} which is, however, a function of V_{DS}. Thus for high dynamic range it would be favorable to bias the device with I_{IS}. Negative bias of V_{GS} will push current through the junction, what has to be avoided. Additionally, it is interesting to

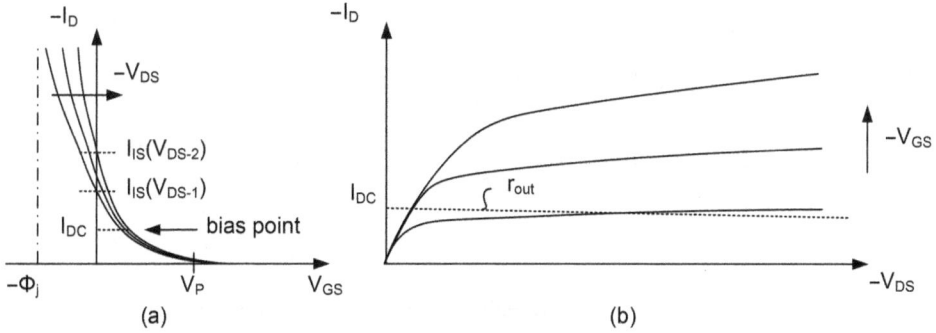

Figure 5.21 (a) transfer characteristic and (b) output characteristic of a p-channel JFET operating as a source follower.

notice that for a high reset potential at the floating diffusion the device already features high linearity when the current source is kept in saturation region. This still holds true for small-signals. For a buried NMOS transistor with a negative threshold voltage that is operating as a source follower this condition would introduce restrictions to the reset potential for the floating diffusion. This is not the case for the JFET implementation. The plotted output characteristic indicates the bias using a current source – indicated by its output resistance r_{out}. The output resistance of the latter and the bias current will define the operating point. Since V_{GS} is approximately kept constant if the drain current of the JFET suffers from low V_{DS} dependence in saturation, the output characteristic and the bias current will be approximately equal for a large voltage range of V_{DS}. This defines the output voltage swing – which, according to the quadratic MOS model, is $V_{GS} - V_P \leq V_{DS}$ – for which the device operates in saturation or, in other terms, the circuit offers high linearity.

Apparently a low pinch-off voltage V_P is important for source follower operation. Since the voltage range for V_{GS} is limited by the built-in potential of the bias junction Φ_j and the pinch-off voltage V_P this may be different for other amplifier configurations. If the doping concentration of the n^+ biasing diffusion is high compared to the n-well that is embedding the p-channel, it can be assumed that the channel is only modulated by the upper junction. A simplifying assumption of an abrupt junction yields

$$V_P = \frac{q \cdot N_A \cdot a^2}{2 \cdot \epsilon_{Si}} - \frac{k_B \theta}{q} \cdot \ln\left(\frac{N_A \cdot N_D}{n_i^2}\right), \tag{5.6}$$

for the pinch-off voltage [SN07], where q is the elementary charge, θ equals the absolute temperature, k_B stands for the Boltzmann constant, N_A equals the p-channel doping concentration, N_D is the n^+ doping concentration, n_i equals the intrinsic carrier concentration, a represents the channel depth and ϵ_{Si} is the permittivity of silicon. Since reasonable doping concentrations for the channel should be lower than the n^+ donor concentration, but higher than the concentration of the well, channel depths of only a few hundreds of nanometres should be chosen to yield a low pinch-off voltage. This

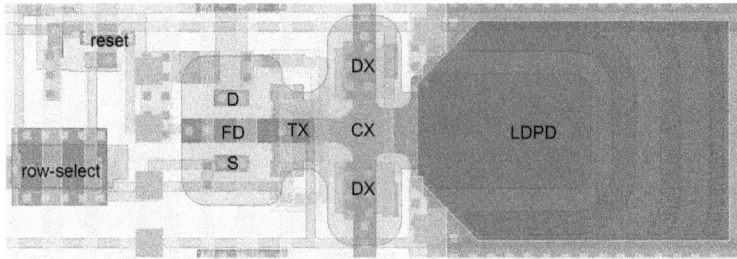

Figure 5.22 Layout of a LDPD with the proposed JFET readout structure.

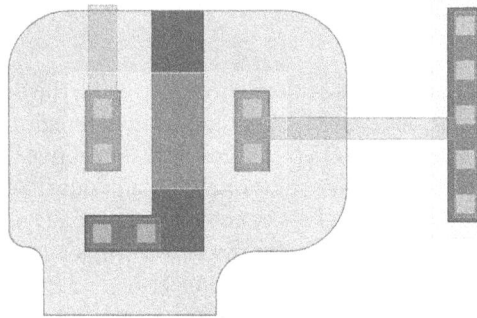

Figure 5.23 Layout of the p-channel JFET test-structure.

predestines an implantation at the far end of the process for a JFET designed for source follower operation.

An exemplary layout of a LDPD designed in a 0.35 μm CMOS process that will be fabricated in the near future is shown in Figure 5.22. The pitch of this test structure is 13.34 μm × 40 μm. Two draining gates (DX) are applied to the collection gate (CX) in perpendicular direction to the charge-transfer direction. The floating diffusion (FD) is then pulled across the p-channel and contacted to a PMOS transistor for reset. For row-select a NMOS transistor was chosen. The fill factor for the designed pitch is approximately 35%. Additionally, it is interesting to note that from the right part of Figure 5.20 it becomes clear that opposite to common APS pixels no positive power supply rail is needed in the pixels since the current source is off-pixel. Thus the fill factor can be increased by elimination of the corresponding interconnect. Alternatively, the space may be used for layouting the remaining interconnects wider and more low-impedance to avoid voltage drops and parasitic coupling.

In order to suppress pixel characteristics and to examine the behavior of the JFET only, the proposed JFET was fabricated and characterized as parallel circuit of 9 single transistors. One of those devices is exemplarily displayed in Figure 5.23. This allows for a simplified characterization of e.g. low leakage currents, which are now multiplied. The data which is presented within this section is rescaled.

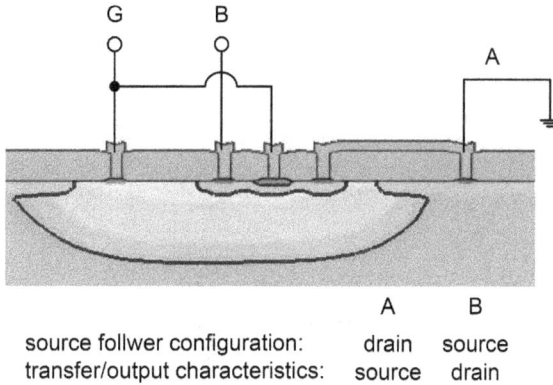

source follwer configuration: drain   source
transfer/output characteristics: source   drain

Figure 5.24 Cross-sectional view of the p-channel JFET as it was simulated with TCAD.

Since the proposed JFET-structure is actually a three-dimensional device in which the channel current propagates orthogonal to the photogenerated charge carrier transport direction (c.f. Figure 5.19), TCAD simulation is rather difficult and time consuming. Thus, to support fundamental understanding and verify hypotheses a two-dimensional variant of the proposed structure was yielded by wrapping the gate around the device as it is depicted in Figure 5.24. This sets the potential of the n-well in which the JFET is embedded. One electrode of the fabricated JFET was short-circuited to the p-substrate. Within a pixel configuration this allows to omit an additional interconnect. However, for characterization of e.g. transfer and output characteristics the other electrode has to be negatively biased with respect to the substrate so that it cannot be shorted to the substrate. The configuration of these two modes is given in Figure 5.24.

In Figure 5.25 various doping concentration profiles are presented that were achieved by variation of the energy and the doping concentration of the channel implant. The implantation was undertaken before the last thermal annealing procedure to make sure the dopants are activated while avoiding their displacement away from the surface by thermal diffusion. Clearly it can be observed that by increasing the energy during implantation, the channel width increases since the maximum of the doping concentration is displaced from the n^+ implant of the controlling junction. In Figure 5.26 electrostatic potential profiles are depicted for the different implantations described before. The Dirichlet boundary conditions were chosen to yield 0 V at all the electrodes of the device.

The transistor should have a low pinch-off voltage to remain in saturation for low $|V_{DS}|$. From the presented results it can clearly be observed that the left column in Figure 5.26 corresponds to a channel implantation with insufficient dose. There the channel is immediately pinched off at $V_{GS} = 0$ V so that no current will flow across the channel. To induce current flow in such a device the junction will have to be forward biased which automatically goes together with an intolerable leakage current through the gate. The mid column corresponds to pinch-off voltage levels in the vicinity of zero, whereas the rightmost column has a dose which allows for current flow at $V_{GS} = 0$ V. The test structures were fabricated and evaluated for varying p-channel

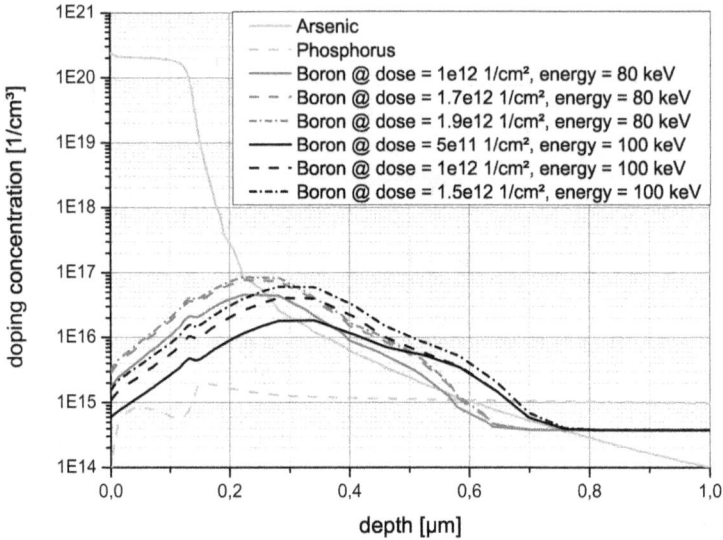

Figure 5.25 Channel doping concentration profiles of the proposed JFET readout structure for various implant doses and energies.

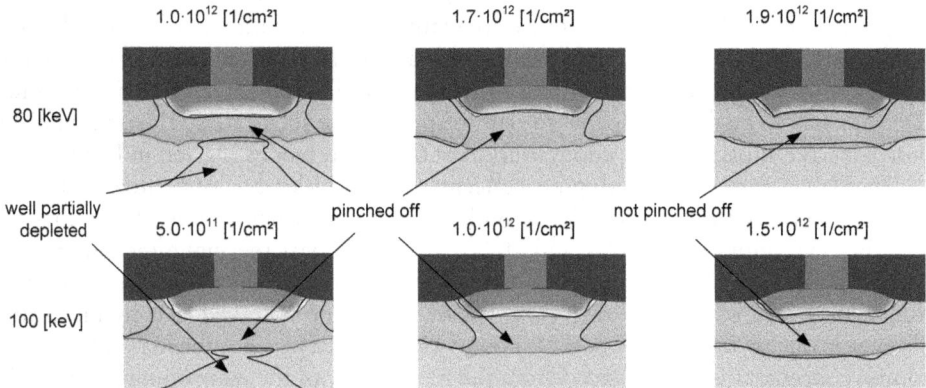

Figure 5.26 Electrostatic simulation results for the proposed JFET readout structure for various implant doses and energies.

and n-well implantations. This was necessary because the TCAD simulations do not yield accurate quantitative results as long as they are not carefully calibrated for the employed process.

In the Figures 5.27–5.30 the measured characteristics of a p-channel JFET with $W = 1.52\,\mu m$ and $L = 1.2\,\mu m$ are displayed. Its channel was implanted with $2 \times 10^{12}\ 1/cm^2$ at $80\,keV$ and embedded within an n-well that was implanted with $5 \times 10^{11}\ 1/cm^2$ at $350\,keV$. As can be seen in Figure 5.28, the transfer characteristic can be divided in three different sections with different behavior. In section I,

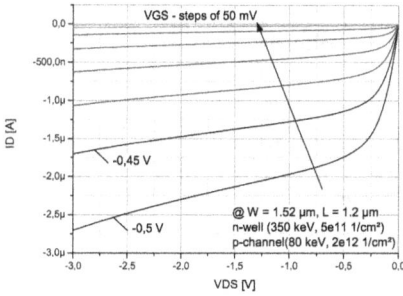

Figure 5.27 Measurements of output characteristics for $W = 1.52\,\mu$m and $L = 1.2\,\mu$m.

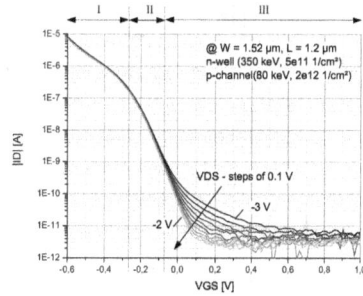

Figure 5.28 Measurements of transfer characteristics for $W = 1.52\,\mu$m and $L = 1.2\,\mu$m.

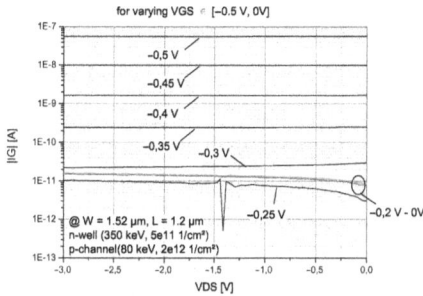

Figure 5.29 Measurements of gate-leakage versus V_{DS} for $W = 1.52\,\mu$m and $L = 1.2\,\mu$m.

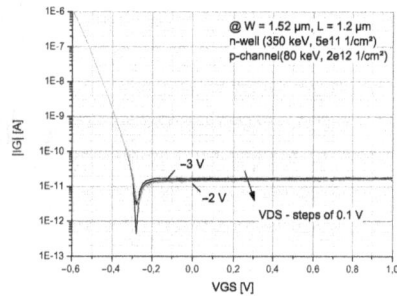

Figure 5.30 Measurements of gate-leakage versus V_{GS} for $W = 1.52\,\mu$m and $L = 1.2\,\mu$m.

which is active for strongly negative V_{GS}, the drain current is much less dependent on variations of V_{GS} than in section II, which might be explained by series resistances or a finite expansion of the channel depth. For some V_{GS} the channel will be fully conducting so that further decrease of the gate-source voltage will only contribute to the current flow by parasitic leakage current e.g. through the gate itself. As it was verified with TCAD simulations, channel length modulation is the dominant cause for the dependence of the channel current with respect to the drain-source potential. Here, the strongly negative bias of the drain region takes part at the depletion region of the channel, which – for long channel transistors – is mostly performed by the gate. This results in a V_{DS} dependent expansion of the depletion zone, reducing the channel length which then results in a higher saturation current (c.f. e.g. [Sin01]). In section III of Figure 5.28 a strong dependence on V_{DS} can be observed. To verify the origin of this phenomenon TCAD simulations have been performed at different biasing conditions. In Figure 5.31 it can be observed that for strong negative bias of the drain-source voltage a current originating from the epitaxial layer is present that

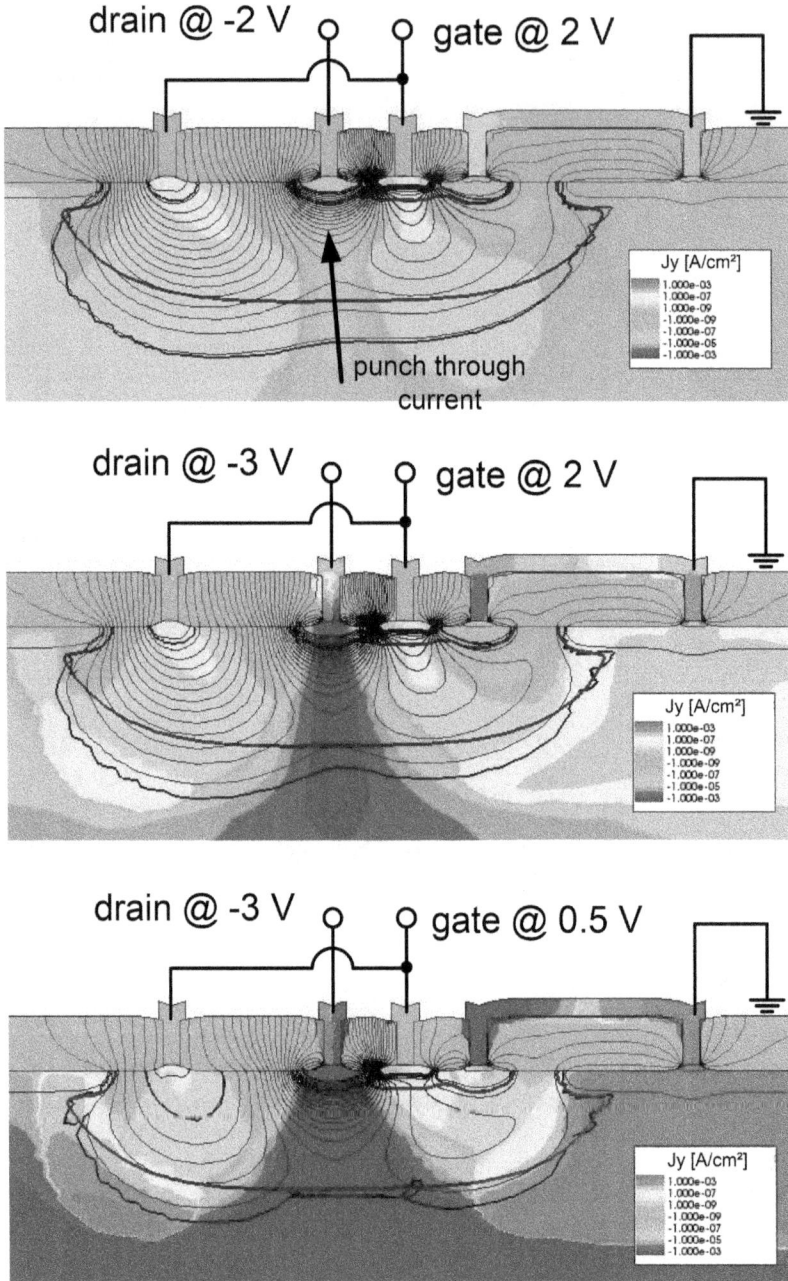

Figure 5.31   Electrostatic simulation result of current densities at different biases for observation of the strong V_{DS} dependence of I_D.

flows through the n-well into the drain region. This is caused by the the depletion zones from the drain/n-well junction and the n-well/p-epitaxial layer junction which unite to one single depletion zone. Comparing the two upper profiles in Figure 5.31 one can observe that due to the stronger negative bias of the drain in the mid plot the current density is higher, what also results in a deformation of the equipotential lines in the vicinity of the current path. Comparing the lower profiles in Figure 5.31 shows that this is of course also dependent on the gate-bulk voltage. Applying a higher V_{GB} will increase the depletion zone what results in a higher resistance and thus a reduced leakage current as it was also observed within the measurements in e.g. Figure 5.28. The leakage path within the real device will differ. This originates from the way the JFET structure was rearranged to allow for two-dimensional simulation. Conversely to the fabricated device presented in Figure 5.23, the structure for TCAD simulation has its gate wrapped around to the left-side (c.f. Figure 5.24). This results in a larger lateral distance from the drain to the epitaxial layer. Observing the source region in Figure 5.24, however, shows that the lateral distance will in reality be much shorter than the vertical. This is caused by the p-type threshold adjustment implantation which is hindering the out-diffusion of the n-well. Thus, in reality the punch-through current will be a lateral phenomenon.

Figures 5.27 and 5.30 present the output characteristics of the device. Since at $V_{GS} = 0$ V current flow is largely dependent on V_{DS} and is small in magnitude, the gate-source junction was slightly forward biased to yield meaningful currents. As can be seen in Figure 5.30, the forward bias of the junction induces a parasitic leakage current through the gate. Two regions can be distinguished what can be explained by having two parasitic leakage components through the gate. The gate-source diode is turned on for $V_{GS} \leq 0$ V and conducts a significantly lower current at $V_{GS} \geq 0$ V when it is reversed-biased. The gate-drain diode, here, always remains reversed-biased. Those two diodes both induce parasitic leakage through the gate – but they do exhibit a different sign. Thus, for approximately $V_{GS} = -0.3$ V there is no DC current observed which cannot be explained by having leakage only through the gate-source diode. Consequently, it can be concluded that there is always a current flowing through the gate which, however, can be compensated at approximately $V_{GS} = -0.3$ V. Unfortunately, from a non-deterministic point of view it is known that both components will exhibit uncorrelated shot-noise components which will alter the devices performance even when there is no DC-current through the gate (c.f. discussions about noise in diodes in e.g. [van86; BS97; Amb82]). Figure 5.29 confirms the behavior observed in Figure 5.30 – the leakage current is hardly dependent on V_{DS} because that junction is reverse-biased, but it is strongly dependent on V_{GS}. In the Figures 5.32–5.35 output and transfer characteristics for varying geometries are depicted.

By comparing Figure 5.27 with Figures 5.32 and 5.33 it becomes clear that I_D below $V_{GS} = -0.3$ V scales more by varying length than by the width. Comparing the transfer characteristic of Figure 5.28 with 5.35 it can clearly be seen that for shorter length the V_{DS}-dependence of I_D which was mostly associated with the punch-through is increased. Interestingly the transfer characteristic for larger widths shows a lower punch-through current component. This is similar to the output characteristics. The scaling of the current for increasing width or length might be explained by having a larger and more homogeneous n-well implant. The higher n-well concentration might help maintaining the predominant modulation of the depletion zone from the gate.

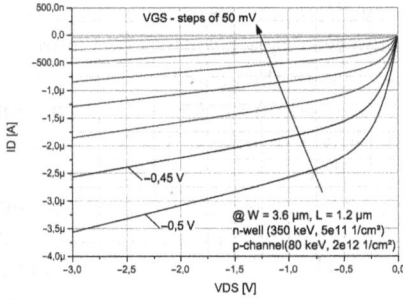

Figure 5.32   Measurements of output characteristics for $W = 3.6\,\mu m$ and $L = 1.2\,\mu m$.

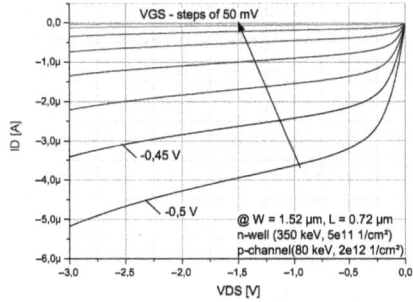

Figure 5.33   Measurements of output characteristics for $W = 1.52\,\mu m$ and $L = 0.72\,\mu m$.

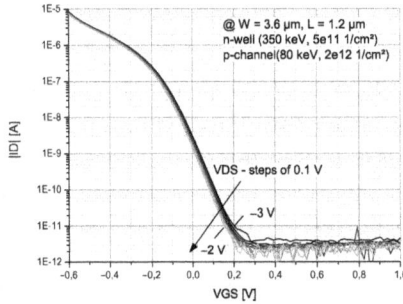

Figure 5.34   Measurements of transfer characteristics for $W = 3.6\,\mu m$ and $L = 1.2\,\mu m$.

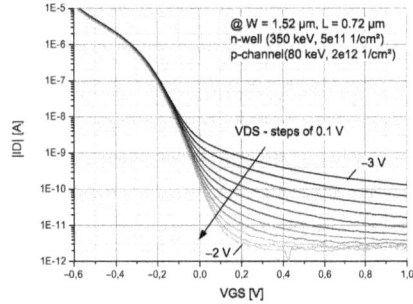

Figure 5.35   Measurements of transfer characteristics for $W = 1.52\,\mu m$ and $L = 0.72\,\mu m$.

This has to be verified in the future by extension of the n-well for constant W/L of the JFET. With two-dimensional simulations, however, this is difficult if not impossible to prove.

In Figures 5.36 and 5.37 the noise performance of the $1.52 \times 1.2\,\mu m^2$ p-channel JFET is presented. It was biased with a rather high current of $I_D = 5\,\mu A$ to enable comparison against the presented NMOSFETs in Figures 4.17–4.18. As can be seen from Figure 5.28, the gate-source junction is slightly forward bias to allow for $I_D = 5\,\mu A$. It can be observed, that the p-channel JFET yields similar performance to the NMOSFET transistors while having much smaller width and length dimensions. This also makes the JFET advantageous for low-light imaging applications, since it can be expected to achieve a low input capacitance compared to the large n-type MOSFET devices and thus a higher conversion gain. This statement has to be treated carefully, since the JFET also necessitates the embedding n-well, resulting in a larger area consumption. In the end real constraints as for instance pixel pitch, integration time, frame rate or SNR for a specific application have to be used for proper comparison.

The performance of the $1.52 \times 1.2\,\mu m^2$ p-channel JFET in source follower configuration is given in Figures 5.38–5.39. Since the pinch-off voltage of the device is in the

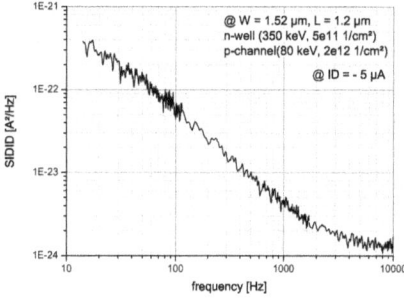

Figure 5.36 Output referred noise current spectral density for $1.52 \times 1.2 \, \mu m^2$ p-channel JFET at $I_D = 5 \, \mu A$.

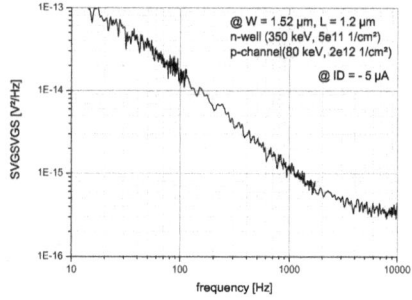

Figure 5.37 Input referred noise voltage spectral density for $1.52 \times 1.2 \, \mu m^2$ p-channel JFET at $I_D = 5 \, \mu A$.

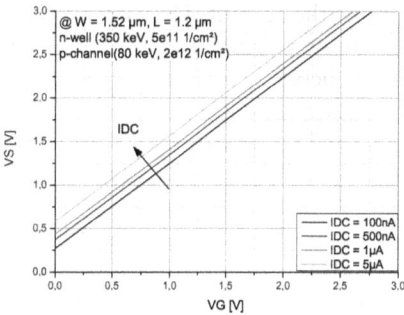

Figure 5.38 Measurements of the transfer function of the $1.52 \times 0.72 \, \mu m^2$ p-channel JFET in source follower configuration.

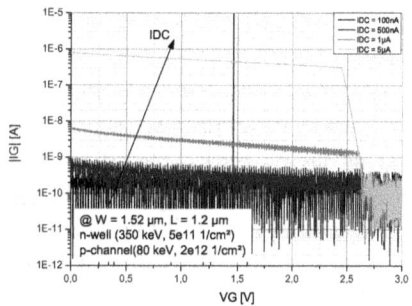

Figure 5.39 Measurements of the gate current of the $1.52 \times 0.72 \, \mu m^2$ p-channel JFET in source follower configuration.

vicinity of zero, the device remains in saturation almost throughout the entire operational range. At higher gate potentials the source potential is no longer increasing. This is caused by the compliance of the current source which was set to 3 V. With increasing bias current the gate-source diode is more forward biased resulting in increasing offset-voltages as can be observed in Figure 5.38 and higher leakage currents as it is depicted in Figure 5.39. The decreasing leakage current at higher V_G is again caused by the compliance setting. Unfortunately, this might not be applicable for imaging purposes since the leakage current will alter the charges on the storage node of the photodetector. To decrease the leakage current while maintaining a proper biasing current as it is necessary to charge the read-line from the source follower output to the signal conditioning circuitry, the intersection current I_{IS} depicted in part a) of Figure 5.21 has to be increased. This is originated in the fact, that the bias current of the source follower defines the slew-rate. The value of the bias current can then be picked such as to operate the JFET in an optimal manner.

Within the Figures 5.40 and 5.41 the effect of a varying n-well dose can be observed. It becomes clear, that for lower dose the pinch-off voltage is shifted to higher

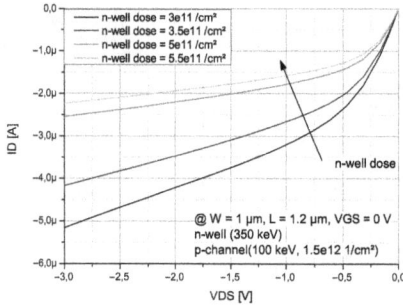

Figure 5.40 TCAD simulation results of output characteristics for varying n-well dose.

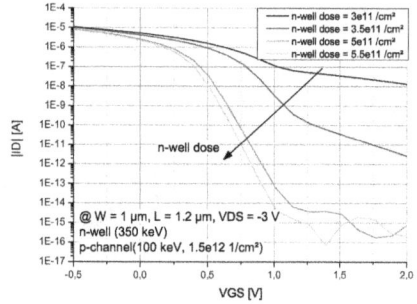

Figure 5.41 TCAD simulation results of transfer characteristics for varying n-well dose.

values and the gate is no-longer capable of controlling the current. The same applies for increasing implant energies for the p-channel as can be observed within Figures 5.42 and 5.43. The difference in the order of magnitude between these two effects is very likely caused by the fact, that by varying the n-well there is still some capability left to control the channel by the upper n^+ diffusion. Increasing the energy of the p-channel implant, however, separates the channel from the controlling n^+ diffusion resulting in a higher pinch-off voltage with the channel length modulation being rarely affected.

In general it is well known, that TCAD simulation rarely give reliable results if TCAD is not properly calibrated to the underlying process (c.f. [Syn09a; Syn09b]). However, as did become clear in this section, it was very well possible to reproduce the phenomena that were observed from the measurement results. The strategy which is used to decide on the future redesign is thus starting from the measurements and using the simulations to decide which parameters have to be changed and in which order of magnitude. The first generation of JFETs exhibited an intersection current I_{IS} in the transfer characteristics which was much too low to allow proper speed while avoiding leakage current through the gate (c.f. Figures 5.21 and 5.28). An increase in p-channel dose would result in an increased drain current but would simultaneously result in a stronger channel length modulation, if the n-well dose is not increased. The n-well dose, however, is defined by an external restriction from the photodetector to which the structure has to be connected in the future. Thus, the p-channel dose should be kept at $2 \times 10^{12} \, \text{cm}^{-2}$.

This results in the implant energy as the only left variable to be rearranged. From Figure 5.43 it becomes clear that an increase from 80 keV to 100 keV increases the pinch-off voltage approximately 400 mV. As can be seen in Figure 5.28 this would result in $I_{IS} \approx 1 \, \mu\text{A}$ which is a tolerable value. The TCAD simulation results of such a device in a source follower configuration are given in Figures 5.44 and 5.45. The offset-voltage can be adequately removed by meaningful biasing currents in the lower micro amperes domain. Furthermore, the linear operational range is very wide and extends almost down to 400 mV for some biasing conditions. As can be observed in Figure 5.45, the finite transconductance of the JFET in combination with its finite output impedance result in the fact that with constant bias current, V_{GS} cannot be

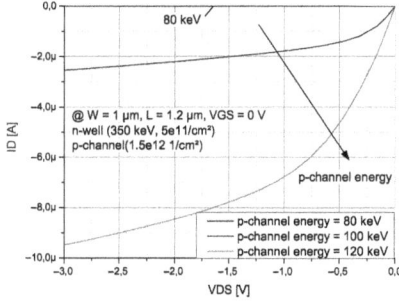

Figure 5.42 TCAD simulation results of output characteristics for varying p-channel energy.

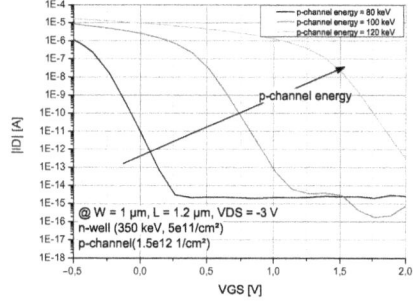

Figure 5.43 TCAD simulation results of transfer characteristics for varying p-channel energy.

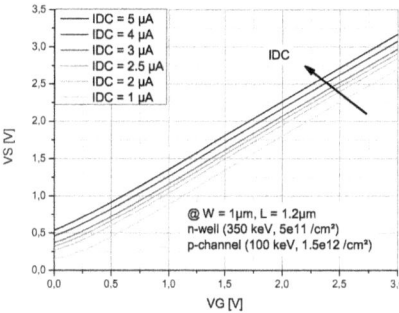

Figure 5.44 TCAD simulation results of the p-channel JFET in source follower configuration – 1.

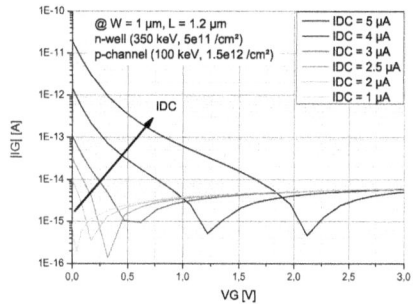

Figure 5.45 TCAD simulation results of the p-channel JFET in source follower configuration – 2.

kept at zero over the entire operational range. There is a transition to be observed from having the gate-source diode slightly forward biased for low V_G towards biasing it reversed for higher V_G. The change from one region to another is, of course, dependent on the bias current.

For better performance the width of the n-well should be increased to ensure a homogeneous doping concentration and to support the capability of the gates to control the channel and to avoid the punch-through effect. The length of the device might be slightly enlarged to avoid channel length modulation. These modifications will hardly affect the overall area consumption of the device and are thus considered to be meaningful. Applying the JFET structure in combination with an LDPD the punch-through phenomenon will always be present, since the LDPD is based on having an n-well that is mostly depleted to allow for fast charge transfer. Thus, an important issue that has to be studied is the impact from the punch-through phenomenon on the leakage current from the gate. If this can be proved not to be an issue, and the proper design of I_{IS} can be established, sensitivity studies should be employed to optimize the biasing current for the purpose of having an optimal signal-to-noise ratio and dynamic range. This might be supported by analytical modeling of the device, which would also allow for circuit level simulation with e.g. SPICE/SpectreRF. The device is then to be studied in combination with the actual photodetector. Finally, studies may be undertaking how to scale the device without intolerably altering the performance and using the least amount of process variations to maintain cost-efficiency.

Chapter 6

On the design of PM-ToF range imagers

This chapter is dedicated to the actual design of low-noise range image sensors. As it has been shown in Chapter 2, the PM ToF principle is advantageous for low-cost precision range measurements in operating ranges from decimetres to metres.

In Section 6.1 the concept of the range imagers that were developed within the framework of this work is presented. Since photon induced shot noise is a major issue for optical range imagers, the physical limitations defined by this phenomenon are derived in Section 6.2. In Section 6.3 the objectives and considerations for the LDPD based ToF range image sensors are presented. Its implementation and evaluation by measurements is given in Section 6.4. From the imperfections within the first generation further considerations were necessary that led to a redesign which is covered in Section 6.5 followed by an outlook on the impact of the remaining imperfections on the systems performance in Section 6.7. Finally an alternative approach to [Sp11a; Sp11b] which aims at better matching performance is presented in Section 6.6.

6.1 BASIC CONCEPT AND CONSTRAINTS

In Figure 6.1 a schematic of the basic concept of the PM ToF principle is depicted. A pulsed light source is collocated aside the range detecting image sensor. The displacement of the light source and the image sensor can be neglected if the object distance is long; in other words if $\alpha_d \approx 0$. The emitted light pulse is diffused to probe a certain field-of-view, reflected by the object under observation, focused on the image sensor by an optical system and detected by the image sensor. The image sensor measures the time-of-flight by the multiple short time integration method (c.f. Section 2.3.3). A controlling unit defined by a *field-programmable gate array* (FPGA) is realized on a *printed circuit board* (PCB) to allow for a flexible stimulation of the range camera. The FPGA is intended to control the pulsed light source and the actual image sensor. To allow for a compensation of delays that may be introduced by routing or signal shaping circuitry, a programmable delay must be implemented in order to properly synchronize the laser and the shutters (cf. Chapter 6.7.4). This can be realized within the FPGA or within an additional delay line.

The timing depicted in Figure 6.2 corresponds to MSI with three shutters that implement the operation of short time integrators: $u_{T_{\text{SW}},\tau}(\tau_{\text{ToF}}) = \int_{\tau}^{\tau+T_{\text{SW}}} x_s(\tau_{\text{ToF}},t)\mathrm{d}t$ (c.f. Section 2.3.3). Here, ideally, shutter TX1 yields $V_{\text{FD1}} \propto u_{T_p,t_3}$, TX2: $V_{\text{FD2}} \propto u_{T_p,t_3+T_p}$

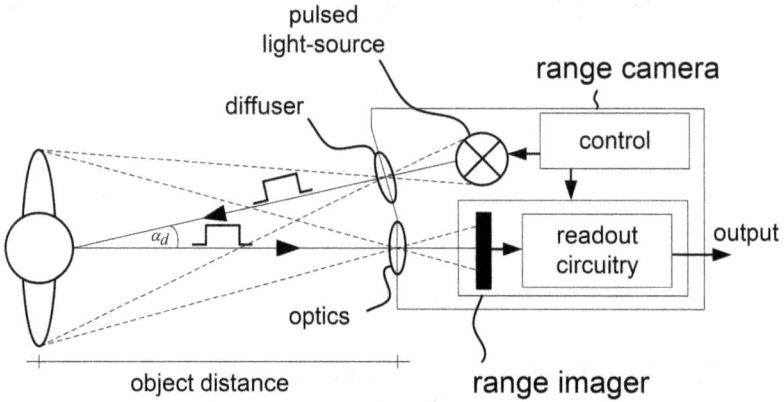

Figure 6.1 Schematic of the basic PM ToF principle (according to [Jer09]).

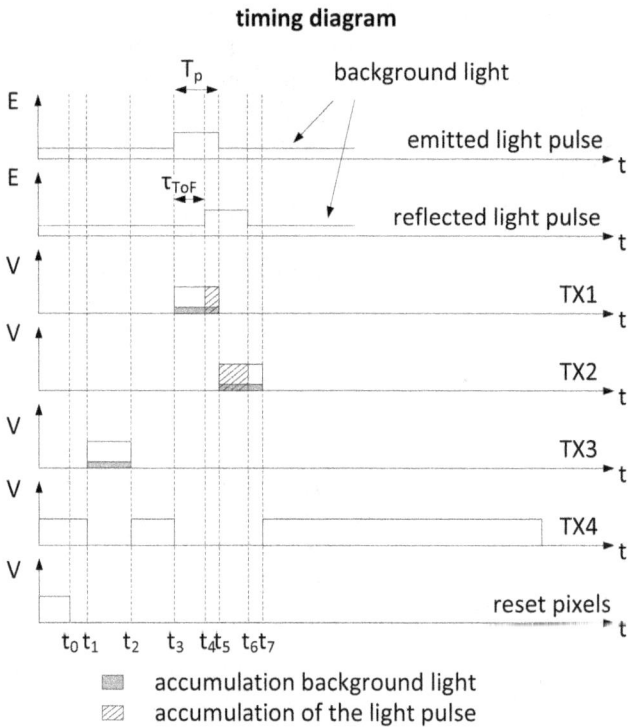

Figure 6.2 Schematic of the basic MSI PM ToF timing.

and   TX3: $V_{FD3} \propto u_{T_p,t_1}$,   so   that   the   object's   distance   can   be evaluated as

$$z = \frac{c}{2} T_p \frac{V_{FD2} - V_{FD3}}{V_{FD1} + V_{FD2} - 2 \cdot V_{FD3}}. \qquad (6.1)$$

Shutter TX3 is intended for the definition of a range independent reference which only accumulates background light, thus no laser is triggered during the corresponding shutter period. TX1 and TX2 correspond to the range dependent signals. TX4 is necessary to drain photogenerated charge carriers during the intervals in which non of TX1-TX3 are selected (cf. Chapter 6.3.2). Conversely to the simple MSI method based on two shutter windows, this ideally allows for elimination of the reflectance and the background light dependence. This, however, becomes difficult since neither does the laser exhibit an ideal rectangular pulse shape, nor does the image sensor perform perfect short time integration. Even if these effects could be neglected, the photon noise of the background light causes a deterioration which is of course dependent on the intensity of the signal, the background light level and the object's position, orientation and reflectance as it will be thoroughly explained in Section 6.2. The presented timing can be executed several times (without reset) in order to accumulate more signal related photogenerated charge carriers. This measure corresponds to time-averaging and thus improves the accuracy.

In a realistic setup the shutters will not perfectly carry out the above mentioned functionality, but will suffer from e.g. non-linearity. If each shutter can be modeled by a map $V_{FDi} = f_{TXi}(u_{T,\tau}(\tau_{ToF}))$ – assuming mutual independence for the sake of simplicity – the object's distance can be extracted by

$$z = \frac{c}{2} T_p \frac{f_{TX2}^{-1}(u_{T_p,t_3+T_p}(\tau_{ToF})) - f_{TX3}^{-1}(u_{T_p,t_1}(\tau_{ToF}))}{f_{TX1}^{-1}(u_{T_p,t_3}(\tau_{ToF})) + f_{TX2}^{-1}(u_{T_p,t_3+T_p}(\tau_{ToF})) - 2 \cdot f_{TX3}^{-1}(u_{T_p,t_1}(\tau_{ToF}))} \qquad (6.2)$$

if f_{TXi} are bijective functions and thus their inverse functions f_{TXi}^{-1} exist. Characterization of the functions f_{TXi} and development of their inverse functions has to be done during calibration (c.f. [MBG11]).

6.2 PHYSICAL LIMITATIONS DUE TO PHOTON INDUCED SHOT NOISE

Considering an ideal range imager, which perfectly implements the concept described in Section 6.1, the remaining limitation will be the photon induced shot noise which limits the accuracy that can be achieved. To derive parameters such as the minimum integration time or accumulation count, respectively, or the dynamic range that has to be covered by the readout circuitry, analysis has to be done to estimate the depth resolution. Several approaches have been presented in the past to provide estimates for range imagers based on CW modulation [Sei07; Sei08; FPR09; Lan00] and pulse modulation ([Elk05; Jer09; Sp10]). All of them have applied Gaussian error analysis (equation 3.62). This, however, is a simplification in terms of that it linearizes the relations, which in general may become inaccurate if the noise is not small compared to the large-signal excitations. A more rigorous approach is the transformation of probability density functions (equation 3.59), which will be presented here.

It is assumed that the photon induced shot noise causes a Poisson distribution for the electron counts that are accumulated in the storage nodes of the photodetector that will transduce these electrons to output signals $V_{FD1} - V_{FD3}$. The Poisson

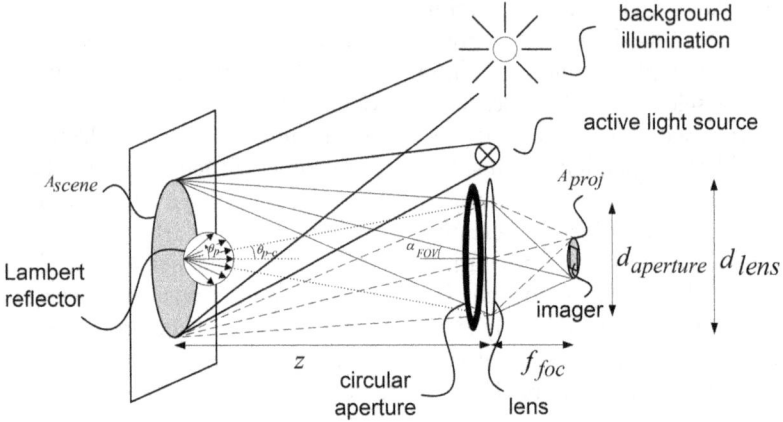

Figure 6.3 Simplified schematic of the range measurement setup.

distribution is fully defined by its expectation $\mathbb{E}(N)$ (c.f. Section 3.3.2). First expressions for $\mathbb{E}(N_{FD1}) - \mathbb{E}(N_{FD3})$ dependent on the physical conditions defined by the measurement setup have to be derived. The spectral irradiance $E_{e\lambda-pixel}(\lambda)$ given in $Wm^{-2}\,nm^{-1}$ at the surface of the detector generates an expected amount of

$$\mathbb{E}(N) = \int_0^\infty \frac{E_{e\lambda-pixel}(\lambda)\eta_{ext}(\lambda)T_{int}A_{pa}\lambda}{hc}\,d\lambda \qquad (6.3)$$

electrons, which have to be kept in a storage node of the photodetector until readout. Here A_{pa} is the photoactive region of one pixel which equals the pixels area A_{pix} multiplied with its fill factor FF, η_{ext} equals the extrinsic quantum efficiency and T_{int} defines the integration time which is set by T_p and varies with τ_{ToF}. $E_{e\lambda-pixel}$ is dependent on the active light source as well as the ambient light conditions, the object's distance, tilt and surface conditions and the optical setup of the camera. A common simplification enabling an eased derivation of $E_{e\lambda-pixel}$ is the assumption of having an object to observe which is orthogonal to the optical axis of the sensor and the light source (c.f. [Lan00; Elk05; Sp10]). Further, it is assumed that the light source generates an illumination in shape of a cone, so that an area of a circle is projected on the scenery. The reflectance is assumed to be only diffusive and is modeled by the relations of a *Lambert reflector* for which the radiant intensity I_e is proportional to $\cos(\theta_p)$, with θ_p being the polar angle. The optical system is assumed to project the very same circular area on the sensor. A schematic of such a range measurement setup is given in Figure 6.3. The spectral radiant flux $\Phi_{e\lambda-scene} = E_{e\lambda-scene}A_{scene}$ appearing at the imaged area of the scenery $A_{scene} = z^2\pi\tan^2(\alpha_{FOV})$ is comprised by the active illumination and the background illumination. Here, z equals the object distance and α_{FOV} defines the plane angle of the field-of-view. The radiant flux is reflected into the half-space according to $d\Phi_{e-\lambda} = I d\Omega_{sa}$ or in spherical coordinates $d^2\Phi_{e-\lambda} = I(\theta_p)\sin(\theta_p)d\theta_p d\phi_a$. Only for polar angles below θ_{p-c} will the reflected flux be focused by the optics onto the actual sensor. Thus, the spectral radiant power appearing at the lens can be evaluated as

$$\Phi_{e\lambda-\text{lens}} = \Phi_{e\lambda-\text{scene}} r_{\text{scene}} \frac{\int_0^{2\pi} \int_0^{\theta_{p-c}} \cos(\theta_p) \sin(\theta_p) d\theta_p d\phi_a}{\int_0^{2\pi} \int_0^{\pi/2} \cos(\theta_p) \sin(\theta_p) d\theta_p d\phi_a}$$

$$= \Phi_{e\lambda-\text{scene}} r_{\text{scene}} \left[\sin(\theta_{p-c}) \right]^2 \tag{6.4}$$

$$\approx \Phi_{e\lambda-\text{scene}} r_{\text{scene}} \left(\frac{d_{\text{aperture}}}{2z} \right)^2, \tag{6.5}$$

with the diameter of the circular aperture d_{aperture} and the reflectance of the surface r_{scene}. Here, the optical aperture may be replaced by the lens diameter d_{lens} if no aperture is used. For the last approximation it was assumed that $\sin(\theta_{p-c}) = \sin[\tan^{-1}(d_{\text{aperture}}/2z)]$ can be approximated by the first element of a Taylor series, thus that $z \gg d_{\text{aperture}}$. It is assumed that the object is focused onto the plane at the focal length f_{foc}. The spectral irradiance impinging on one pixel can be evaluated by $E_{e\lambda-\text{pixel}} = \Phi_{e\lambda-\text{lens}}/A_{\text{proj}}$ which yields

$$E_{e\lambda-\text{pixel}} = \frac{\Phi_{e\lambda-\text{lens}} \tau_{\text{lens}}}{f_{\text{foc}}^2 \pi \tan^2(\alpha_{\text{FOV}})} \tag{6.6}$$

$$= \frac{r_{\text{scene}} \tau_{\text{lens}}}{4 f_{\text{foc}}^2 \pi \tan^2(\alpha_{\text{FOV}})} \frac{d_{\text{aperture}}^2}{z^2} z^2 \pi \tan^2(\alpha_{\text{FOV}}) \left[E_{e\lambda-\text{bg}} + \frac{\Phi_{e\lambda-\text{as}}}{z^2 \pi \tan^2(\alpha_{\text{FOV}})} \right] \tag{6.7}$$

$$= \frac{\Phi_{e\lambda-\text{as}}}{4z^2} \frac{r_{\text{scene}} \tau_{\text{lens}}}{f_{\#}^2 \pi \tan^2(\alpha_{\text{FOV}})} + \frac{E_{e\lambda-\text{bg}}}{4 f_{\#}^2} r_{\text{scene}} \tau_{\text{lens}} \tag{6.8}$$

with τ_{lens} being the lens transmittance, the f-number $f_{\#} = f_{\text{foc}}/d_{\text{aperture}}$, the spectral irradiance of the active light source $E_{e\lambda-\text{as}} = \Phi_{e\lambda-\text{as}}/A_{\text{scene}}$ and the background illumination given as a spectral irradiance $E_{e\lambda-\text{bg}}$. It was assumed that the diffuser homogeneously spreads the source flux across A_{scene}.

If it is assumed that the background illumination is due to the sun and that the attenuation of its irradiance due to the atmosphere can be neglected, its spectral irradiance appearing at the scene is given by

$$E_{e\lambda-\text{bg}}(\lambda) = \frac{2\pi h c^2}{\lambda^5 \left[\exp\left(\frac{hc}{k_B \theta_{\text{surface-sun}} \lambda} \right) - 1 \right]} \frac{r_{\text{surface-sun}}^2}{d_{\text{earth-sun}}^2} \tag{6.9}$$

what corresponds to a modeling of the sun as a black body so that Planck's law can be applied. As described in Chapter 2, range detectors often use optical filters to limit the amount of background illumination. If the bandwidth of such an optical filter is narrow, the wavelength dependent sensitivity of the actual photodetector can be assumed constant. For very narrowband filters temperature regulated light sources may become necessary, so that a drift of the emission frequency can be neglected.

Given a setup with ideal optical filters that allow for illumination within $\lambda_l \leq \lambda \leq \lambda_u$, the expectation of the amount of photogenerated charges which are transferred to the storage nodes can be expressed as

$$\mathbb{E}(N_{TXi}) = \int_{\lambda_l}^{\lambda_u} \frac{r_{scene} \tau_{lens} \eta_{ext} A_{pix} FF\lambda}{4hcf_\#^2} \left[\frac{\Phi_{e-L} T_{int-TXi}(z)\delta(\lambda - \lambda_L)}{\pi z^2 \tan^2(\alpha_{FOV})} \right.$$
$$\left. + E_{e\lambda-bg}(\lambda) T_{para-int}(z) \right] d\lambda \qquad (6.10)$$

with the total radiant flux Φ_{e-L} emitted at the wavelength λ_L and shutter dependent functions $T_{int-TXi}(r)$ and $T_{para-int}(r)$. Assuming ideal rectangular shaped pulses as indicated in Figure 6.2, these functions can be expressed as

$$T_{int-TX1}(z) = N_{accu} \cdot \left(T_p - \frac{2z}{c} \right) + \beta_1 N_{accu} \frac{2z}{c} \qquad (6.11)$$

$$T_{int-TX2}(z) = N_{accu} \cdot \frac{2z}{c} + \beta_2 N_{accu} \cdot \left(T_p - \frac{2z}{c} \right) \qquad (6.12)$$

$$T_{int-TX3}(z) = \beta_3 N_{accu} T_p \qquad (6.13)$$

$$T_{para-int}(z) = N_{accu} T_p + \beta_4 (\Delta T - N_{accu} T_p) \qquad (6.14)$$

where ΔT is the total observation time window, N_{accu} is the accumulation count and $\beta_1 - \beta_4$ are device dependent constants that model the parasitic accumulation in time frames, where the photodetector's storage nodes should ideally be light insensitive. For simplicity, here, these parameters are assumed to be zero.

Typical parameters of a range measurement camera are: $\Phi_{e-L} = 75$ W, $\lambda_L = 905$ nm, $\lambda_l = 820$ nm, $\lambda_u = 920$ nm, $T_p = 30$ ns, $A_{pix} = 40$ μm × 40 μm, $FF = 38\%$, $\alpha_{FOV} = 15°$, $f_\# = 0.95$ to 1.4, $f_{foc} = 25$ mm and $\eta_{ext} = 3.6$–5%. These correspond to the demonstrator camera which was included in Table 2.1 and the embedded range detector given in Table 2.3. The extrinsic quantum efficiency can be estimated from TCAD simulations and estimates of the photodetector's storage capacitance. The latter can also be estimated from TCAD simulations for the diffusion nodes themselves or from specific capacitance per area/perimeter values as they can be extracted from separate test structures (c.f. [Du09a; Sp10]). Interconnect metal-to-metal capacitances can be estimated by e.g. Diva®[1]. For the results that will be presented in in this section τ_{lens} and r_{scene} are assumed to equal 1, $f_\# = 0.95$ and $\eta_{ext} = 5\%$.

The statistical description is done via vanilla Gaussian error analysis and the more rigorous transform of the probability functions according to the rational function

$$z = \frac{T_p c}{2} \frac{N_{FD2} - N_{FD3}}{N_{FD1} + N_{FD2} - 2N_{FD3}}. \qquad (6.15)$$

---

[1]Diva® is a layout and verification tool from Cadence Design Systems, Inc. that also implements parasitics extraction.

Gaussian error analysis yields a standard deviation (c.f. Equation 3.62)

$$\sigma_z = \sqrt{\text{var}(z_r)}$$

$$= \frac{T_p c}{2} \frac{\sqrt{(N_{FD2} - N_{FD3})^2 N_{FD1} + (N_{FD1} - N_{FD3})^2 N_{FD2} + (N_{FD1} - N_{FD2})^2 N_{FD3}}}{(N_{FD1} + N_{FD2} - 2N_{FD3})^2}.$$

$$(6.16)$$

Employing Equation 3.59, $F_{p-z_r}(z)$ is expressed as

$$F_{p-z_r}(z_r) = \iiint_{D_z} f_{p-N_{FD1-r}, N_{FD2-r}, N_{FD3-r}}(N_{FD1-r}, N_{FD2-r}, N_{FD3-r}) dN_{FD1-r} dN_{FD2-r} dN_{FD3-r}$$

$$(6.17)$$

$$D_z = \left\{ N_{FD1-r}, N_{FD2-r}, N_{FD3-r} \left| \frac{T_p c}{2} \frac{N_{FD2-r} - N_{FD3-r}}{N_{FD1-r} + N_{FD2-r} - 2N_{FD3-r}} \le z_r \right. \right\}. \quad (6.18)$$

The three shutters TX1 − TX3 are assumed to be independent, thus

$$f_{p-N_{FD1-r}, N_{FD2-r}, N_{FD3-r}}(N_{FD1-r}, N_{FD2-r}, N_{FD3-r})$$

$$= f_{p-N_{FD1-r}}(N_{FD1-r}) f_{p-N_{FD2-r}}(N_{FD2-r}) f_{p-N_{FD3-r}}(N_{FD3-r}). \quad (6.19)$$

The integral in Equation 6.17 is solved numerically. For that, exception handles have to be embedded, since the mapping defined in Equation 6.15 can yield results that are not within $[0, cT_p/2]$. For some tuples $(N_{FD1-r}, N_{FD2-r}, N_{FD3-r})$ Equation 6.15 may not even be defined, namely when $N_{FD1-r} + N_{FD2-r} - 2N_{FD3-r} = 0$. This results in heavy deteriorations of the measurements for cases when the background illumination dominates over the active illumination which This becomes visible in probability density distributions $f_{p-z_r}(z_r)$ for strong background illumination in combination with long distances.

In Figure 6.4 exemplary probability density functions $f_{p-z_r}(z_r) = dF_{p-z_r}(z_r)/dz_r$ at an object distance of 1 m are given. In this picture the non-symmetry is clearly visible. For numerical evaluation of this phenomenon, which is introduced by the non-symmetrical Poisson distribution and the non-linear mapping of the distance $z(N_{FD1}, N_{FD2}, N_{FD3})$, two parameters are introduced:

$$\sigma_{pos} = \sqrt{\int_{\mathbb{E}(z_r)}^{\infty} [z_r - \mathbb{E}(z_r)]^2 f_{p-N_{FD1-r}, N_{FD2-r}, N_{FD3-r}}(N_{FD1-r}, N_{FD2-r}, N_{FD3-r})} \quad (6.20)$$

$$\sigma_{neg} = \sqrt{\int_{-\infty}^{\mathbb{E}(z_r)} [z_r - \mathbb{E}(z_r)]^2 f_{p-N_{FD1-r}, N_{FD2-r}, N_{FD3-r}}(N_{FD1-r}, N_{FD2-r}, N_{FD3-r})}. \quad (6.21)$$

It is also interesting to note that without background illumination the shape of the PDF tends to larger values of the depth, while with background illumination this changes. It becomes obvious how significantly the background light degenerates the

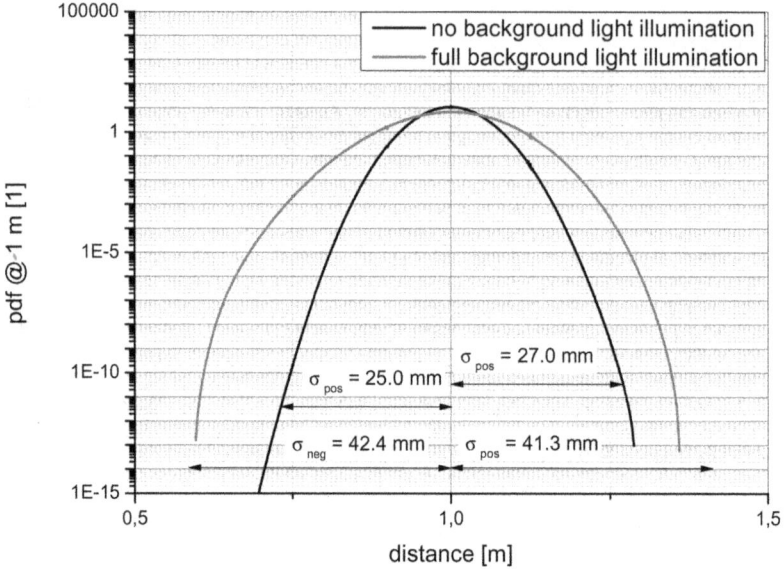

Figure 6.4 Probability density functions of the depth at 1 m object distance with and without parasitic background illumination from the sun.

image quality. The standard deviation changes from approximately 40 cm towards approximately 60 cm. In Figure 6.5 the standard deviation as a function of the object's distance is depicted. The graphs demonstrate the impact of the background light illumination on the accuracy – especially for long distances, where the amount of signal generated charges is decreasing. Furthermore, this plot demonstrates the underestimation of the distance uncertainty due to photon induced shot noise for long distances, where the denominator of equation 6.15 easily becomes close to zero. Nevertheless, for short distances Gaussian error analysis is in good agreement to the results obtained by transformation of the PDFs according to the non-linear function $z(N_{FD1}, N_{FD2}, N_{FD3})$.

To yield a sufficient accuracy in the appearance of harsh ambient light conditions, multiple accumulations can be performed to increase the amount of signal generated charge carriers and thus the resolution. PDFs evaluated at an object distance of 1 m are depicted for varying accumulation counts in Figure 6.6. Additionally, the standard deviations are given. Clearly, it can be observed that the accuracy significantly improves for higher N_{accu}. From the presented data, it can be extracted that an increase of N_{accu} by a factor of hundred corresponds to a decreased standard deviation by a factor of ten. This can also be extracted from Gaussian error analysis. If $N_{FD1} - N_{FD3}$ all scale by N_{accu}, σ_z scales by $\sqrt{N_{accu}}$ (c.f. equation 6.16).

Figure 6.7 and Figure 6.8 present the accuracy given as standard deviation and relative error $\sigma_z(z)/z$, respectively, for varying accumulation count.

To provide a sufficient accuracy of e.g. 5% with a fixed accumulation count – here 1000 – a high full-well capacity of 1×10^8 e$^-$ becomes necessary. This can be observed in Figures 6.9 and 6.10. Those figures present the amount of photogenerated charge

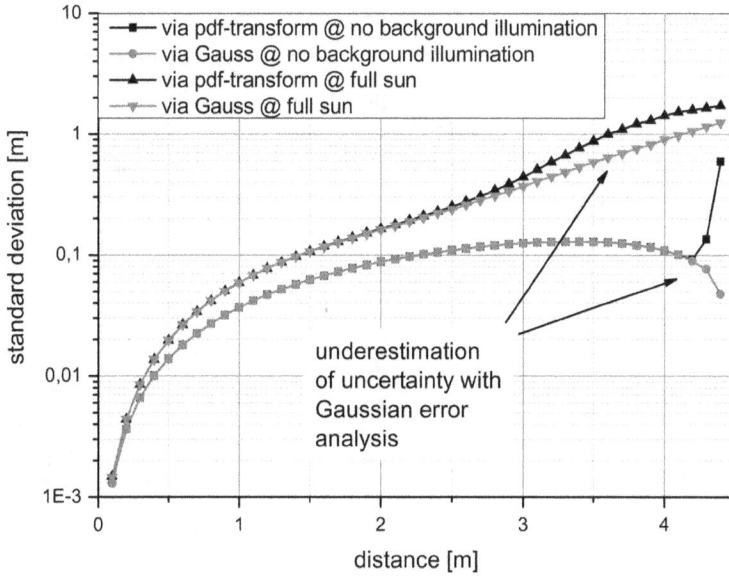

Figure 6.5 Standard deviation as a function of the object distance with and without parasitic background illumination from the sun.

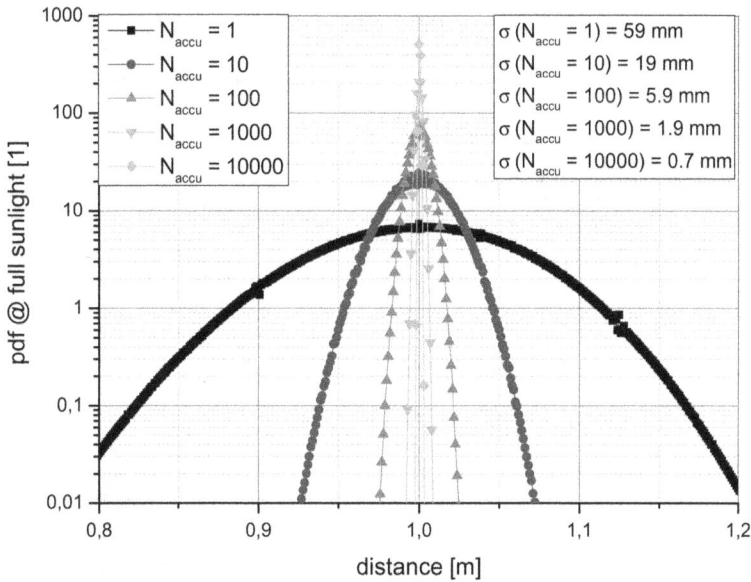

Figure 6.6 Probability density functions of the depth at 1 m object distance with parasitic background illumination from the sun for varying numbers of accumulation.

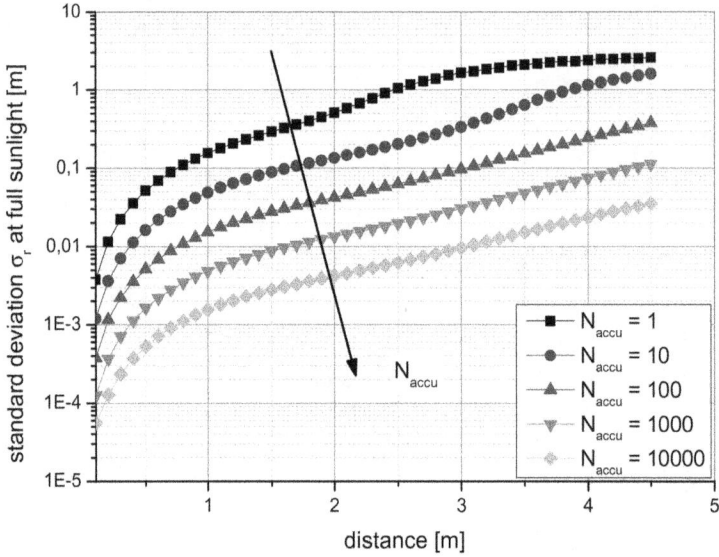

Figure 6.7 Standard deviation as a function of the object distance with background illumination from the sun at varying accumulation count.

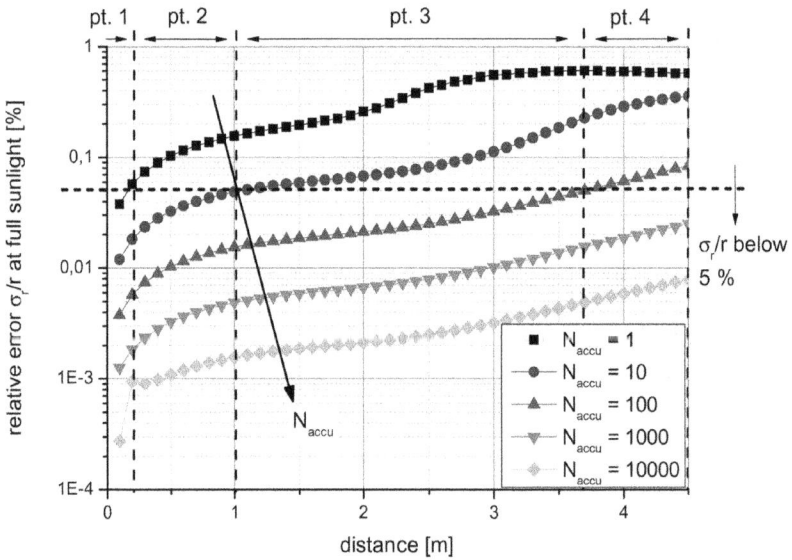

Figure 6.8 Relative error as a function of the object distance with background illumination from the sun at varying accumulation count.

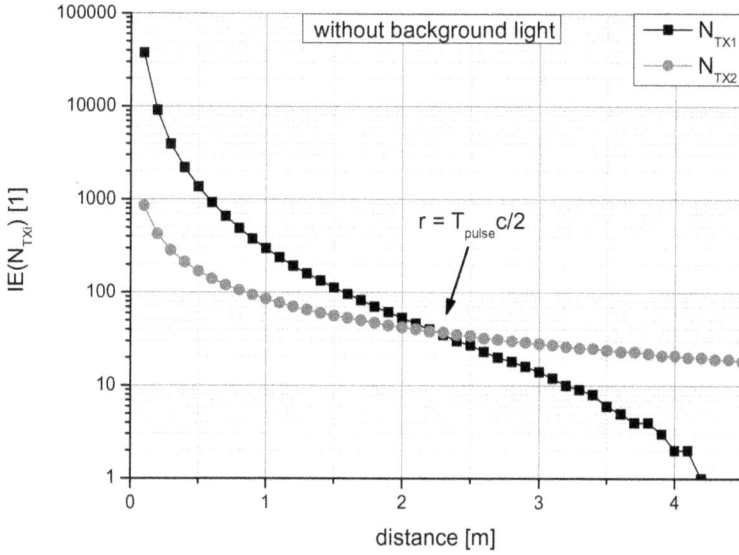

Figure 6.9 Stored photogenerated electron count as a function of the object distance without background illumination from the sun.

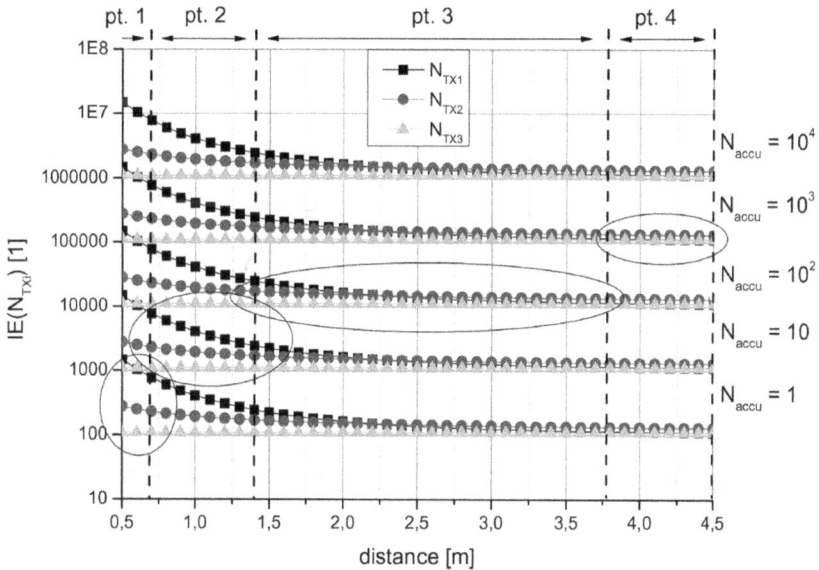

Figure 6.10 Stored photogenerated electron count as a function of the object distance with background illumination from the sun at varying accumulation count.

carriers, that have to be detected and stored to yield appropriate depth resolutions, with and without parasitic background illumination and for varying accumulation count. However, such high full-well capacities are rather impractical, since they limit the minimum pixel pitch and thus for instance the lateral spatial resolution, fill-factor or indirectly also read noise etc. that can be realized. An additional problem that is arising here, is that given a constant accumulation count the dynamic range provided by the storage node to guarantee the depth resolution is also to be provided by the readout circuitry.

An alternative approach is to extend the dynamic range by multiple acquisitions with different integration times or accumulation count[2]. The total range $[0, T_p c/2]$ can be separated in several regions, for which different accumulation counts are used. With for example four measurements with e.g. $N_{accu} \in \{1, 10, 100, 1000\}$ that together form a single frame, the dynamic range can be extended by keeping the storage node capacitance comparatively low.

If the object's distance is long and the background light illumination is high, many photons have to be collected to provide a sufficient resolution. While for high background light illumination and an object in the vicinity of the sensor, the signal generated charge carrier counts will be rather high what may push the storage node into saturation for high N_{accu}. Thus, after all acquisitions are done, the range extraction has to be done with the highest N_{accu} that does not push any of the shutters $N_{FD1} - N_{FD3}$ into saturation. From Figure 6.10 it becomes clear, that the full-well capacity that has to be provided is basically determined by the distance range with the longest distance (pt. 4). In this example for instance approximately $4 \times 10^5 \, e^-$ have to stored.

6.3 DESIGN OBJECTIVES AND CONSIDERATIONS

In this section the design objectives and trade-offs that have to be made are discussed. Afterwards, using the defined constraints the photodetector used to perform the MSI based demodulation for the PM ToF principle (c.f. Section 2.3.3) is chosen and the architecture of the demonstrators that were developed within the framework of this work are explained. This is followed by a schematic of the sensor system architecture and a short review on the employed CMOS process.

6.3.1 Design objectives

To properly understand the trade offs in a PM ToF range camera system, a fictitious specification is given in Table 6.1. The distance range is typical for ToF cameras as explained in Chapter 2. The laser pulse length determines the maximum measurement range. The chosen value of 30 ns corresponds to a measurement range up to 4.5 m. Though this is larger than the specified range of 2 m, this is considered as meaningful, since such modules are widely available. As has been explained in Sections 2.3.3 and 5.2.2 the laser pulse length should match the shutter length to avoid parasitic accumulation of background illumination.

---

[2]There are many other methods known on how to extend the dynamic range (c.f. e.g. [Dar12]). However, their treatment is beyond the scope of this work.

Table 6.1 Fictitious specification of a range camera.

imager resolution	128 × 96
distance range	0.15 to 2 m
reflectance range	0.1 to 1
accumulation count	1, 100, 400 and 1000 (dynamic)
background illuminance	0 to 100 klx
optical bandpass filter	880 to 930 nm
total transmittance of optics	0.9
field of view	15° × 15° (circular)
f-number	0.95
duty cycle of the laser	1/1000
radiant flux of the laser	4 × 75 W
laser wavelength	905 nm
laser pulse width	30 ns
shutter width	30 ns
frame rate	≥20 fps
precision	≤1.7 cm (1σ)

Nevertheless, since the time interval of 30 ns is not necessary to resolve the 2 m range, the shutter length may be reduced in future designs. The background illumination is specified in lx, since this is usually the case (c.f. Section 2.4). Unfortunately, this is not a proper physical quantity since it relies on the spectral sensitivity of the human eye, but does not say anything about the spectral properties of the light sources present in the scene. To allow for a better comparison within this work it is assumed that ambient illumination has a spectral radiant flux distribution according to Planck's law as a model for the emission of the sun acting as a black body at surface temperature of $5777\,K$ at a distance of 1.496×10^{11} m from the earth which has a radius of 6.955×10^8 m (c.f. Equation 6.9). This way defined specifications for the illuminance can be unambiguously translated into irradiances by weighting with the luminosity function (c.f. e.g. [PPBS08]). Absorption of radiant flux due to the earth atmosphere is neglected because it is strongly dependent on humidity and air pressure (c.f. Chapter 1). Furthermore, any attenuation arising from the change of the relative position to the sun according to the sun path is neglected.

It is thus assumed that the irradiation from that ambient light source is orthogonally impinging on the scenery and there is no attenuation due to the atmosphere. Since all the described phenomena basically reduce the ambient illumination the model can be considered a worst case estimation which typically overestimates the reality. For applications where the setup is operated in a foggy ambiance or for indoor applications where artificial illumination is used, the assumption of direct sunlight can be an overestimation of several orders of magnitude. For space applications, however, this is what the setup will have to withstand in order to properly resolve distances. The specified illuminance of 100 klx corresponds to approximately 85% of the full solar spectrum impinging orthogonally onto the surface of the atmosphere of the earth. It is assumed that a device which properly operates in such harsh conditions is feasible for a diversity of applications, e.g. for industrial, scientific or automotive applications (c.f. Chapter 1). Reflectance is, of course, a wavelength dependent parameter. Assuming though that the optical bandpass filter properly attenuates outside the passband, it

Table 6.2 Fictitious specification of a range imager.

storage node capacitance	10 fF
pixel size	$40 \times 40\,\mu m^2$
fill-factor	38%
extrinsic quantum-efficiency	$\geq$5% at 905 nm
linear output voltage range	$\geq$2 V
dark current of the storage nodes	$\leq$25 000 e$^-$/s
transfer time	$\ll$30 ns
readout time	$\leq$10 ms
read noise	$\leq$1 mVrms

can be assumed to be a constant value. The specified range corresponds to reflectance values ranging from dry asphalt with pebbles to snow (c.f. [Lan00]). Together with the precision[3], large field-of-view, the reflectance range and the frame rate, the constraints for the active light source are rather tight. There is a strong need for an active light source with a high radiant flux to generate enough signal-related photons to yield an appropriate precision. Alternatively, pixel parameters such as pixel size or quantum efficiency can be increased. This, however, is rather difficult and comes at an expense of lower speed or crosstalk, as will be discussed later on. To ensure that the sensor is not saturating at settings with high reflectance at short distances while performing well at low reflectance at long distances, employing adjustable multiple accumulation count settings is meaningful.

This results in a loosened constraint for the dynamic range of the image sensor. The random error is chosen to be less than 1.7 cm, because if it can be assumed that the PDF can be roughly estimated by a Gaussian distribution, the probability of having an error below 5% of the measurement range is approximately 99.7%. In a real application additional constraints as for instance cost, calibration effort etc. may be introduced, but are neglected in this work. Furthermore, it is assumed that all specifications have to be met under all circumstances. Since this might not be possible at all for different applications, parameters can be mutually traded against each other to allow for adequate solutions. These specific cases, however, have to be dealt with extensive use of optimization and are neglected within this rather fundamental work.

The model which was explained in the former section can be employed to allow for verification of the reasonableness of the given range camera setup. Specifications for the image sensor which were found to meet the specifications of the range camera are given in Table 6.2.

The pixel pitch was chosen as a compromise between charge transfer speed and photoactive area. The fill-factor is a value that actually resulted from the other constraints during layout of the pixel. The size of a storage node is a typical value. As shown in the former section, its size can be decreased if the use of multiple accumulation

---

[3]According to [Sta94] *precision* describes the random error from a measurement setup, whereas *trueness* describes the systematic measurement error. Together they result in the term *accuracy*. Insufficient trueness can be compensated by adjustment of the measurement setup (provided the measurement is bijective) whereas the precision ultimately limits the measurement accuracy.

counts is accepted, what is assumed during this fictitious case study. In reality, however, each accumulation count setting has to be calibrated resulting in increased cost and processing time and performance. Abandoning multiple accumulation counts though results in extreme constraints for noise performance and dynamic range of the image sensor. As will be shown later on, the impact of the readout noise on the precision is rather low because multiple accumulation counts are allowed. This results in a rather loose constraint of the read noise of 1 mV standard deviation. Read noise, here, is understood as the noise affecting the signal from the source follower input to the output of the image sensor. Other phenomena such as reset noise or dark current shot noise are treated additively. The extrinsic quantum efficiency is a rough guess made from large area test structures. The specification for the linear output voltage swing is a typical value for the input voltage range of analog-to-digital converters. Linearity errors, here, are defined as the maximum distance of an LQ-fit of a straight line with degrees of freedom for offset and gain to the transfer characteristic of the readout circuitry. The charge-voltage conversion also introduces some non-linearity which can, however, not be easily predicted without extensive use of TCAD and electromagnetic parasitic extraction tools. Non-linearity is though not very important, since it is dealt with during calibration later on. The major deviations from the simple relation $z \propto (V_{FD2} - V_{FD3})/(V_{FD1} + V_{FD2} - 2V_{FD3})$ are arising from the laser pulse shape, the photodetector and the optics. Apart from the last point these statements will be proven by measurements within the later text. The task of resolving these insufficiencies by calibration, however, is beyond the scope of this work. The dark current is an important parameter for sceneries at long distance, low reflectance and strong background illumination. Here, high accumulation counts have to be employed to meet the specification for the precision. Since the time interval in which these accumulations can be performed is ultimately limited by the duty cycle of the active light source, 1000 accumulations need approximately 30 ms. Within this time interval dark current can significantly affect the accuracy of the measurements. The duty cycle also affects the maximum frequency of tolerable ambient light sources according to Nyquist-Shannon's sampling theorem and the non-ambiguity range of range measurements. For the presented constraints a Nyquist frequency of approximately 16 kHz results for the tolerable ambient light. The non-ambiguity range amounts to 0 to 4500 m which is a factor of $\text{dutycycle}^{-1} + 1$ larger than the resolvable range of maximal 0 to 4.5 m. Considering the usage of multiple accumulation counts and the 30 ms which become necessary for $N_{\text{accu}} = 1000$, a pixelclock in the MHz-domain becomes necessary to yield a frame rate of 25 to 30 fps. This can be easily realized by use of multiple output channels and/or the embedding of on-chip ADCs. Since this is though considered as a detail which results in much design effort but does not contribute to the fundamental understanding of the range imager itself, this constraint is loosened. A readout time of approximately 10 ms is considered meaningful for the demonstrators presented within this work. To ensure a "sufficiently" large amount of charges being transferred from the photoactive region towards the storage node within the proposed 30 ns, "transfer times" of photogenerated charge carriers should be "much shorter" than the shutter length. This non-properly defined constraint makes it very difficult to understand the limitations and to design a system in order to meet specifications as will be examined in Sections 6.5 and 6.7. There, a model for the photodetector is demonstrated that defines an effective charge-transfer measure, resembles measurement

Table 6.3 Range precision of the proposed imager limited by the photon shot noise and based on the model of an idealized imager from Section 6.2 according to Table 6.1 and 6.2.

accumulation count		1	100	400	1000
frame rate @ 10 ms readout time		99.7 fps	76.9 fps	45.5 fps	25 fps
precision 1σ	$z = 15$ cm, $r_{scene} = 100\%$ & $E_{V-bg} = 0$ lx	0.3 cm	–	–	–
	$z = 2$ m, $r_{scene} = 100\%$ & $E_{V-bg} = 0$ lx	12 cm	1.2 cm	0.6 cm	0.1 cm
	$z = 2$ m, $r_{scene} = 10\%$ & $E_{V-bg} = 0$ lx	38 cm	3.8 cm	1.9 cm	0.4 cm
	$z = 15$ cm, $r_{scene} = 10\%$ & $E_{V-bg} = 100$ klx	1.1 cm	0.1 cm	0.1 cm	0.1 cm
	$z = 2$ m, $r_{scene} = 10\%$ & $E_{V-bg} = 100$ klx	43 cm	4.3 cm	2.1 cm	1.3 cm

results and thus allows to accurately predict the systems' performance in a meaningful manner.

The results of the estimation of the range precision by the model presented in Section 6.2 is given in Table 6.3 for a selected set of sceneries. For a scenery containing short object distance and high reflectance, the proposed detector goes into saturation at all settings of the accumulation count except for $N_{accu} = 1$. Here, the precision for this specification is worst at long object distances and low reflectances. Additional ambient illumination hardly affects the accuracy. This demonstrates that the major limitation in this case study is arising by the amount of photogenerated charge carriers introduced from the active light source. By allowing for a combination of the accumulation counts 1 and 1000, the specification for the precision seems to be met. Embedding a multitude of accumulation counts to form one single frame with extended dynamic range can be performed by subsequent storage of the particular "sub-frames" and evaluation of saturation effects to yield the largest non-saturated values; because these are associated with the best accuracy. Unfortunately, at the vicinity of the according thresholds the amount of accumulated signal may largely differ what results in jumps in the signal-to-noise ratio (c.f. e.g. [Dar12]). Intermediate regions in the dynamic range for which the signal-to-noise ratio is less than one are also known as *dynamic range gaps*. In a proper design it has to be examined if the designed accuracy meets the specifications in the vicinity of those SNR jumps. A simple way to avoid large jumps is not to use largely varying accumulation counts. That way resulting differences are kept minimal. Nevertheless, this is traded against calibration effort and frame rate.

To study the impact of read noise and dark current the model from Section 6.2 is slightly modified. This is best done in the storage nodes referred charge domain. Here, reset noise, dark current shot noise and read noise can be added according to Equation 3.11. For the sake of simplicity, here Gaussian error analysis is applied. Assuming the mapping $z \propto (N_{FD2} - N_{FD3})/(N_{FD1} + N_{FD2} - 2N_{FD3})$ remains a proper estimate of the distance although the signals are affected by the deterministic component of the dark current, the variance can be estimated by

$$\sigma_z^2 = \left| \frac{\partial z}{\partial N_{FD1}} \right|^2 \sigma_{N_{FD1}}^2 + \left| \frac{\partial z}{\partial N_{FD2}} \right|^2 \sigma_{N_{FD2}}^2 + \left| \frac{\partial z}{\partial N_{FD3}} \right|^2 \sigma_{N_{FD3}}^2, \qquad (6.22)$$

where,

$$\sigma_{N_{FDi}}^2 = N_{FDi-photo} + \frac{N_{accu}}{f_{rep}} J_{dark} + \frac{k_B \theta C_{SN}}{q^2} + \sigma_{read}^2 \left| \frac{C_{SN}}{q \cdot A_{read}} \right|^2 \qquad (6.23)$$

was used. Here, $N_{FDi-photo}$ is the photogenerated charge carrier count kept within the storage node at FDi, J_{dark} is the dark current expressed in electron counts per time interval, C_{SN} is the sense node capacitance, A_{read} is the amplification from the source follower input to the analog output of the imager and σ_{read}^2 is the read noise referred to the output of the sensor in Vrms. A more proper treatment of this problem will be demonstrated in Section 6.5. The above model assumes that the reset noise $k_B \theta C_{SN}$ is not compensated by e.g. correlated double sampling. This is caused by the need of a global shutter mode for the PM ToF principle. A rolling shutter mode is in principle possible and would also result in loosened constraints for e.g. the shutter drivers, but the emitted laser pulse counts would have to be increased according to the number of rows, what would clearly result in a low frame rate due to the constraints for eye safety and accuracy. Implementing an analog CDS stage for global shutter mode as it was presented in Section 5.3 would necessitate feedback capacitances for each storage node of the imager. This would result in a large high impedance node between the sampling capacitance and the feedback capacitances which would be very prone to any neighboring disturbances due to high parasitic capacitances which act as parasitic sample capacitors. Such a setup might easily amplify small disturbances over the actual signal level. The alternative approach of not storing the actual reset levels of each storage node, but performing the subtractions $V_{FD1} - V_{FD3}$ and $V_{FD2} - V_{FD3}$ would not improve the noise performance because the reset levels of the different storage nodes have to be assumed as being uncorrelated. One might motivate the subtraction as being necessary because they have to be performed during calibration anyway. This might though not be meaningful, since the differing shutter outputs might be affected from mismatch or non-linearity so that a compensation before the actual subtraction can become meaningful. Digital CDS is not considered within the framework of this work because it would imply a high circuit level complexity.

Evaluation of the the data from Table 6.2 with the modified model yields the results presented in Table 6.4. Comparison of this table with Table 6.3 demonstrates, that the read noise substantially affects the signal at sceneries containing long object distance, low reflectance and high ambient illumination. The precision for those cases, however, was inadequately disturbed by the photon noise anyway. At higher accumulation count good precision is obtained. There the difference in the random error is mainly determined by the dark current noise component. From these two tables the advantage of having multiple accumulation count settings becomes clear. If a design would require a single setting for accumulation counts as it might originate from frame rate or calibration effort purposes, one would have to realize $N_{accu} = 1000$, since the

Table 6.4   Range precision of the proposed imager limited by the total noise performance and based on the model of an idealized imager from Section 6.2 according to the specification in Table 6.2.

accumulation count		*1*	*100*	*400*	*1000*
frame rate @ 10 ms readout time		*99.7 fps*	*76.9 fps*	*45.5 fps*	*25 fps*
precision 1σ	$z = 15$ cm, $r_{scene} = 100\%$ & $E_{V-bg} = 0$ lx	0.8 cm	–	–	–
	$z = 2$ m, $r_{scene} = 100\%$ & $E_{V-bg} = 0$ lx	72 cm	1.4 cm	0.6 cm	0.4 cm
	$z = 2$ m, $r_{scene} = 10\%$ & $E_{V-bg} = 0$ lx	710 cm	8.1 cm	2.6 cm	1.4 cm
	$z = 15$ cm, $r_{scene} = 10\%$ & $E_{V-bg} = 100$ klx	7.5 cm	0.1 cm	0.1 cm	0.1 cm
	$z = 2$ m, $r_{scene} = 10\%$ & $E_{V-bg} = 100$ klx	710 cm	8.3 cm	2.8 cm	1.5 cm

accuracy is otherwise not met. This would imply the desire for a much higher sense node capacitance to allow for measurements at short distances and high reflectance. Unfortunately that increases the uncertainty arising from the readout circuitry substantially as can be seen in Equation 6.23. As has been described during the survey through state-of-the art low-noise concepts for CMOS APS in Chapter 5, the reset noise is typically the dominating factor for the noise performance. This would result in the need of digital CDS or active reset topologies (c.f. Section 5.2.3) which is, however, beyond the scope of this work.

6.3.2   Photodetector selection

As has been described in the previous section, the photodetector that has to be implemented has to have a comparatively large photoactive area of $0.38 \cdot 40 \times 40 \mu m^2$ in combination with a charge transfer time significantly shorter than 30 ns, a storage node capacitance of approximately 10 fF and a low dark current. Furthermore, the photodetector should be able to demodulate signals by means of short time integration. Standard p/n-junction photodiodes typically have a much higher storage node capacitance and exhibit a high dark current. Also they do not allow for charge transfer into light-insensitive storage nodes. Thus they are not feasible for PM ToF purposes. A PPD implementation might meet all the specification but that for charge transfer speed. The LDPD as advancement of the PPD, however, is a candidate which is more likely to meet all constraints. Similar to the PPDs there is also the flexibility of connecting multiple storage nodes to a single photoactive region which is advantageous in terms of resulting photoactive area. The trade-off that is inherently done here, is discussed more thoroughly in Section 6.6. In the remaining part of this section, the concept of the ToF-LDPD is presented.

Part a) in Figure 6.11 displays a cross-sectional view along the axis I-I of the LDPD and part b) shows a top view perspective in which all cuts are depicted. The photoactive region is connected to three storage (FDi) and one draining node (DD).

a) cross-sectional view I-I of a ToF LDPD

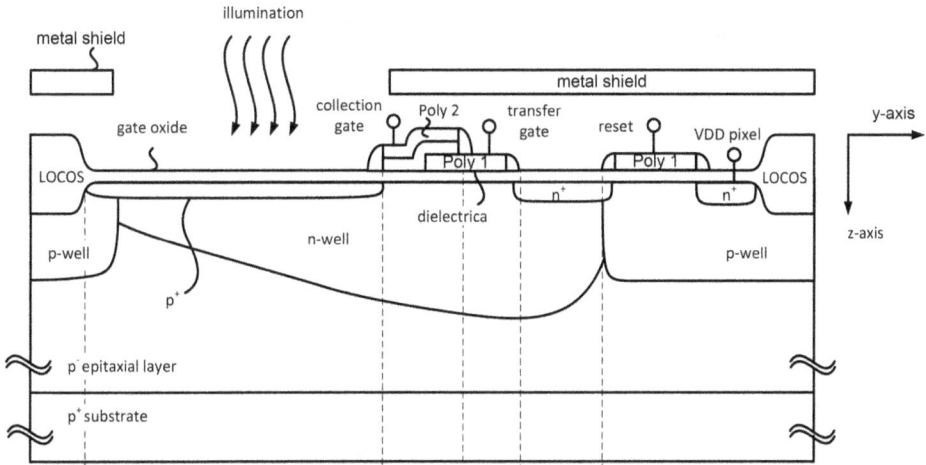

b) top view of a LDPD

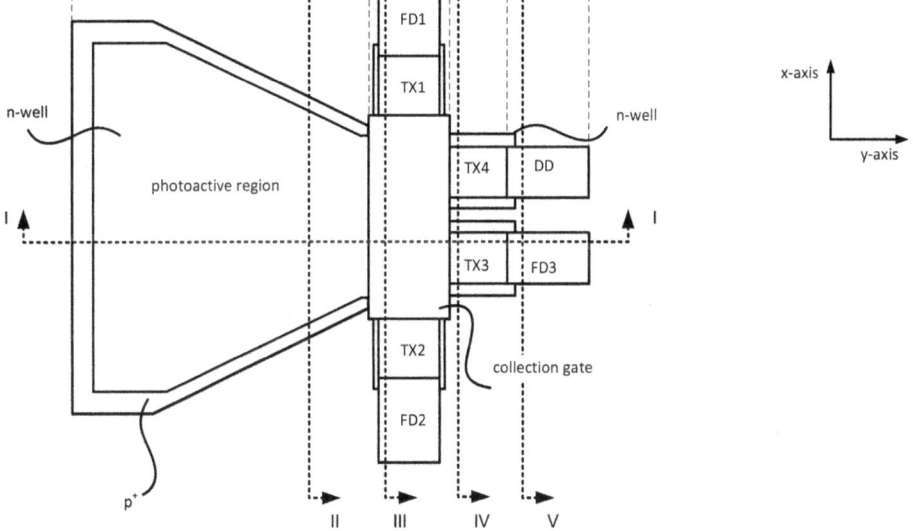

Figure 6.11 Schematics of the LDPD (part one).

The electronic separation of those nodes to the photoactive area is realized by transfer gates (TXi) that are connected to a collection gate (CX). When the laser impinges at the photoactive region, electron-hole pairs are generated and separated by the electric field. Electrons are then accelerated towards the potential maximum, that is defined by either a storage or a draining node, while the holes are drained by the substrate connected to the ground potential. A doping gradient is formed to further accelerate photogenerated electron-hole pairs from the photoactive region towards the readout nodes. The LDPD is operated in a way that the region, in which electrons are transferred towards the storage or draining nodes, always remains fully depleted[4]. For that, proper voltage levels have to be applied at CX and the transfer gates.

To understand the operation of the device, comparison with the basic timing diagram depicted in Figure 6.2 is necessary. A reset is defining the initial state (t_0). In parallel to the emission of the laser pulse a shutter pulse (TX1), that has the same width as the laser pulse, is activated (t_3–t_5). When the reflected laser is impinging at the detector (t_4), charge is accumulated in the floating diffusion (FD1)(t_4–t_5). The delay, which is proportional to the time-of-flight and thus to the distance, has impact on the effective integration-time of the transferred photogenerated charges. After the end of the first shutter pulse a second shutter is activated (TX2)(t_5), so that the remaining charge, that represents the second part of the laser pulse, is integrated at the according storage node (FD2)(t_5–t_6). A third storage (FD3) node is implemented for accumulation of the background light. This may be triggered (TX3) before the emission of the laser pulse to suppress the parasitic tail of the laser (t_1–t_2). Between these shutter windows photogenerated charge should be drained (TX4) to eliminate blooming effects. Contrary to the CW principle, light is only accumulated during the short shutter windows (c.f. Section 2.3). Thus much more laser power might be used in those short time frames to increase the SNR significantly, while the eye safety is still guaranteed. This principle thus provides good robustness against background illumination.

Figure 6.12 presents the other cuts, as they are indicated in Figure 6.11. The cut depicted in part a) is orthogonal to the cut II which is depicted in part a) of Figure 6.11 and placed in the photoactive region. It can be seen, that the doping profile of the n-well forms a gradient to define a potential maximum in the middle of the photoactive region along the II-II cutline which is increasing along the I-I cutline towards the collection gate. Furthermore, the depleted well is dislocated from the LOCOS isolation by the p^+-pinning layer to prevent dark current and noise which can be caused by the traps located at the Si-SiO_2-interface. Part b) presents a cross-sectional view of the device along the III-III cutline to show the opposing diffusions FD1 and FD2. Due to processing conditions, minimal gaps have to be introduced between structures, so that the collection gate can become relatively large to enable the connection to all four transfer gates. From part b) in Figure 6.12 it becomes clear, that this can cause a longer transfer time, since the doping profile of the n-well underneath the collection gate is rather flat. Part c) depicts the insulation region between the n-well implants connected to the two neighboring diffusions FD3 and FD4. To allow for proper insulation between these paths, a p-well implant is introduced. Part d) shows the last cross-sectional view through the ToF-LDPD within the separation of the two diffusions FD3 and FD4 is presented.

A disadvantage of the implementation is the fact that the n-well has contact to the Si-SiO_2-interface so that generation/recombination can occur which is associated with flicker and RTS noise (c.f. Sections 3.3.5 and 3.3.6). At the sidewalls the well may be separated by additional acceptor implants. Furthermore, as discussed in Chapter 5,

---

[4]Actually, the region underneath the transfer gate is not necessarily fully, but rather partially depleted in practice as it is otherwise difficult to ensure that the transfer gates can properly switch OFF.

a) cross-sectional view II-II of a ToF LDPD

illumination

LOCOS

gate-oxide

p^+

n-well

p^- epitaxial layer

p^+ substrate

y-axis

z-axis

b) cross-sectional view III-III of a ToF LDPD

metal shield

FD1   TX1   Poly 2   TX2   FD2

LOCOS   Poly 1   Poly 1   LOCOS

n^+ n^+

gate oxide   dielectrica   n-well

p-well   p-well

p^-

p^- epitaxial layer

p^+ substrate

c) cross-sectional view IV-IV of a ToF LDPD

metal shield

TX4   TX3

gate-oxide   LOCOS

p-well   n-well   p-well   n-well

p^- epitaxial layer

p^+ substrate

d) cross-sectional view V-V of a ToF LDPD

metal shield

DD   FD3

gate-oxide   LOCOS

n^+ n^+

p-well   p-well

p-well

p^- epitaxial layer

p^+ substrate

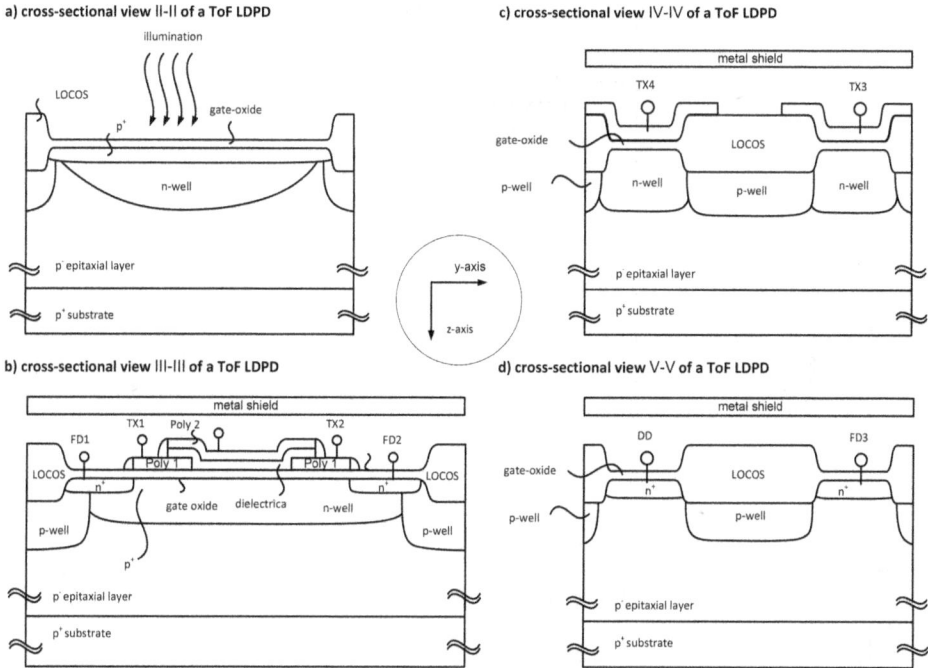

Figure 6.12 Schematics of the LDPD (part two).

interaction with traps located at the Si-SiO$_2$-interface may be avoided by cycling of the device between inversion and accumulation. If the transfer gates are biased with negative potentials during non-transfer phases, interface traps can be emptied and blocked by holes, so that they remain empty during transfer time when the device transit to inversion.

6.3.3   Sensor system architecture

A simplified block diagram of the range detecting image sensor is given in Figure 6.13. The core of the image sensor is defined by the pixel matrix that transduces and accumulates the impinging light pulses. The imager is intended to operate in global (i.e. synchronous) shutter mode. Due to the tight constraints defined by the ToF operation, this, however, has to be carefully implemented as will be described in Sections 6.3.4 and 6.5.

For the global shutter a well-synchronized distribution of the clock for the shutter becomes necessary. Therefore, clock trees and appropriately designed high-speed drivers were implemented. Additionally to the need for the shutter switch drivers, drivers for the distribution of the reset and row-select signals become necessary. The readout of the signals that are directly demodulated by the photodetectors can be done for instance by standard 5T readout scheme which is explained in Section 5.1 or a 4T scheme as it will be explained in Section 6.4. The biasing of the source followers can be done at a column-level. As indicated, further signal conditioning circuitry can

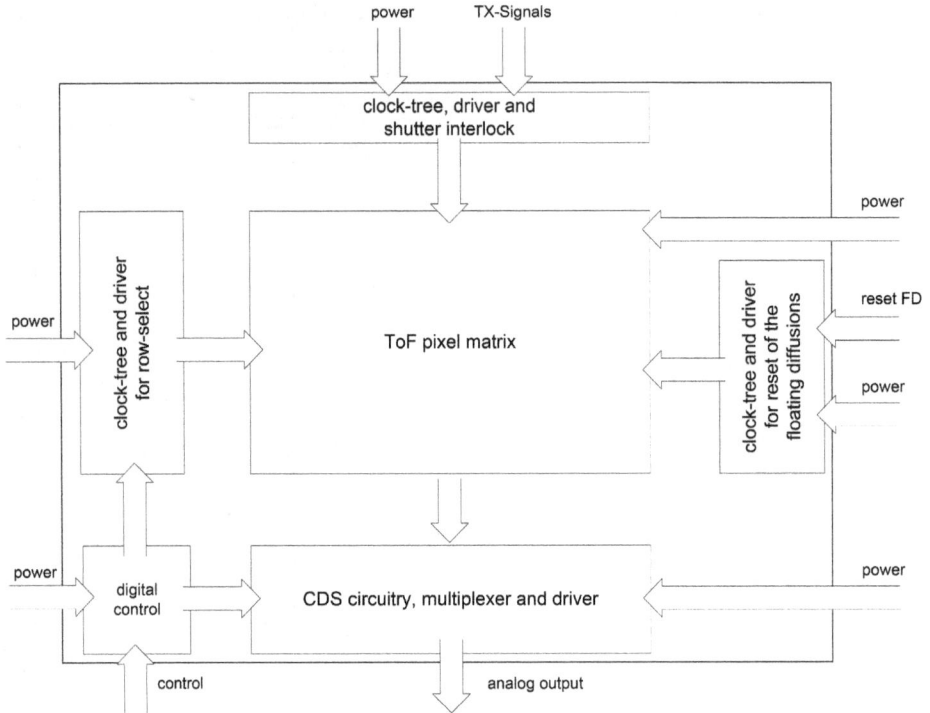

Figure 6.13 Block diagram of a range imager based on PM ToF.

be implemented as well, but can also increase complexity for e.g. the calibration. For instance a switched capacitor circuit can be implemented to directly allow the subtractions given in Equation 6.1. This, however, is not meaningful if the real signals $V_{FD1} - V_{FD3}$ have a non-linear relation to the illumination or, in other words, if they deviate from $u_{T,\tau}(\tau_{ToF})$. A switched-capacitor circuit which carries out the operations $V_{FD1} - V_{FD3}$ or $V_{FD1} + V_{FD2} - 2 \cdot V_{FD3}$ exhibits a higher temporal noise level, since reset, photodiode and readout noise are not correlated. If possible though, the signals should be converted to the digital domain as soon as possible, in which signal conditioning, such as correlated double sampling, can be performed.

6.3.4 Fabrication technology

As has been pointed out in the introductory section of Chapter 4, a 2P4M 0.35 μm CMOS process has been employed for the designs that have been developed within the framework of this work. The process was investigated for its feasibility for optoelectronic applications in general (c.f. [Du09a]) and range imaging applications in particular (c.f. [Sp10]). This work further extends these investigations by studies of the noise performance of a large variety of components (c.f. Chapter 4) for general purposes and components for imaging purposes (c.f. Chapter 5). To extend the capabilities of the process by the high-speed photodetector referred to as LDPD, an additional

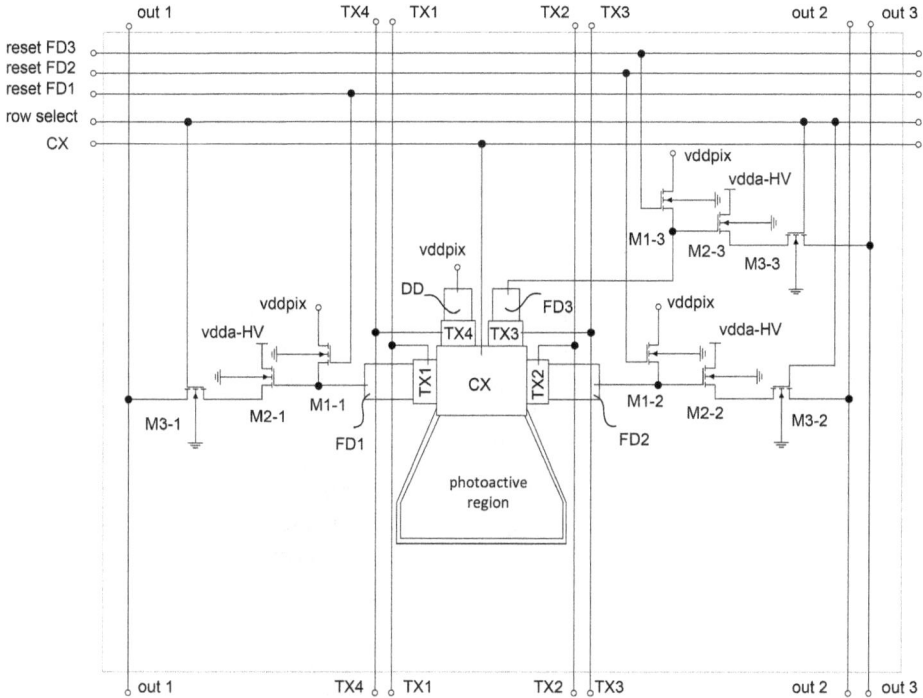

Figure 6.14 Schematic of the ToF-LDPD pixel.

implant has been introduced to create the n-well with the built-in potential gradient (c.f. Chapter 5).

6.4 DETECTOR DESIGN AND EVALUATION[5]

The presentation of the design of the imager presented in Figure 6.14 in this section is reduced to the components that bear main impact on the performance of the range image camera. The blocks digital control, analog multiplexer and output buffers are standard circuits that affect the overall performance very little. The CDS circuitry was not used throughout the characterization due to the reasons explained in the preceding section.

---

[5]The content presented in this section was developed within a collaboration of the departments *Optical Sensor Systems, Integrated Circuits and Systems* and *CMOS Technology and Devices* of the Fraunhofer Institute for Microelectronic Circuits and Systems. The author contributed to this work with the design of the readout chain from the storage nodes to the input of the CDS circuitry which can be bridged to directly connect the column data-line via multiplexer to the analog output buffer, the drivers for the signals which stimulate the pixel matrix and the layout of the pixel matrix and driving circuitry. The content presented in this section resulted in the publication [Sp09].

6.4.1   Readout circuitry

The readout circuit contains a plain 4T APS circuitry implemented once per storage node per pixel, as it is depicted in Figure 6.13.

The positive power supply voltages of the APS (vddpix and vdda-HV) were chosen to 5 V to allow for higher electric fields and thus faster charge transfer within the photodetector. Furthermore, this way the voltage swing and thus the dynamic range are extended. The pixel pitch of 40 μm did not allow the integration of a p-type MOSFET for reset (RST) function. An NMOS transistor was thus chosen which also inherently provides anti-blooming functionality, because it conducts current in OFF-state when the potential of the storage node becomes lower than $-V_{TH-M1}$. To extend the dynamic range by means of realization of a large voltage swing, the RST-signal which is applied to M1 during reset is set to 8 V. This was found to allow charging the storage node up to 5 V without pushing M1 into weak-inversion, so that soft-reset is avoided (c.f. Section 5.2.3). The row select transistor (M3) shares one diffusion with the source follower (M2) to yield a higher fill-factor. Assuming an I_D referred current noise power spectral density $S_{IDID} = 4k_B\theta g_m\gamma_e + \frac{K_f I_D^{\alpha_f}}{C_{ox}W_{eff}L_{eff}\nu^{\gamma_{flicker}}}$ for M2 and a resulting output resistance of $r_{out} = 1/(g_m + g_{mb})$ which is assumed to dominate over the output resistance of the current load, the voltage noise power spectral density referred to the output node of the source follower is calculated to

$$S_{V_{SF-out}V_{SF-out}} \approx \frac{4k_B\theta g_m\gamma_e}{(g_m + g_{mb})^2} + \frac{\dfrac{K_f I_D^{\alpha_f}}{C_{ox}W_{eff}L_{eff}\nu^{\gamma_{flicker}}}}{(g_m + g_{mb})^2}, \tag{6.24}$$

where $g_{mb} = \partial I_D/\partial V_{BS}$ was used. Since a large width simultaneously allows for a low flicker noise component and small output resistance a rather high W/L was of 8/1.2 was chosen for M2. The length corresponds to the minimum gate length for 5 V design rules. This was possible because the constraints for read noise are comparatively low (c.f. discussion in Section 6.3.1). The bias current was chosen to 50 μA which is also tuned for fast readout speed of the pixels due to the good slew rate and tolerable noise performance. The simulated DC voltage transfer functions of the readout circuitry are given in Figure 6.15. Here, V_{in} is the potential at the gate of the source follower and V_{out} is the analog output of the chip. Corner simulations have been performed to study parameter mismatch of the circuitry. According to the specifications given in Table 6.2 the design objectives of the output referred linear voltage range is 2 V. To study the linearity the output referred voltage was evaluated at $V_{in} = 5$ V. The range from that value to 2 V below that value was fitted to the affine linear mapping $V_{out} = gain \cdot V_{in} + offset$ with the LQ-method. The maximum absolute distance between the fit and the simulation results was normalized to the 2 V and plotted in Figure 6.15. The linearity error was evaluated to be below 1% for all corner and temperature settings. The offset which was obtained by the affine linear fit, was found to vary about 190 mV at 27°C and 300 mV over the entire temperature range. The slope of the transfer function was found to be very insensitive to process and temperature variations. Its value is approximately 2/3 and it varies less than 1.5% at 27°C and less than 3.6% over the entire temperature range. This can be explained by the bulk effect of M2. This dominates the output impedance of the source follower configuration and is proportional to the

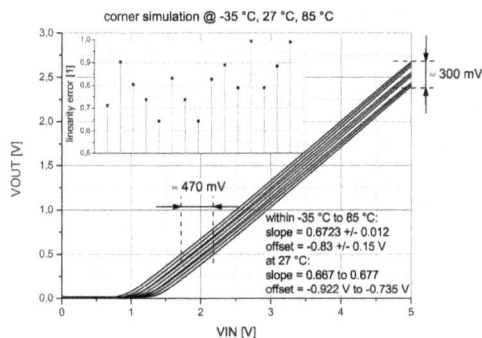

Figure 6.15 DC voltage transfer characteristics of the readout chain.

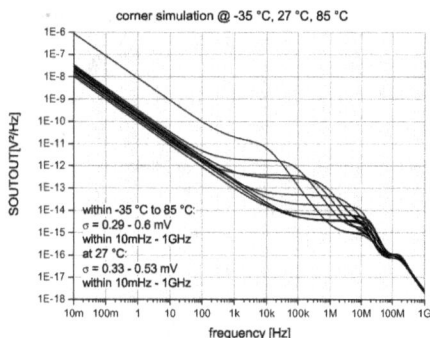

Figure 6.16 Noise characteristics of the readout chain.

transconductance of M2 so that the source follower amplification is hardly sensitive to variations of W/L of M2, the output impedance of the current sink, or the bias current. Assuming the simple Level 1 model for M2 and operation of all devices in strong inversion and saturation, the differential amplification, i.e. small-signal voltage gain, is found to equal

$$\frac{dV_{out}}{dV_{in}} = \frac{1}{1 + \dfrac{K1}{2\sqrt{\phi_S + V_{out}}}} \approx 2/3 \tag{6.25}$$

with $K1 \approx 1.4\,V^{0.5}$, $\phi_S \approx 0.7\,V$ and an output voltage of 0.7–2.5 V.

Figure 6.16 demonstrates the simulated noise power spectral density of the readout circuitry modeled as an LTI system and verified using small-signal noise analysis. Integration over the frequency range from 10×10^{-3} to 1×10^9 Hz yields standard deviations of the output signal below 1 mV for all corner and temperature settings. In the actual application parasitic capacitances, the packaging and the actual load limit the bandwidth, what will result in a reduced noise power exhibition. Nevertheless, the sampling process also mixes the undersampled noise processes, to that proper calculus has to be employed when accurate estimations of the noise performance have to be made (c.f. Section 3.2). As has been discussed in the former section, the impact from the circuit noise on the range measurement accuracy is rather limited, so that the increased effort for proper analysis was not necessary (cf. Chapter 6.3.1).

6.4.2 ToF-LDPD design

The ToF-LDPD itself was designed according to the concept in Section 6.3.2. Contact to LOCOS or the gate-oxide is avoided by separation with p-well implants and the p^+-pinning layer. The doping of the n-well was realized by one additional mask. The implantation gaps were designed such as to yield a doping gradient from the photoactive region towards the collection gate after the thermal processing steps as it was proposed in [Mer95] (c.f. [Du10; MFH10]). The n-well was adjusted to yield a proper

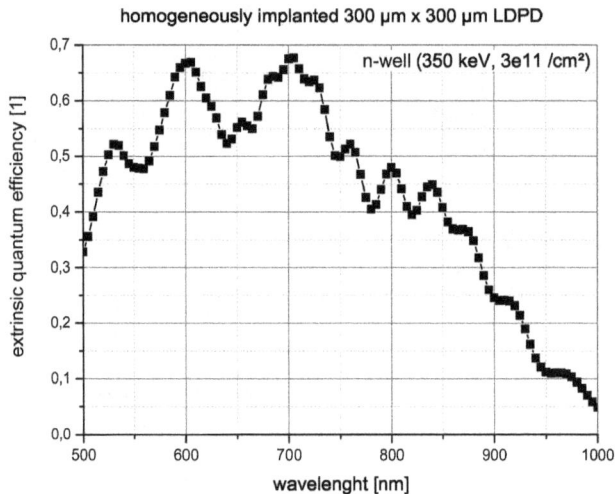

Figure 6.17 Extrinsic quantum efficiency of a homogeneously implanted $300 \times 300 \,\mu m$ n-well implanted with dose of 2.8×10^{11} 1/cm^2 and energy 350 keV underneath the p$^+$-pinning layer.

compromise between charge transfer speed and the capability of properly switching ON and OFF the current paths between photoactive area and the storage or draining nodes (c.f. Figure 6.18). The quantum efficiency was not a major design objective. In Figure 6.17 the extrinsic quantum efficiency of a homogeneously implanted n-well sandwiched between the p$^+$-pinning layer and the p-type epitaxial layer is given. The dose and energy are in vicinity of those from the fabricated n-well which is used for the ToF-LDPD. The measured extrinsic quantum efficiency is much above the 5% given in the specification. It is believed, that the lateral gradient in the doping concentration does not affect this parameter severely enough to fall out of the specification. In the upper part of Figure 6.18 the shutter switch is OFF. The potential gradient is properly increasing towards the collection gate where the charges will be gathered until the shutter gate is turned on and the potential maximum will be defined by the floating diffusion, which is depicted in the lower part of Figure 6.18. Within the actual application, however, one node is always connected to the active region (c.f. Figure 6.2). During the first phase a photogenerated electron e.g. at 'A' travels towards the potential maximum underneath the collection gate along the indicated path. A potential barrier is introduced between the potential maximum and the storage node by setting the potential at the transfer gate (TX) to a 0 V. In the transfer phase the barrier is eliminated by switching TX to 3.7 V.

An increasing dose of the n-well yields a deeper n-well due to the out-diffusion during thermal processing steps. This results in a higher quantum efficiency for longer wavelengths, but also negatively affects the capability of controlling the current flow by the transfer gates. The same applies for higher n-well implant energies. Since the electric field along the indicated cutlines 'A–B' constantly accelerates charge carriers, there exist bijective relations between the traveled distance of charge carriers along

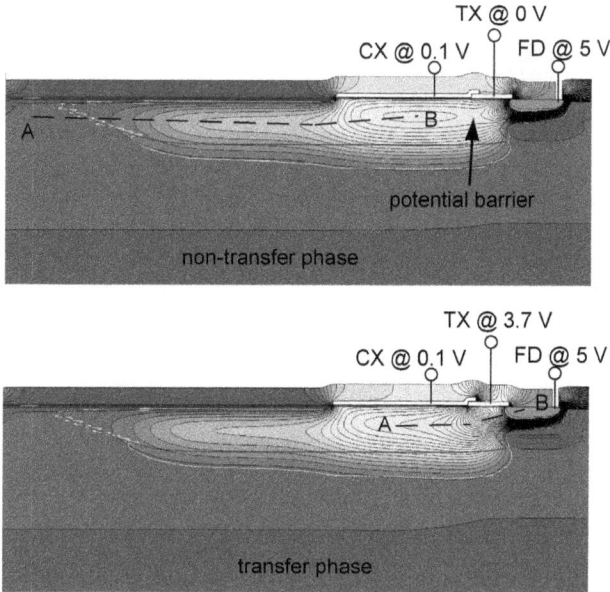

Figure 6.18 Electrostatic potential profile of the proposed ToF-LDPD for an n-well implant dose of 2.8×10^{11} 1/cm^2 and energy 350 keV.

'A–B' and the corresponding elapsed time. Thus the ordinary differential equation $v_x(t) = \mathrm{d}x(t)/\mathrm{d}t$ can be rearranged to

$$t = \int_{x_0}^{x'} \frac{1}{v_x(x')} \mathrm{d}x' \qquad (6.26)$$

in the one-dimensional space domain. Here, x is the space coordinate along the trajectories described by $-\mathrm{grad}(\phi)$ which correspond to the paths that are indicated in Figure 6.18.

'A' describes a point which is far away from the collection gate and the storage nodes. Thus it might be assumed that the proposed approach yields an overestimation of the transfer time. Extraction of the electrostatic potential profile along the paths 'A–B' allows for evaluation of the velocity by $v_x(x) = \mu_e \cdot \mathrm{d}\phi(x)/\mathrm{d}x$. Substitution into Equation 6.26 and integration along the cutlines 'A–B' yields $t(x)$ as depicted in Figures 6.19 and 6.20. In Figure 6.19 the transfer time for travelling from the photoactive region towards the collection gate is given, whereas Figure 6.20 shows the transfer time charge carriers need for propagation from underneath the collection gate to the potential maximum in the floating diffusion after the transfer gate has turned on. The presented approach resulted in an estimated worst case transfer time of approximately 10 ns. This corresponds to the best trade-off, which allowed for proper speed and capability of controlling the current paths. The n-well implant dose of 2.8×10^{11} 1/cm^2 and energy of 350 keV were thus chosen.

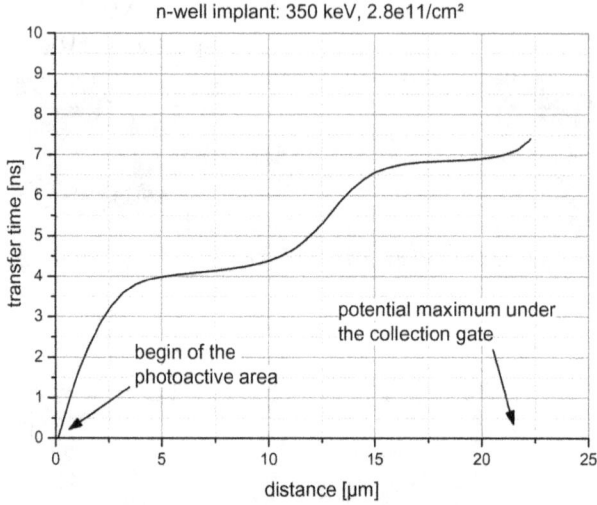

Figure 6.19 Estimation of the worst case transfer time from the photoactive area to the collection gate.

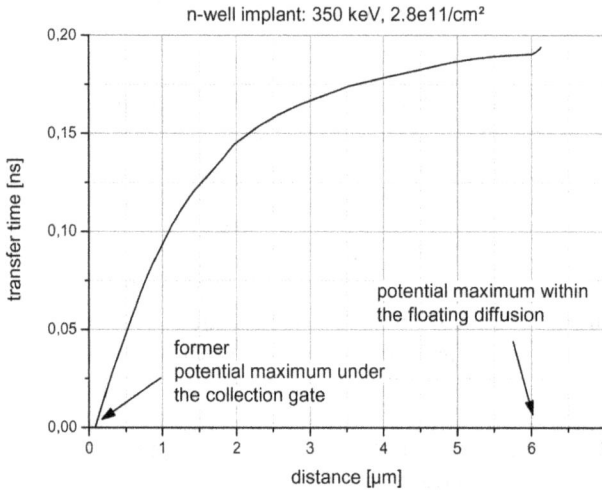

Figure 6.20 Estimation of the worst case transfer time from the collection gate to the floating diffusion.

6.4.3 Evaluation of the first generation LDPD based PM-ToF imager

Figure 6.21 depicts a microphotograph and the layout of the ToF-LDPD of the fabricated demonstrator. As it is indicated in the microphotograph, a large CDS block was implemented. In order to allow for realization of averaging by multiple accumulation.

ON THE DESIGN OF PM-TOF RANGE IMAGERS

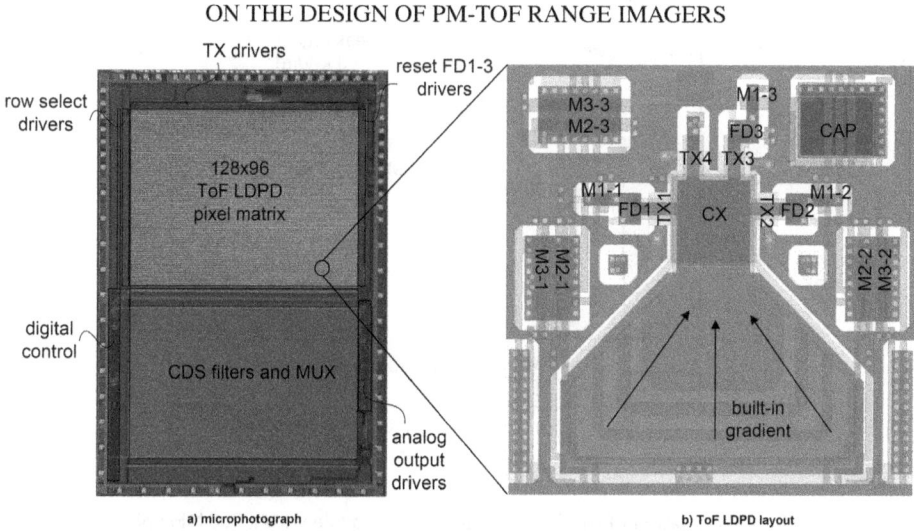

Figure 6.21 Microphotograph and pixel layout of the first generation ToF-LDPD range imager.

Since it was verified that the ToF-LDPD provides sufficient charge handling capabilities, these accumulations were carried out in the photodetector itself to avoid additional noise contribution from the readout circuitry.

For the sake of simplicity, the ToF-LDPD demonstrator was characterized using an emulating test principle in which scaling of the impinging radiant flux due to spherical broadening is separated from scaling due to the actual time-of-flight (c.f. Equations 6.10–6.14). The measurement setup contains a laser pulse which is fixed at a distance of 40 cm from the sensor and arranged along its optical axis (c.f. Appendix B.2). The time instance at which the laser pulse is emitted can be controlled by a laser delay line that allows for time increments down to 0.25 ns. The irradiance can be attenuated employing neutral density filters. The peak irradiance per pulse at the sensor was varied between 3 to 6000 Wm^{-2}. The irradiance level of 6000 Wm^{-2} was achieved by reducing the distance between laser module and imager to 30 cm. The shape of the employed laser pulse has been characterized with a *HCA-S-SI-SMA* photoreceiver which has a bandwidth of 200 MHz. The full width at half maximum of the laser pulse is approximately 31 ns as depicted in Figure 6.22. The presented shape clearly deviates from the ideal rectangular pulse shape. The rise and fall transitions appear similar to relaxation processes as they occur for instance in simple RC-low pass filters. The phase response of the photoreceiver was not available to the author. However, assuming a single pole low pass filter with a characteristic time constant of $1/2\pi \cdot 200$ MHz $= 0.8$ ns a rise time $t_{10-90} \approx 1.8$ ns can be estimated. Since this is much shorter than the observed transition time, it can be concluded that either the photoreceiver is not properly modeled by a single pole low pass filter or that the laser pulse is truly significantly deviating from the ideal pulse shape. Assuming the latter, it is interesting to note that approximately 110 ns elapse before the laser pulse is finally decaying below 1% of its peak value.

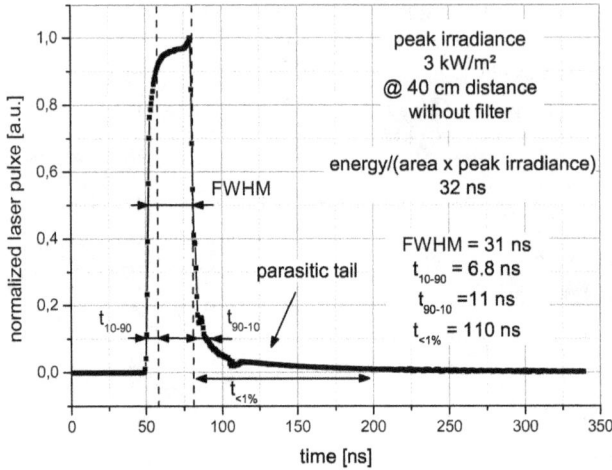

Figure 6.22   Characterization of the 905 nm-laser pulse module with a 200 MHz photoreceiver.

Figure 6.23 presents the response functions of the emulated time-of-flight principle. The axis of ordinates presents the measured output voltage for varying time shifts given on the abscissa. Since the radiant flux is not affected by spherical broadening at differing object distances, the emulated ToF principle is similar to the convolution of the shutter impulse function and the laser pulse. This statement is, however, to be treated carefully, since convolution applies only to linear time-invariant systems. If the ToF-LDPD would carry out an ideal short time integrator operation and the laser pulse would have a perfectly rectangular shape, the response function should correspond to the functions depicted in Figure 2.7. The deviation from the ideal shape can thus be caused by two phenomena, imperfections of the laser shape and of the shutter function. The impulse response function of an ideal short time integrator to the measured laser pulse is given in Figure 6.24. The presented data is clearly differing from the ideal shape. Nevertheless, it also becomes clear that the shutter performance itself is not yet optimal. Another imperfection that can be observed is mismatch between the three shutters. This phenomenon will be investigated more thoroughly in Section 6.6.

The data presented in Figure 6.23 were measured at an irradiance of $3\,kW/m^2$, a shutter window width of 75 ns and a laser pulse width of 30 ns. A significant non-linear relation between variations of the irradiance and resulting output voltage was observed as depicted in Figure 6.25. Here, the output voltage levels of FD1 from response functions to the emulated ToF principle as depicted in Figure 6.23 were evaluated for their maxima. These maxima were plotted for different irradiance levels and shutter widths while the laser pulse width was kept constant. Interestingly, increasing of the shutter width at constant laser pulse width resulted in significant improvements of the linearity. This observation leads to the conclusion that the transfer process of photogenerated charges into the storage nodes is intolerably slow at low irradiances

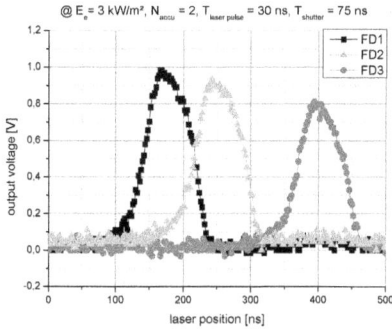

Figure 6.23 Response functions to the emulated ToF principle.

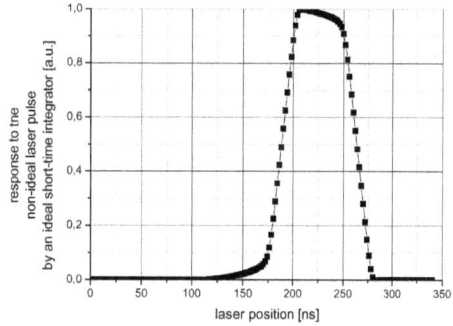

Figure 6.24 Response of an ideal short time integrator to the laser shape given in Figure 6.22.

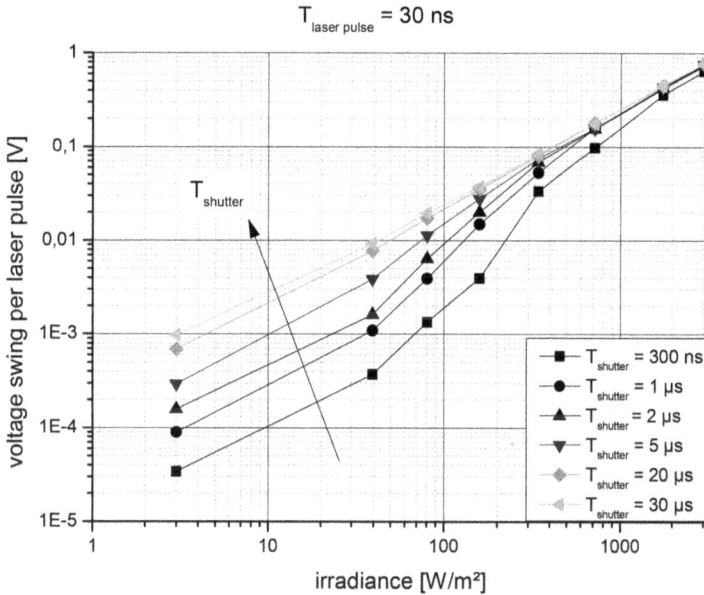

Figure 6.25 Output voltage characteristics for varying shutter pulse width at constant laser pulse width.

in the presented design. The origin of this insufficiency and how the performance can be improved is investigated in the next section. Since the observed transfer speed is far from the specifications given in Table 6.2 no further characterization of the range imager will be presented.

6.5   SPEED CONSIDERATIONS FOR LDPD BASED TOF IMAGE SENSORS[6]

6.5.1   Design considerations for charge transfer speed improvement

In the former section it was concluded that the first ToF-LDPD demonstrator suffered from insufficient charge transfer speed – especially at low irradiances. From further analysis of the previous design two possible causes for this phenomenon emerged.

On the one hand, finite element simulations of the lateral drift-field photodetector showed that the n-well that defines the actual charge transfer, exhibits a concentration gradient in the wrong direction due to its geometrical design in the region of the transfer gates. In Figure 6.11 it is depicted that the n-well that forms the photoactive area and propagates charges towards the storage nodes, has an increasing doping concentration along the I-I cutline and has a maximum below the collection gate. Considering for instance the region in part b) where the n-well extends towards TX3, it can be observed that the expansion along the x-axis, which is orthogonal to the charge transport direction, is small compared to the dimensions of the n-well underneath the collection gate.

This results in the described lower doping concentration and thus in an electric field pointing in the wrong direction. The definition of a minimum area of the implantation window of the n-well implant in order to assure that there are no concentration gradients in the vicinity of the transfer gates can circumvent this bottleneck. In Figure 6.26 it is indicated how the n-well dimensions were rearranged to yield proper charge transfer. Along the cutlines I-I and II-II it can be seen that the n-well regions below TX3 and TX4 were joined in order to avoid further increase of the collection gate area. The separation of the neighboring storage nodes has then to be accomplished by channel stop implants at the surface of the silicon layer as can be seen in Figures 6.26 and 6.27. The later demonstrates this functionality by means of TCAD simulation results.

Another issue derived from simulations at circuit level including parasitics of the shutter path, was the insufficient distribution of the shutter signals across the pixel matrix. This distribution in the former sensor was limited by off-chip buffering of the power supply voltage. Since the operation of PM ToF range imagers relies on determination of time differences between emitted and received laser pulses, the shutter timing control plays an extremely important role, because it is used to define the integration time windows for the photodetectors serving as pulse receivers. A delay in the signal distribution across the imager can be compensated by calibration of the sensor output signals as long as it is much shorter than the shutter window. Nevertheless, propagation delays of driving stages and signal propagation across interconnects of

---

[6]The redesign was carried out within a collaboration of the departments *Optical Sensor Systems* and *CMOS Technology and Devices* of the Fraunhofer Institute for Microelectronic Circuits and Systems. The author contributed to this work during discussions on pixel level and by the error tracking and redesign on circuit level. The content presented in this section resulted in the publication [Sü13].

(a) top view of a LDPD

(b) cross-sectional view I-I of a ToF LDPD

(c) cross-sectional view II-II of a ToF LDPD

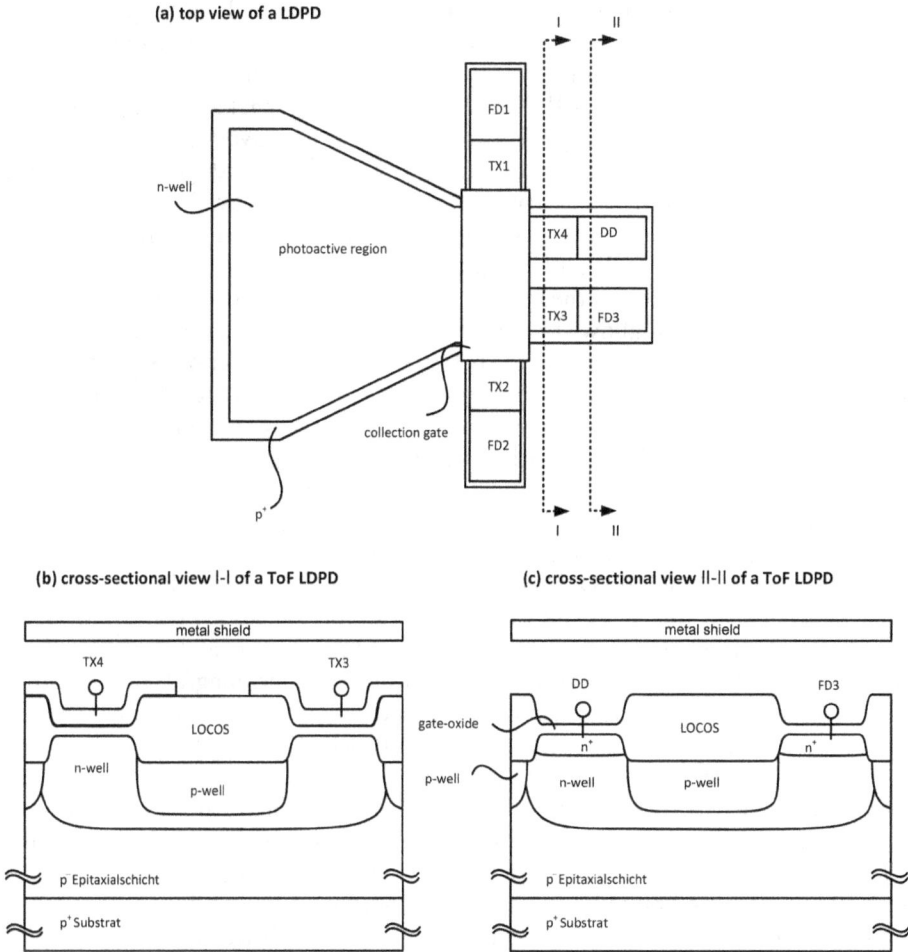

Figure 6.26 Schematic of the pixel redesign for faster charge transfer.

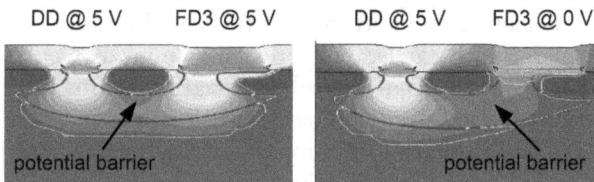

Figure 6.27 Electrostatic potential profiles along the cutline II-II from Figure 6.26.

some millimetres can easily amount to some nanoseconds. In [Sü13] a comparison with state of the art high-speed image senors is given. CW ToF imagers are often operated at frequencies between 20 to 50 MHz (e.g. [KKK12; PMP12; SMP11]). In CW imagers based on the mixing or the correlation method, in principle only the sinusoidal fundamental tone is needed for demodulation (c.f. Section 2.3.2). CW demodulation based on the sampling method, however, is prone to generate high harmonics which degrade the performance in a similar way as the limited bandwidth degrades the PM ToF performance. A global shutter 400×256 pixel imager with pulse lengths shorter than 50 ns is reported in [THK13]. Nevertheless, the chip package that is used comprises 424 pads to achieve the performance and was not available. Unfortunately, very little literature about the design of symmetrical global shutter generation for high frequencies, as they become necessary for PM ToF imagers, is available.

It will now be demonstrated how a low-power, high-speed symmetrical shutter control circuitry for PM ToF can be implemented, to yield a good background light suppression and a high ratio of the resulting effective pulse width to the intended shutter width. Additionally, it allows for variable voltage levels of the shutter, what yields the necessary flexibility to introduce the best potential gradient into the photodetector during charge transfer operation. Furthermore, it will be explained how an accidental short-circuiting of the storage nodes can be prevented by the implementation of a logical circuit into the high-speed path. After explanation of the used approach simulation results are presented, followed by measurement results of the proposed PM ToF principle based 128×96 pixel range imager.

To start with it, it must be stated that in practice, the pulses that trigger the transfer gates of the PPD/LDPD detectors are usual far from ideal. These pulses might for example overlap due to delays or disturbed shape. Since this could introduce a parasitic sensitivity or in worst case even cause a short circuit, it should be avoided. High-speed signals always tend to suffer from jitter, thus a meaningful solution for this problem would be to probe the signal of the preceding shutter period and activate the switching procedure only after a certain value has been reached. For this, a probe of a well defined worst-case column in combination with a logical circuit and a global driving stage might be one solution. However, for a symmetrical signal distribution the sensed worst case signal has to be applied to a clock tree. The signal propagation of these stages would thus directly result in a bad ratio of the effective resulting pulse width and the intended pulse width. Thus this approach either suffers from a column level mismatch or a non-sufficient effective shutter width. A better approach is to probe the preceding shutter signal for each column. The logic can then be implemented in one of the last stages of the driver to provide a sufficiently large shutter width. Otherwise the time that elapses between the probe and the actual switching process might be insufficiently long. For pulses as short as 30 ns this could be a serious problem.

The shutter generation is carried out for each column. A CQFJ84 package, that provides only 84 pads, was chosen to embed the imager. Since many pins were reserved for different functionality and the PCB board was chosen to be as simple as possible, most of the buffering had to be done on chip. Thus capacitors are integrated on-chip to provide the charge, that is necessary for the compensation of the switching processes in order to prevent spikes at the supply rails and for stabilization of the power supply. To provide a low impedance connection to the drivers, that actually charge the load, the capacitors are located as close as possible to the load.

In Figure 6.28 the schematic of the global shutter control circuitry is shown. The four master high-speed shutter signals TX1-4 propagate through the clock-tree that provides a symmetrical clock distribution across all columns of the range imager. Exemplarily, a driving stage that comprises an interlock circuit that prevents any

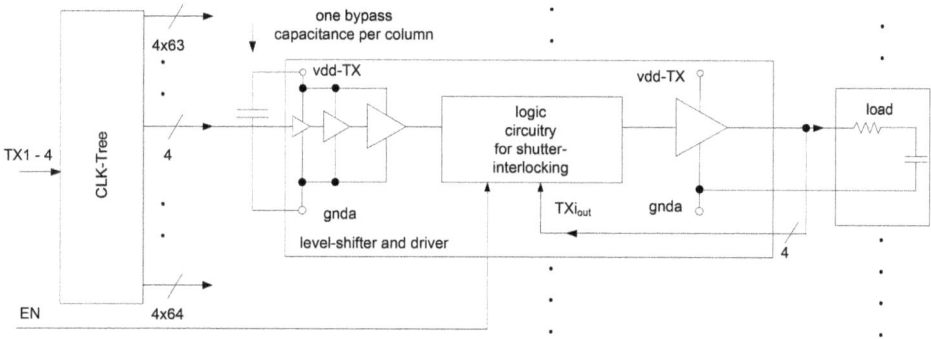

Figure 6.28 Architecture of the shutter generating circuitry.

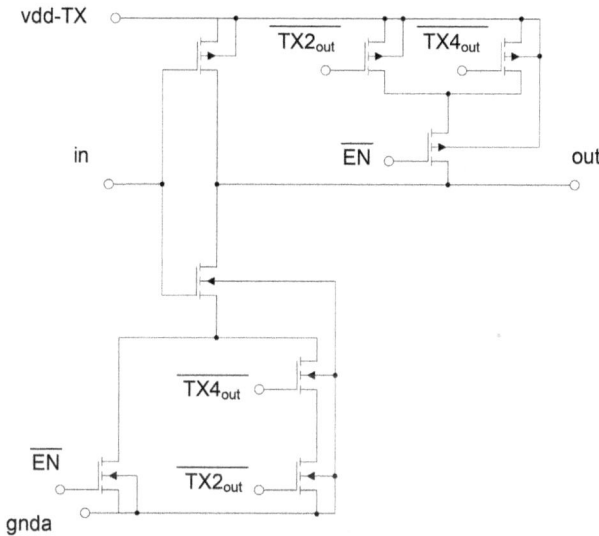

Figure 6.29 Interlock circuit for shutter switch TX1.

overlap between various TX signals is depicted. To yield a sufficient ratio of effective to intended pulse width, the interlock circuit is inserted before the last driving stage. This has the advantage that only very little signal propagation has to be done, from detecting the OFF-state of the preceding shutter and the actual switching process for the following shutter pulse. For verification, the interlock may be disabled by setting the enable input (EN) to the ground potential. The drivers were designed to be sufficiently fast for positive power supply voltage levels of 1.7 to 3.7 V.

In Figure 6.29 the interlock circuitry for shutter switch TX1 is presented. If EN is set to LOW, the circuit operates like a standard inverting driving stage. If, on the

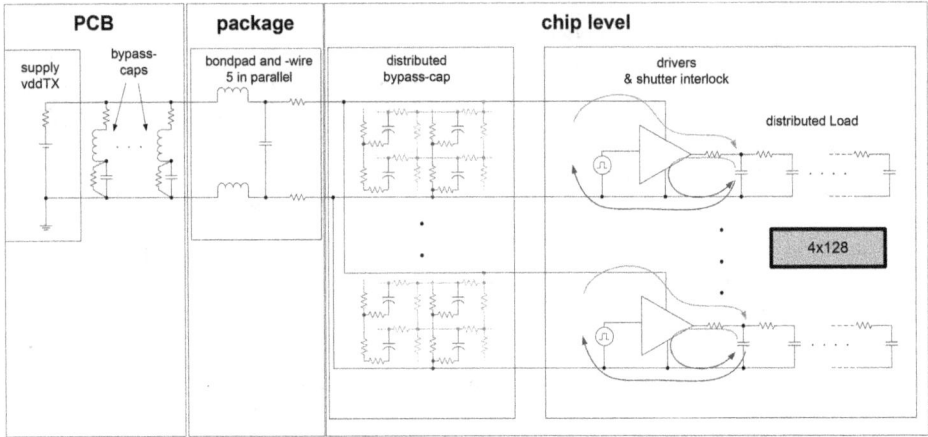

Figure 6.30 Simplified parasitic model.

contrary, EN is set to HIGH, the circuit operates as a digital circuit with the Boolean operation *inhibition*. Additionally, it offers the necessary driving capability. According to Figure 6.2 it was necessary to protect TX1 against TX2 and TX4, TX2 against TX1 and TX4, while for TX3 there was only need for protection against TX4.

Figure 6.30 presents a simplified schematic of the circuit model, with which the operation was verified. It is emphasized how the voltage distribution is undertaken, so that rise times below 5 ns become possible. After switching the driving stages to HIGH, a current is delivered from the on-chip capacitor, that was designed to yield a maximal voltage spike on the positive power rail of 10%. If the driving stages are switched OFF, the load is discharged through the pull-down path of the amplifier. The wiring was carefully done to avoid an impermissible large ground-bounce during that process. After one switching process, the capacitor had to be charged again within the shutter period to provide enough voltage swing for the next switching process.

For that, carefully routed wires were used in combination with 5 pads for the on-chip supply voltage for transfer gates vddTX and 5 pads for analog ground, to yield a low-impedance connection to the bypass and buffering capacitors on the PCB. The large on-chip capacitors employ polysilicon to gate oxide (GOX) to n-well capacitances, since they provided the highest specific area capacitance available. However, this type of capacitor suffers from a high series resistance. To minimize this effect, a design strategy involving the use of 4 metal layers was developed that is depicted in Figure 6.31. Additionally, the signal wires for the shutters were routed through these connections, while parasitic coupling was avoided with the introduction of rather large gaps. The simulation model comprises a parasitic model for the PCB, the bypass capacitors, the package, a distributed model for the on-chip capacitance, all 128×4 drivers and a distributed model for the load (c.f. Figure 6.30).

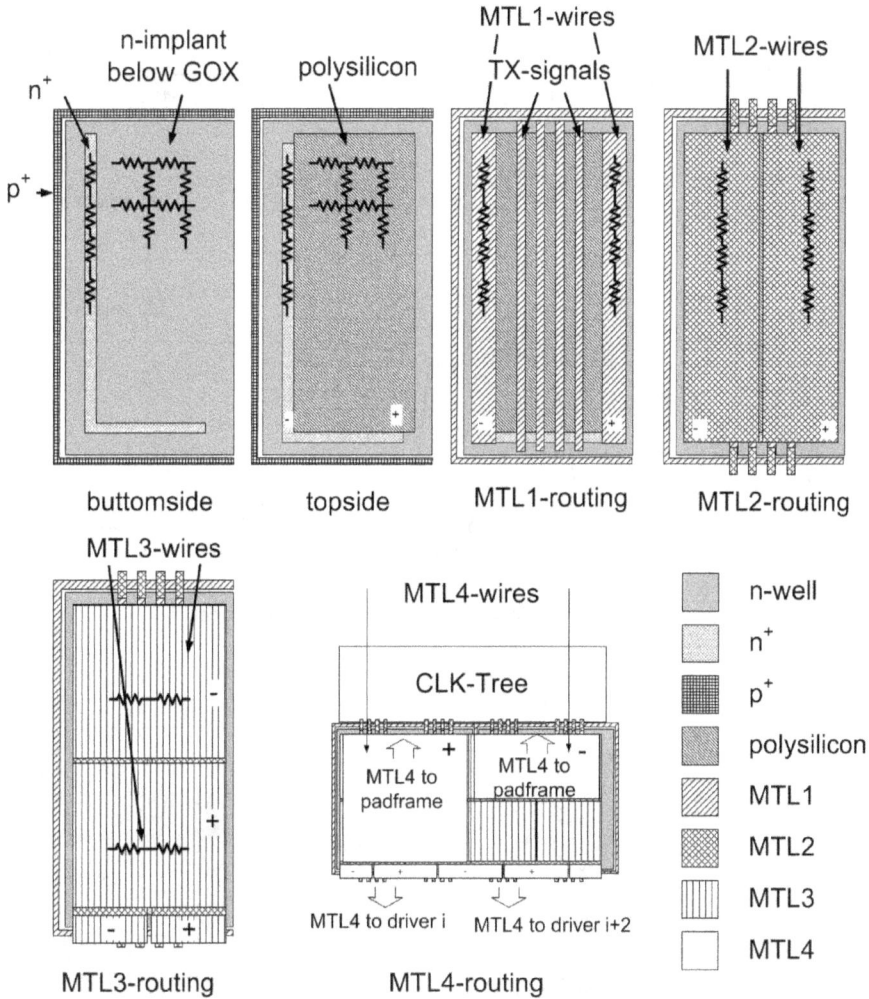

Figure 6.31 Layout concept for the buffer capacitor and the interconnects.

In Figure 6.32 the simulation results can be observed. It is demonstrated that although an overlapping is induced by the stimulating shutter signals, an overlapping active period of much less then 5 ns has been yielded, which – considering the expected charge transfer times in the order of several tens of nanoseconds – is expected to have negligible impact on the overall performance. Furthermore, with an effective shutter width of approximately 27 ns compared to the ideal 30 ns, a considerably good ratio was achieved. The skew was observed to be less than 1 ns across the columns, while the propagation across the rows yielded values below 3 ns. The maximum peak current consumption was 100 mA, while the average was below 2 µA for 100 fps and one accumulation per frame.

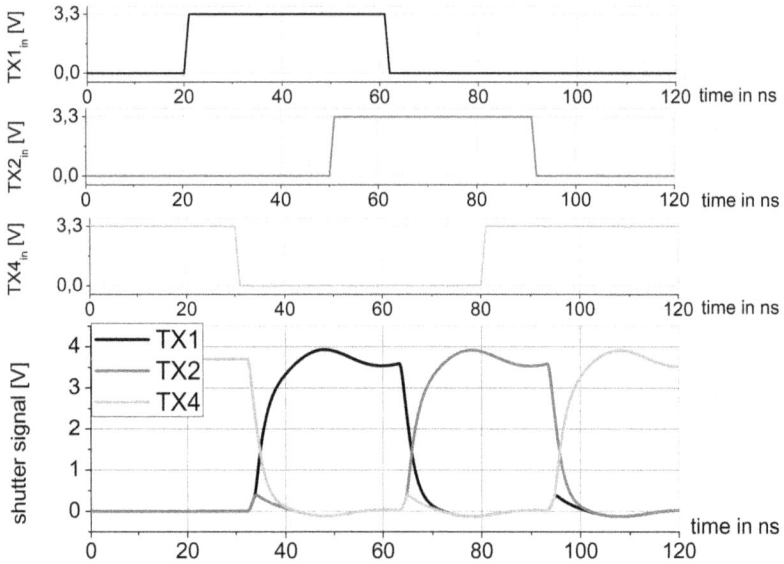

Figure 6.32   Simulation results of the parasitic model.

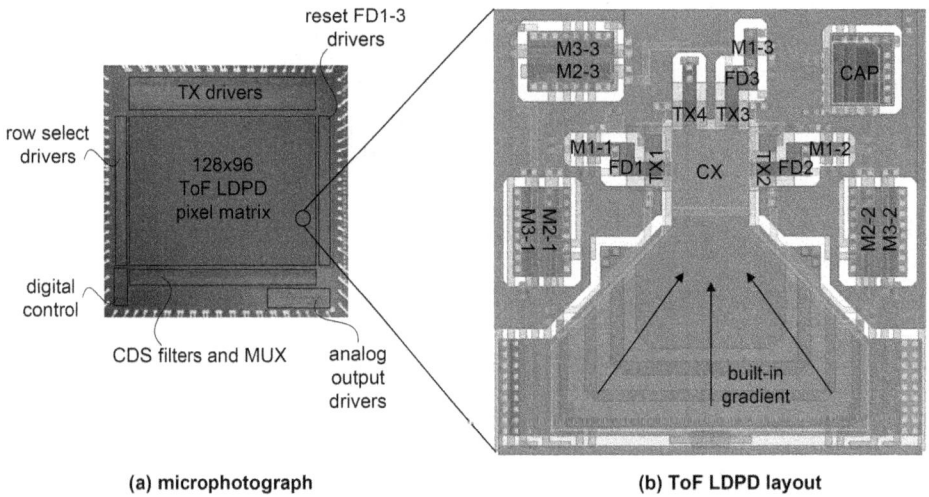

(a) microphotograph

(b) ToF LDPD layout

Figure 6.33   Microphotograph and pixel layout of the second generation ToF-LDPD range imager redesign.

6.5.2   Evaluation of the second generation LDPD based PM-ToF imager

Figure 6.33 presents the microphotograph and the layout of the redesigned ToF-LDPD. The large CDS block from Figure 6.21 is not integrated anymore, because the LDPD

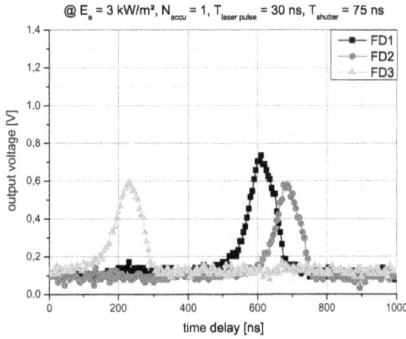

Figure 6.34 Shutter response to the emulated ToF principle at 3 kWm².

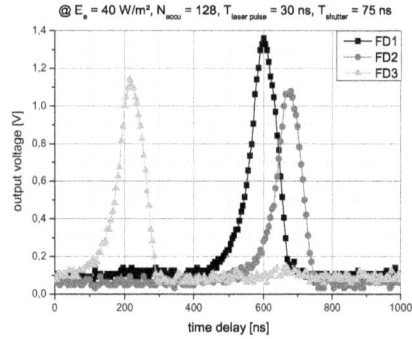

Figure 6.35 Shutter response to the emulated ToF principle at 40 W/m².

provided sufficient charge handling capability for multiple accumulations within the photodetector itself.

6.5.2.1 Shutter performance

For verification of the shutter performance the emulated ToF principle from Section 2.3.3 was employed. From the comparison of the Figures 6.34 and 6.35 it can be observed, that the shutter functionality is given at low and high irradiance levels. From Figure 6.36 this becomes even more evident. Here, the shutter width was varied for constant laser pulse widths. Comparison with the previous design proves that the redesign demonstrates a significantly improved shutter performance (c.f. Figure 6.25). Whereas the previous design yielded a linear output to irradiance characteristic at shutter widths larger than 20 μs, good linearity is now obtained at shutter widths down to 75 ns. This method relies on the fact that for all observed shutter pulse widths, the output voltage level should ideally be equal since $T_{\text{shutter}} \gg T_{\text{laser-pulse}}$. Thus, for verification of linearity for shutter widths shorter than 75 ns another laser module with shorter pulse width would have to be used (c.f. Figure 6.22).

In Figure 6.38, the response of the ToF-LDPD to the laser pulse given in Figure 6.22 is depicted. It can clearly be seen that for high irradiance levels, the shutter properly resembles the ideal response to the laser pulse. For low irradiance levels, however, the response still deviates from the ideal. This cannot be observed from Figure 6.36, since firstly it is a log-log plot and secondly only maxima of the emulated ToF principle were evaluated. The actual shape of the response was not taken into account. Better measures of the quality of the shutter response could thus for instance be defined as

$$m_{\text{lin}-1} := \int_{-\infty}^{\infty} \left(f_{\text{shutter}}(\tau) - f_{\text{ideal}}(\tau) \right)^2 \, d\tau \tag{6.27}$$

$$m_{\text{lin}-2} := \max_{\tau} \left(f_{\text{shutter}}(\tau) - f_{\text{ideal}}(\tau) \right), \tag{6.28}$$

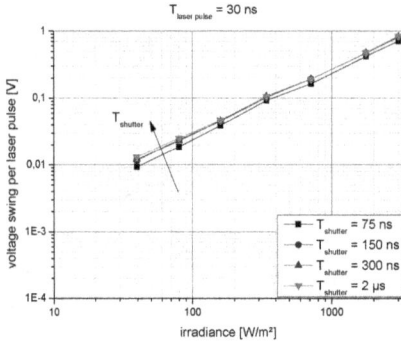

Figure 6.36 Output voltage characteristics for varying shutter pulse width at constant laser pulse width.

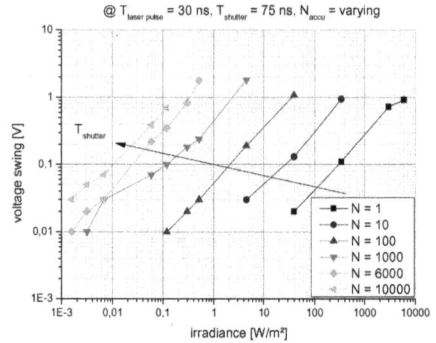

Figure 6.37 Output voltage characteristics for various accumulation counts and irradiance levels.

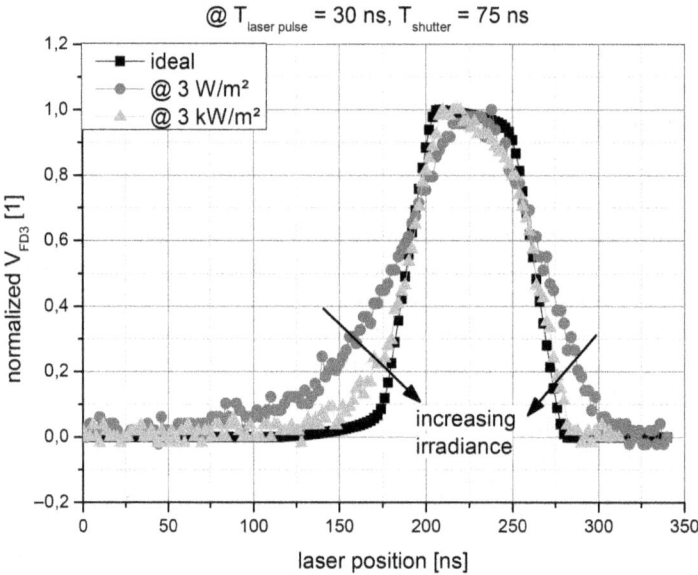

Figure 6.38 Comparison of the shutter response realized in the redesign to that of an ideal short time integrator.

were τ is the time shift defined by the delay line, $f_{shutter}$ is the response function of the shutter to the emulated ToF principle and f_{ideal} is the ideal response. The proposed measures originate from regression theory. Clearly, m_{lin-1} is similar to the norm that is employed in the *least-squares method*, whereas m_{lin-2} originates from the *minimax-method*.

It can be concluded that charge transfer is an irradiance dependent phenomenon. Simple analysis based on electrostatic potential profiles as it is described in Section 6.4.2

is not accurately describing the charge transfer speed, since the variation of the electrostatic potential due to photogenerated charge carriers is not taken into account. Furthermore, the two-dimensional profiles, that were demonstrated for the previous design do not model effects that are dependent on the third space dimension. The small implantation gaps in the vicinity of the transfer gates cannot be properly taken into account in the two-dimensional cut along the charge transfer direction. Another difficulty which may, however, be circumvented by an increased calibration effort, is that the doping profiles might differ from reality. Especially for the relatively lowly doped n-well this can cause significant mismatches between simulations and reality. Charge transfer is not an instantaneous phenomenon. Terms such as *complete charge transfer* have to be treated very carefully. As will be pointed out in Section 6.7 a proper physical model of the charge transport is necessary. It allows for accurate predictions of the imperfections of the shutter response and can also predict their impact on the actual application – the time-of-flight measurement.

6.5.2.2 Noise performance[7]

The readout circuit presented in Section 6.4.1 was evaluated for FPN and temporal noise performance. To allow for proper comparison between the simulation results and the measurements, the input of the system was defined as the voltage difference between the gate-potential of the source follower and the ground potential, whereas the output at which measurements have been carried out was defined by the analog output of the imager which was sampled with the ADC from the camera. To achieve this, the reset transistor was turned ON in order to allow for the definition of the FD-potential by vddpix. Since the reset transistor is turned on simultaneously for 12288 pixels, the resulting capacitance at the FD node is substantially increased. All measurements were carried out in the voltage domain, so that the high input capacitance corresponds to a significantly attenuated input noise level because firstly the high capacitance results in a low thermal noise level $k_B\theta/C$ and secondly filters noise components from the source itself.

As presented in Section 6.4.1, the expected variation of the offset was -0.735 to -0.922 V as predicted from corner simulations. The characterization of 12288 pixels from one imager demonstrated variations within -0.78 to -0.745 V what differs from the simulation results (c.f. Figure 6.39). This might be caused by the way the corner parameters of the simulation setup were chosen. It can though also be originated in the fact that the measurements were done at camera level. Thus more offset might be introduced from e.g. the analog-to-digital converter. The measured range for the gain was 0.667 to -0.674 (c.f. Figure 6.40), what fits the predicted simulation results from Figure 6.15. It is interesting to note, that offset and gain are both approximately distributed in a Gaussian manner. Since the mismatch at least for the offset is rather severe, it has to be compensated for during calibration.

The temporal noise component is firstly tested for stationarity. For each frame the sample variance was calculated by ensemble statistics. The resulted sample variances have then be compared between the subsequent frames. The temporal sample variance of the ensemble sample variance is then used as a measure for stationarity. Ideally,

[7]The author wants to acknowledge Adrian Driewer for the execution of the measurements.

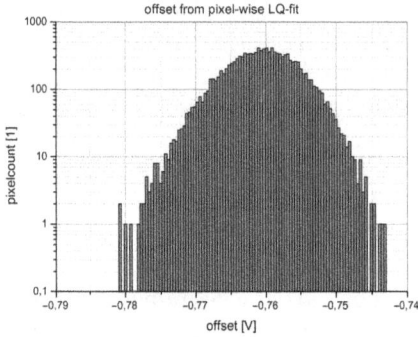

Figure 6.39　Histogram of the offset distribution of one imager.

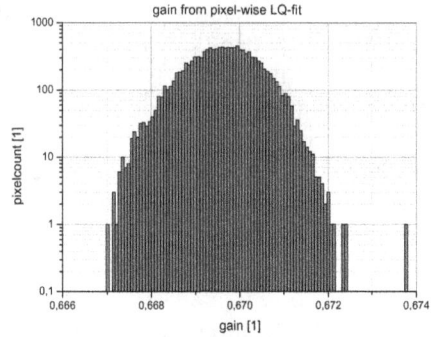

Figure 6.40　Histogram of the gain distribution of one imager.

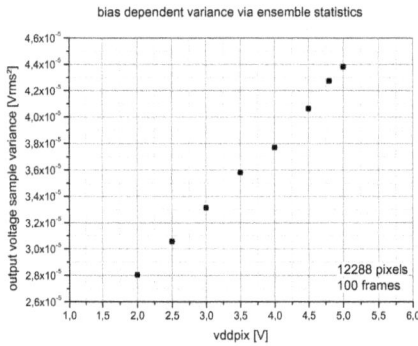

Figure 6.41　Stationarity test of the temporal noise – part 1.

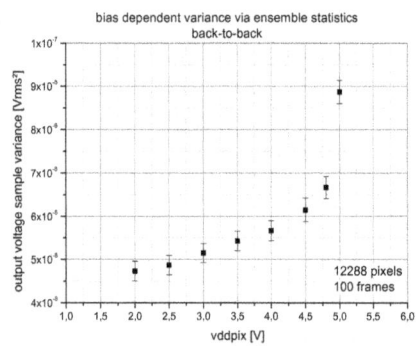

Figure 6.42　Stationarity test of the temporal noise – part 2.

each frame yields the same sample variance so that temporal statistics of the frame statistic would not show significant differences. However, when the random processes are non-stationary differences in the sample variance of each frame are expected to become visible.

In Figure 6.41, the error bars that are used to indicate the sample standard deviation (temporal statistics) of the sample variance within one frame (ensemble statistics) indicate that temporal fluctuations of the ensemble statistics of single frames are negligible – the bars are barely visible due to the very low fluctuation of the sample variance measure. This is though rather misleading because ensemble statistics do not simply evaluate the temporal noise but the fixed-pattern noise as well. This might be resolved by pixel-wise fitting of e.g. offset and gain. Unfortunately, it was observed, that the linearity error of these fits is still significantly higher than the actual temporal noise component. Taking the difference of two subsequent frames, however, properly eliminates the fixed-pattern noise within the pixels as it is proposed in the *photon transfer method (PTM)* (c.f. [Ass10]). This can be observed in Figure 6.42, for which two subsequent frames were subtracted (back-to-back) followed by the described statistical

variance via temporal statistics

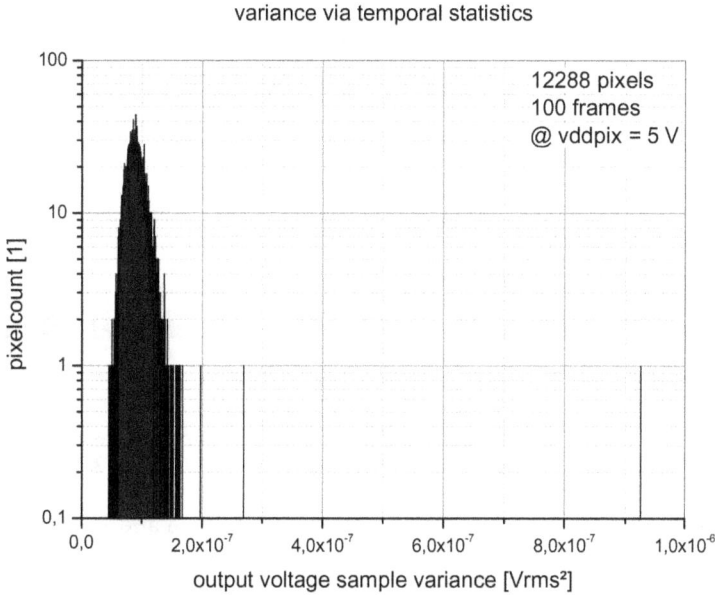

Figure 6.43 Ergodicity test of the temporal noise – part I.

analysis. Typically it is assumed that two subsequent frames are totally uncorrelated, so that the subtraction increases the sample variance by a factor of two. As proposed by [Ass10] this factor is considered and compensated for within the analysis[8]. The error bars now indicate a significantly larger temporal variation of the ensemble sample variance than before. Furthermore, it has to be noted, that the order of magnitude of the noise is also much smaller than before. The temporal standard deviation of the ensemble sample variance is negligible compared to the temporal mean value of the ensemble sample variance. Thus it can be assumed that the noise processes are in first order stationary. Interestingly, it can be observed that the output voltage reference is dependent on the vddpix potential and is significantly higher at 5 V. LTI based noise simulations verified that this substantial increase of the noise originates from the reset transistor which is no longer in linear region and contributes to the noise performance by its flicker noise component. Since the noise performance is in spec, further investigations of this phenomenon were not done within the framework of this work.

To analyze the ergodicity of the temporal noise, temporal statistics are compared to the ensemble statistics. Figure 6.43 presents the histogram of the pixel-wise temporal analysis of the sample variance. Most of the pixels exhibit a sample variance in the order of approximately 0.1×10^{-6} Vrms^2 for vddpix $= 5$ V. Some pixels also demonstrate significantly more noise. This will be investigated later on. Figure 6.44 shows the

---

[8]This problem is very similar to the considerations made in Section 5.3. There it was demonstrated, that the assumption of having totally uncorrelated samples is not always justified and may lead to errors for the predicted noise.

variance via ensemble statistics

12288 pixels
100 frames
@ vddpix = 5 V

pixelcount [1]

output voltage sample variance [Vrms²]

variance via ensemble statistics
back-to-back

12288 pixels
100 frames
@ vddpix = 5 V

pixelcount [1]

output voltage sample variance [Vrms²]

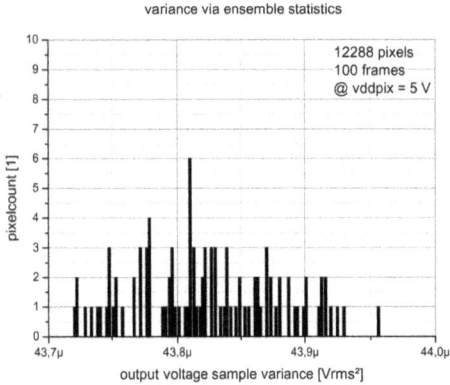

Figure 6.44 Ergodicity test of the temporal noise – part 2.

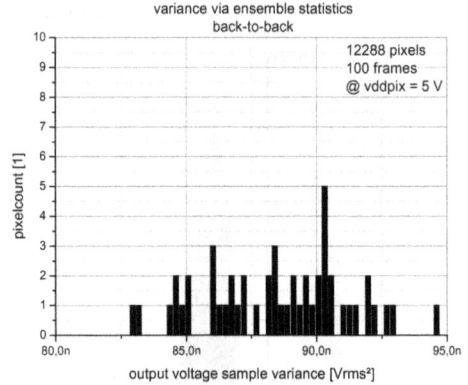

Figure 6.45 Ergodicity test of the temporal noise – part 3.

histogram of the direct pixel-wise ensemble sample variance and Figure 6.45 presents the ensemble sample variance of the subtraction of two subsequent frames (back-to-back). The latter is very similar to the transient statistics so that firstly, the above statement of significantly affecting direct ensemble statistics by *fixed pattern noise (FPN)* is justified and secondly it is demonstrated that in first order it can be assumed that the temporal noise is ergodic. Unfortunately, the comparison allows only for weak comparisons, because the frame count is significantly lower than the pixel count.

A typical method to analyze hot-pixels within an imager is to observe the temporal sample variance for each pixel as it was depicted in Figure 6.43. The presentation of the results though, usually slightly differs. Often this phenomenon is presented as logarithmic plot of $1 - \mathrm{CDF}_{sample}(\mathrm{var}_{sample})$ as it is depicted in Figure 6.46. Most pixels exhibit a sample variance of approximately 1×10^{-7} Vrms². Clearly 4 pixels demonstrate significantly more temporal noise than the others out of which one has a sample variance of almost 1×10^{-6} Vrms². This is almost a factor of ten higher than the noise exhibited by the majority of the pixels. To finalize this survey, a more meaningful measure of the temporal noise is depicted in Figure 6.47. The measure that is given here is yielded by evaluation of the before mentioned cumulative distribution functions such as to provide a worst-case value of the sample variance which is not exceeded by 99.7% of the pixels. Since analysis was performed with 12288 pixels, the few *hot pixel* candidates are not taken into account and thus do not disturb the performance measure.

6.5.2.3 Camera level verification[9]

The sensor was further employed in a 3D range camera consisting of an FPGA, a peak 4×75 W laser module emitting 45 ns pulses at 905 nm with a repetition rate of 8 kHz. The used optics form a cone-shaped field-of-view of 30°. First calibration results show

[9]The author wants to acknowledge Adrian Driewer for the execution of the PTM measurements and TriDiCam for the range measurements.

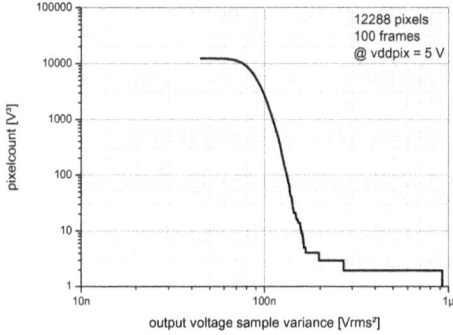

Figure 6.46 Hot pixel analysis.

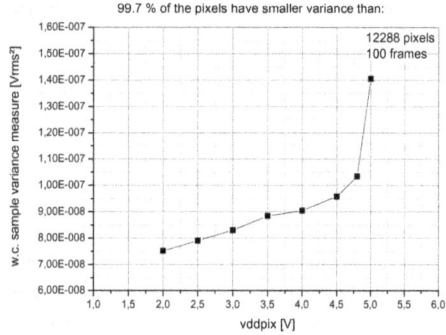

Figure 6.47 Trustworthy estimates of the sample variance.

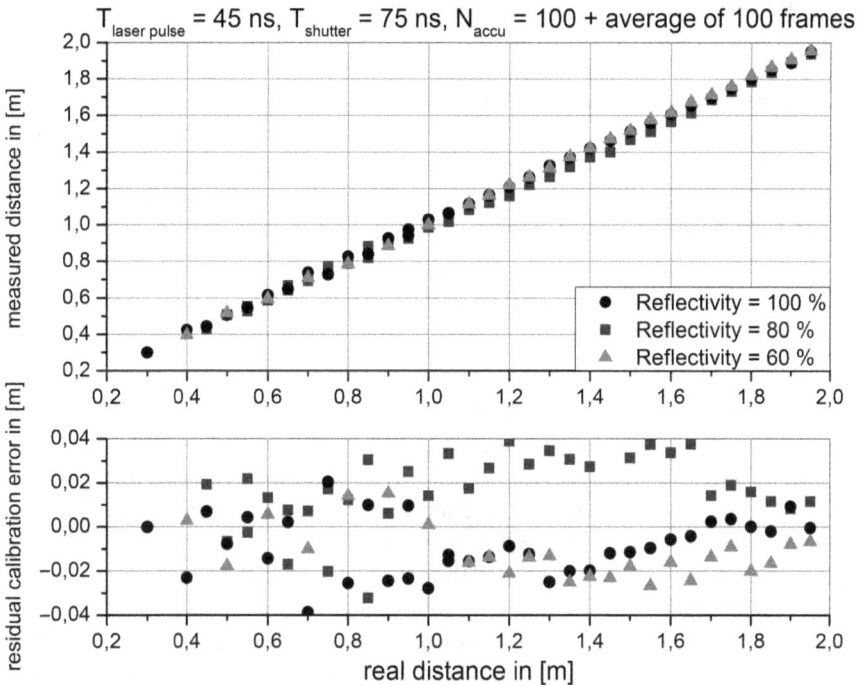

Figure 6.48 First calibration results.

performance in the centimetre regime as shown in Figure 6.48. The measurement setup which was employed here, comprises a linear-motion bearing on which the camera is mounted in order to properly set defined distances between the range camera and a planar object which is orthogonal to the optical axis. The planar object can be covered by different wallpapers, so that defined reflectances are obtained. The presented range measurements were carried out for three different reflectances. The calibration was performed in a least-squares manner with the multivariate polynomial fitting function

$$z_{\text{fit}} = \sum_{m+n \leq o} \alpha_{m,n} V_{\text{FD1}}^m V_{\text{FD2}}^n, \tag{6.29}$$

where o defines the order of the multivariate polynomial fitting function. The results obtained and presented in Figure 6.48 have been achieved with $o = 10^{10}$. The residuals are in the cm-domain. In the future optimization capabilities of the calibration procedure, characterization of the impact from ambient illumination and a larger reflectance range on the precision have to be investigated. Section 6.7 will present this on a theoretical level. Verification based on actual range measurements, however, is beyond the scope of this work.

Additionally to the actual range measurements that have been performed on camera level, PTM measurements have been carried out (c.f. Appendix C). The definition of the linearity measure differs from the definition from Section 6.3 (c.f. [Ass10]). Still, it is considered meaningful in order to allow for comparison and is thus given here. It has to be noted, that the capacitance actually does not result from the PTM method but can be calculated by combination of the PTM results with the measurement results that have been employed for the noise analysis. By turning the reset transistor ON the potential on the storage node can be set with the vddpix potential. This allows the characterization of the transfer characteristics which moreover allows to translate the observed output variation to a variation in the voltage domain at the storage node which can then be employed to estimate the sense node capacitance:

$$C_{\text{SN}} = \frac{\Delta N_{e^-} q \cdot A_{\text{read}}}{\Delta V_{\text{out}}}. \tag{6.30}$$

Here, ΔN_{e^-} is the difference of the photogenerated charge carrier count which can be determined from the PTM method, ΔV_{out} is the corresponding output voltage difference, q is the elementary unit and A_{read} is the small-signal amplification from the input of the source follower transistor input to the output. The read noise also differs from the definition from Section 6.3. There, read noise was defined as the temporal noise component between source follower input and analog output. The PTM method, however, determines the read noise within the actual application. Thus the floating diffusion is separated from the vddpix potential before readout by switching the reset transistor OFF. This introduces for instance a reset noise component to the read noise which was not compensated by CDS in this application (c.f. Section 5.2.3). Fortunately, addition of the reset noise to the read noise from Figure 6.47 yields a comparable result to the read noise observed from PTM measurements:

$$\sqrt{0.14 \times 10^{-6}\,\text{Vrms}^2 + A_{\text{read}}^2 \frac{k_B \theta}{C_{\text{SN}}}} \approx 0.55\,\text{mVrms} \approx 0.61\,\text{mVrms}. \tag{6.31}$$

---

[10]The choice of the order has to be justified by cross-validation in the future in order to exclude the possibility of overfitting.

Table 6.5 Extract of PTM measurements of the second generation LDPD based PM-ToF range camera.

extrinsic quantum efficiency	14% (@ 525 nm)
linearity error	1.6%
read noise	610 µVrms
sense node capacitance	12 fF
linear output voltage range	1.8 V
dark current of the storage nodes	15 000 e⁻/s

The parameters for dark current, sense node capacitance and read noise are in-spec.

The extrinsic quantum efficiency from the PTM is referred to the pixel area and thus differs from the definition used in Section 6.2 where it was referred to the photoactive area. Using the latter definition it would equal ≈37%. Unfortunately, no continuous light source with wavelength of 905 nm was available, so that the extrinsic quantum efficiency has to be estimated indirectly. This can be done by employing the characteristics from Figure 6.36 that resulted from stimulation with the laser pulse given in Figure 6.22. The photon count within one laser pulse can be estimated by

$$N_{\text{photon/puls}} = \frac{\lambda \int_0^\infty E_e \cdot f_{\text{laser}}(t) \cdot A_{\text{pix}} FF \, dt}{hc}, \tag{6.32}$$

with the irradiance level E_e, the normalized laser shape function f_{laser}, the pixel area A_{pix}, the fill-factor FF, Planck's constant h and the velocity of light c. The integral $\int_0^\infty f_{\text{laser}}(t) dt$ was solved numerically to 32 ns and is displayed in Figure 6.22. Using the PTM measurements in combination with the transfer function from the source follower input to the output, the resulting stored photogenerated charge carrier counts can be calculated by

$$N_{e^-/\text{puls}} = \frac{C_{\text{SN}} \cdot \Delta V_{\text{out}}}{q \cdot A_{\text{read}}}, \tag{6.33}$$

with the sense node capacitance C_{SN}, the output voltage ΔV_{out}, the elementary unit q and the amplification from the sense node towards to output A_{read}. Taking the ratio of the photogenerated charge carrier count and the impinging photon count the extrinsic quantum efficiency is yielded. In Figure 6.49 this is evaluated for all results from Figure 6.36. It can be observed, that the quantum efficiency for the transfer time 75 ns is below the others, what can be explained by not having collected all photogenerated charge carriers within the shutter width. Since the tail of the laser is approximately 110 ns long, this shutter length setting is not considered to be meaningful for estimates of the extrinsic quantum efficiency. The other values are fluctuating in a seemingly random manner.

Figure 6.50 thus depicts the histogram for shutter lengths larger than 75 ns. The estimated mean extrinsic quantum efficiency amounts to 35%. This is a surprisingly high value, considered that the extrinsic quantum efficiency at 525 nm is 37% which is just slightly higher. Actually, it is expected to yield a significantly lower extrinsic

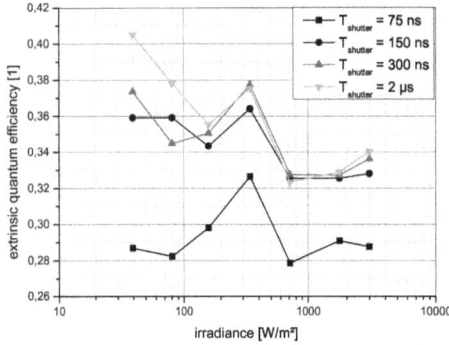

Figure 6.49 Estimation of the quantum efficiency at 905 nm – 1.

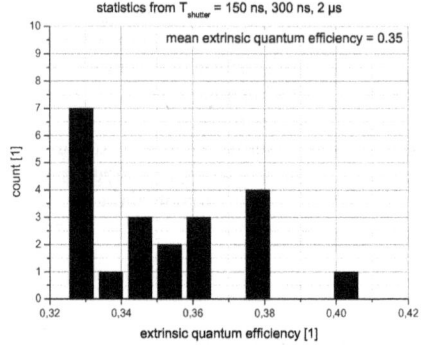

Figure 6.50 Estimation of the quantum efficiency at 905 nm – 2.

quantum efficiency at 905 nm because the effective absorption depth according to Beer-Lambert law is much larger at 905 nm what should affect the extrinsic quantum efficiency of the LDPD even more than a standard photodetector due to the doping gradient. Another effect that can be observed throughout the extrinsic quantum efficiency estimates, is a strong correlation at for instance the irradiance level 340 W/m^2, that is substantially shifting the average to higher values. Since this correlation is not expected from purely random measurement errors, it is assumed that a systematic error is present. This can likely be introduced by the neutral density filters which can be affected by e.g. aging effects. To yield proper estimates a calibrated light source becomes necessary.

Another interesting phenomenon that affects the range measurement accuracy, is that the floating diffusions are still sensitive to illumination when the shutter gates are turned OFF. This can be observed by comparison of the output characteristics for e.g. continuous illumination as depicted in Figure 6.51. The presented data was calculated by subtracting the output signals from their reset values. The reset value of FD2 was slightly deteriorated so that the response had to be offset corrected.

The parasitic sensitivity is actually a slightly non-linear function of the irradiance that also differs depending on which node is currently selected as can be seen in Figure 6.52, where the quotient of the intended signal in the activated shutter to the parasitic signal in the deactivated shutter is given. This probably originates in the complex geometry of the region in which the storage nodes are connected to the collection gate (c.f. Figures 6.11 and 6.26). For the sake of simplicity, however, it may be assumed that the parasitic sensitivity relative to the sensitivity of the activated shutter amounts to approximately 1%. This is a proper overestimation until the selected shutter goes into saturation.

6.6 MATCHING CONSIDERATIONS

The previous design was based on the fast transfer of photogenerated charges from one single photoactive region into three floating diffusions (FD) or one draining diffusion.

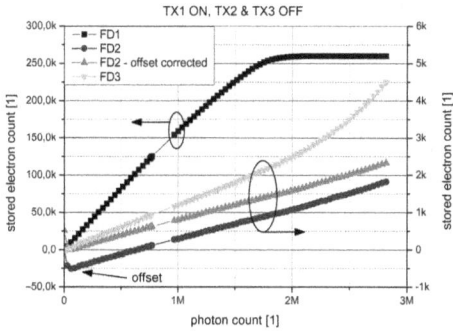

Figure 6.51 Parasitic sensitivity at 525 nm – 1.

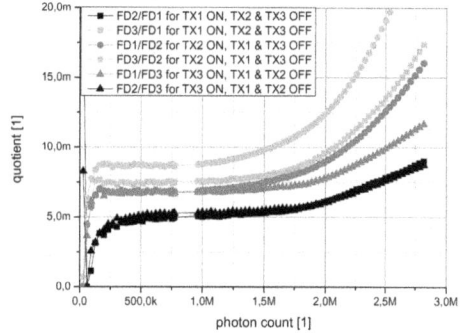

Figure 6.52 Parasitic sensitivity at 525 nm – 2.

To sufficiently separate those nodes a rather large collection gate (CX) is necessary. For a fast transfer this might result in a bottle-neck, because the potential gradient under the CX might become insufficiently steep. Furthermore, the connection of the four transfer gates and respective storage or draining nodes at one single CX results in a mismatch due to differing orientation and parasitic capacitances e.g. to routed metal interconnects or diffusions of neighboring devices. These mismatches have to be compensated by calibration, what complicates on-chip calculation of the range. Unfortunately, the mismatch can also affect the precision, if parameters such as sense node capacitance or charge transfer capability are severely distorted as it was observed in the previous designs (c.f. Figure 6.35). Especially the performance of the shutter switch TX1 was heavily differing from the remaining shutters. This can, of course, be circumvented if the accumulation of the ambient light level is not undertaken with use of TX1 but only TX2/TX3. This can be employed by rearrangement of the shutter sequence and by making two subsequent images. Compared to the actual concept, the resulting frame rate is lower by a factor of two. Worse than the reduced frame rate, however, is that the accumulation of the ambient light within an additional frame is prone to aliasing. The specified readout time was 10 ms, what resulted in a frame rate of 20 fps since a combination of the accumulation counts 1 and 1000 is employed. According to the Nyquist-Shannon sampling theorem, aliasing will occur at ambient illumination frequencies higher than 5 fps due to the reduced frame rate of 10 fps. Compensation of ambient light from fluorescent lamps which typically flicker at 100 to 120 Hz would thus not be possible. Advanced fluorescent lamps with high-frequency ballasts can shift the flicker frequencies in the kHz-domain making it even more difficult to compensate the ambient light. In the original concept the resulting Nyquist frequency amounts to approximately 16.7 kHz because the sampling rate is defined by the laser repetition rate, which is much higher than the actual frame rate.

In this section an alternative approach for a ToF-LDPD is presented, that aims at a better matching performance and a higher charge transfer speed. With this concept it is possible to properly carry out the MSI PM ToF principle with simultaneous accumulation of the ambient light level.

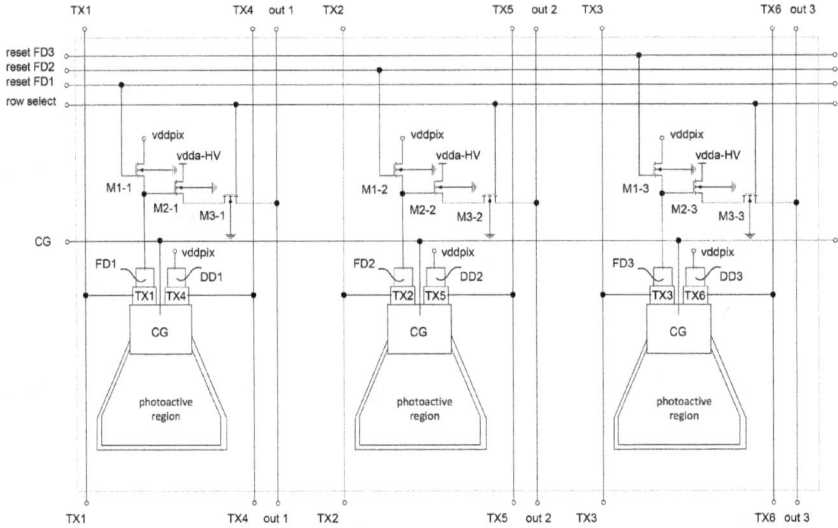

Figure 6.53 Schematic of the proposed ToF-LDPD.

6.6.1 Alternative ToF-LDPD concept

The fixed pattern noise (FPN) of pixel matrices in the mm^2-domain is ultimately limited by doping concentration gradients or mismatches in the geometry of e.g. diffusions, metal interconnects or polysilicon structures as they can originate for instance from photo-lithography, annealing or etching. This can result in mismatches in e.g. quantum efficiency, sense node capacitance or threshold voltage of the source follower transistors or more severely the charge transfer process as described in the former text. As can be seen in Figure 6.53, contrary to the previous designs, a pixel architecture for PM ToF application is proposed here, that implements one photoactive area for each storage node. This enables a better matching, as will be presented later, since orientation, routed metal interconnects and separation to different structures like e.g. transistors are identical for each node. Additionally to the improved matching performance, the area of the collection gate can be minimized, what results in an improved charge transfer. The use of subpixels for matching and speed enhancement is traded against photoactive area, that is approximately 3 times smaller than in the previous device. Since the amount of photogenerated charges is in first order proportional to the photoactive area, the separation results in $\sqrt{3} \approx 4.8$ dB less SNR due to shot noise. Another disadvantage is the degraded *modulation transfer function* (MTF) which might, however, be resolved with complex optics at the cost of additional 4.8 dB loss in SNR. The basic operation differs only slightly from the previous design, as is visualized in Figure 6.54. Here, the difference is that six shutter signals have to be implemented. The charge transfer is carried out either towards one storage node (t_3–t_5 for FD1, t_5–t_7 for FD2 or t_1–t_2 for FD3) or to its respective draining node.

In the following the important details of the pixel layout, that is depicted in part b of Figure 6.56 will be stressed out. The proposed pixel has an area of $13.34 \times 40\,\mu m^2$

timing diagram

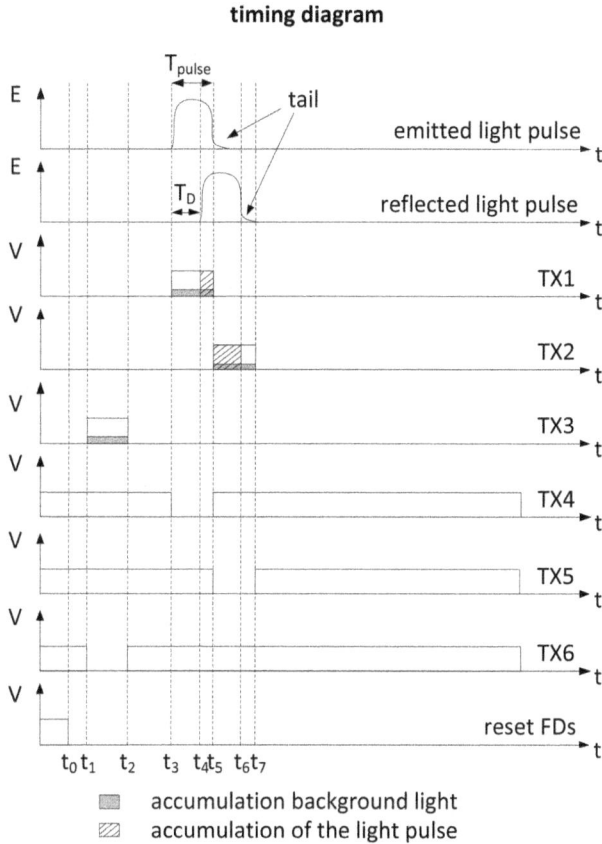

Figure 6.54 Basic ToF timing diagram for the matched architecture.

per sub-pixel, with a fill factor of 38% each. The collection gate area was minimized to yield a maximum transfer speed. This was done by connecting the transfer gates and their respective storage or draining nodes diagonally to the collection gate. To achieve a homogeneous doping concentration in the channel, the width of the n-well has to be sufficiently large. In general, the n-well should avoid the contact to Si-SiO$_2$ interfaces to prevent RTS and flicker noise. The geometry of the transfer gates should be designed in a way to achieve the full depletion of the well during reset or draining operation and to properly separate the storage nodes by potential barriers introduced by the depleted well (c.f. Figure 6.27). The diagonal connection of the transfer gates to the CX have the advantage of an inherently long distance of the floating diffusions and their respective draining nodes. To improve this further, p-type implantations can be used for separation. To avoid parasitic crosstalk between the subpixel and the draining node of the neighboring pixel, a small region with gate-oxide is used to allow for a deeper p-type implantation, that separates the respective n-wells. Since a contact through these thin oxides is not necessary it can be designed smaller than design rules actually allow. The polysilicon transfer gates are designed to

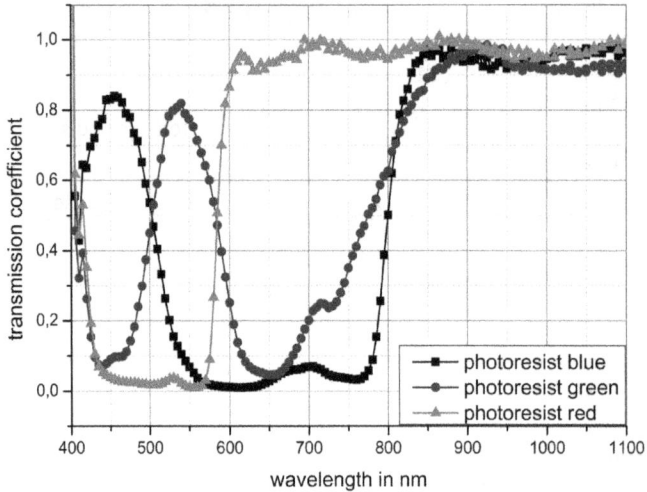

Figure 6.55 Exemplary transmittance characteristics of colour filters.

sufficiently overlap their respective channels. This avoids parasitic leakage currents. The explained details cannot be verified by standard parasitics extraction tools and should thus be verified by technology CAD tools. Since the draining node is almost permanently conducting the background illumination towards the respective positive power supply rail, a wide power supply interconnect is necessary. To avoid peaks that may occur due to large dynamic variations in the illumination, the remaining space in the pixels is used for poly1-poly2-metal1-capacitors. Since the positive power supply rail for the source follower conducts a DC current, it might also be designed wider, so the source follower remains in saturation in the worst case and electromigration can be neglected. The restrictions for the other interconnects are rather loose, since they either lead only low currents or may even conduct no DC current at all. However, for large-scale imagers capacitive crosstalk may influence the performance significantly, what is though neglected here.

In addition to the details, that concern the design for ToF applications, it is interesting to note that the proposed architecture is predestined for the integration of color and range detection in one single pixel. This is known as *red-green-blue-range imaging* (RGBZ). Color filters are available, that have a high transmittance for the infrared range as can be seen in Figure 6.55[11]. By structuring the color filters in a barrel manner, a RGBZ imager is obtained. For its operation it is only necessary to modify the timing, since for color imaging other integration times may be necessary. However, the three colors can be accumulated in parallel. For range detection the timing presented in Figure 6.54 can be applied. The proposed approach has the advantage of a better MTF compared to [KYO12], where RGB pixels are separated from an additional range

[11]The author wants to acknowledge Melanie Jung for the data of the transmittance from the color filters.

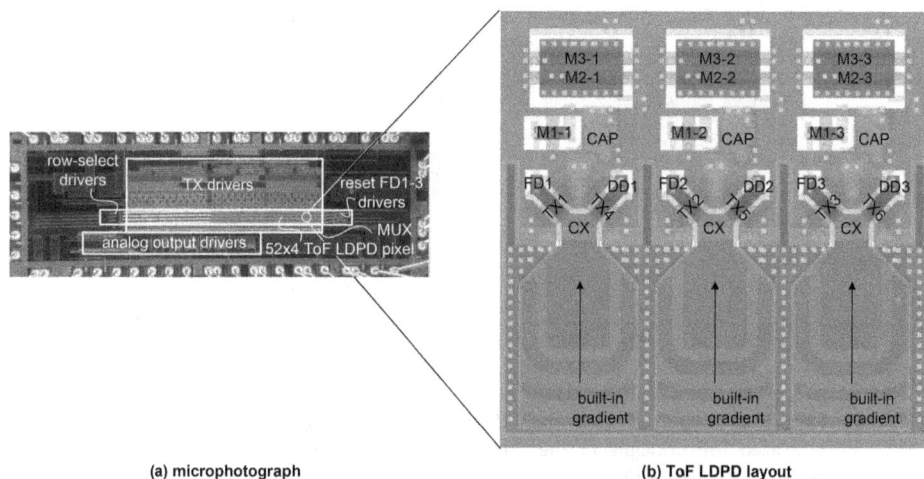

(a) microphotograph   (b) ToF LDPD layout

Figure 6.56 Microphotograph and pixel layout of the third generation LDPD based PM-ToF imager.

pixel. The approach presented in [KKK12], however, is quite similar to this one. Here 8 pixels, that may work in color mode, can be binned to yield range measurements by the continuously modulated principle (CW).

6.6.2 Evaluation of the third generation LDPD based PM-ToF imager[12]

Figure 6.56 presents the microphotograph of the fabricated test imager and the layout of the proposed ToF-LDPD. The test imager comprises 13 mini test matrices each consisting of 4×4 pixels. In this section the best ToF-LDPD variant is described and evaluated by measurements.

6.6.2.1 Shutter performance

In Figure 6.57 the good matching of the three subpixels can be observed. Here, the response to a laser pulse with a peak irradiance of $3\,W/m^2$ was characterized by application of the standard timing, that is presented in Figure 6.54 while the laser pulse was shifted in time by the delay line. To amplify the signal sufficiently, 1024 accumulations have been carried out. Later on, the data was shifted in the time domain to allow for a better evaluation of the performance. In Figure 6.58 the output voltage as a function of the irradiance level is depicted. As can be observed, the pixel has a high linearity over several orders of magnitude. For a good resolution of this graph, however, depending on the actual irradiance varying accumulation count settings were employed.

---

[12]The author wants to acknowledge Stefan Bröcker for the design of the PCB that was used throughout the evaluation.

@ E_e = 3 W/m², N_{accu} = 1024, $T_{laser pulse}$ = 30 ns, $T_{shutter}$ = 30 ns

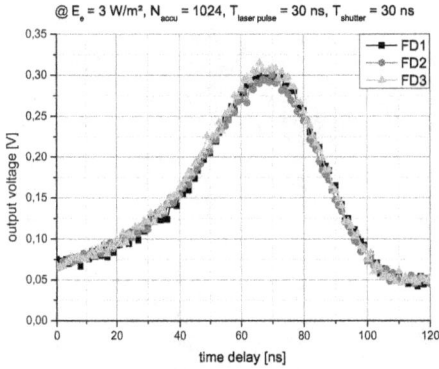

Figure 6.57 Response characteristics to the emulated ToF principle of the three different storage nodes FD1-3.

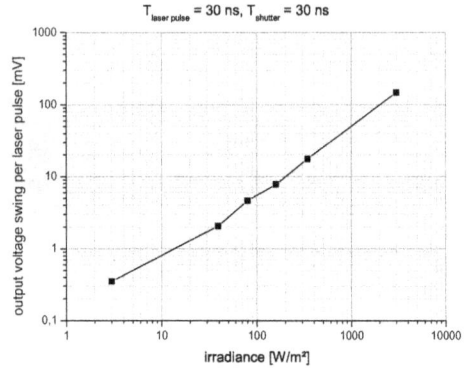

$T_{laser pulse}$ = 30 ns, $T_{shutter}$ = 30 ns

Figure 6.58 Output voltage characteristics as a function of the irradiance.

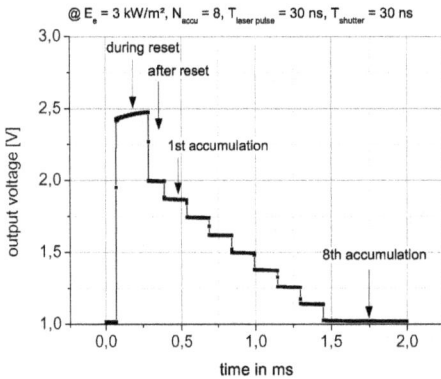

@ E_e = 3 kW/m², N_{accu} = 8, $T_{laser pulse}$ = 30 ns, $T_{shutter}$ = 30 ns

Figure 6.59 Detector response with 8 accumulations in time domain.

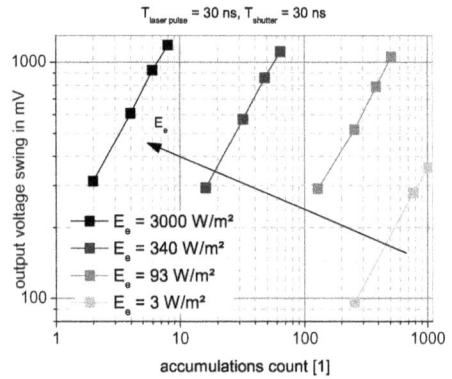

$T_{laser pulse}$ = 30 ns, $T_{shutter}$ = 30 ns

Figure 6.60 Output voltage characteristics for various accumulation counts and irradiance levels.

Figure 6.59 depicts the response of the photodetector to the laser pulse in the time domain for 8 accumulations. This graph has been recorded with an oscilloscope. The linearity of the output voltage swing with respect to the accumulation count can be seen, but it is better illustrated in Figure 6.60. Here, the linearity with respect to the accumulation count over several orders of magnitude can be observed for 4 different irradiance levels.

The differing shapes of the response to the emulated ToF principle for varying accumulation count and irradiance levels can be found in Figures 6.61–6.64. The response of a perfect short time integrator to the non-ideal laser pulse, that is depicted in Figure 6.22 is also displayed. The measured responses are similar to the ideal response. The remaining mismatch may be explained by the non-ideal stimulation of the transfer

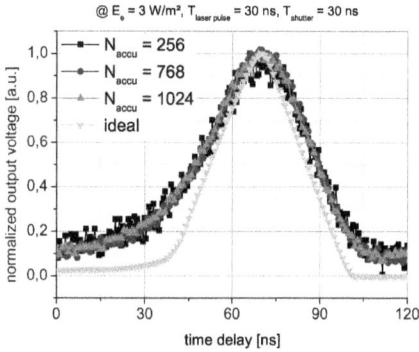

Figure 6.61 Response to the emulated ToF principle at an irradiance level of 3 Wm^{-2}.

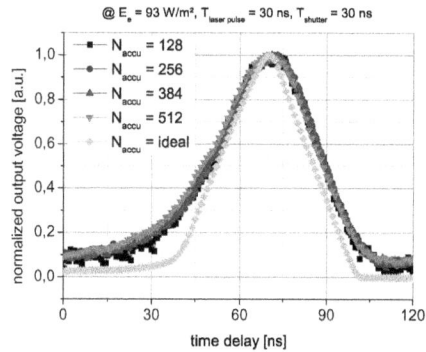

Figure 6.62 Response to the emulated ToF principle at an irradiance level of 93 Wm^{-2}.

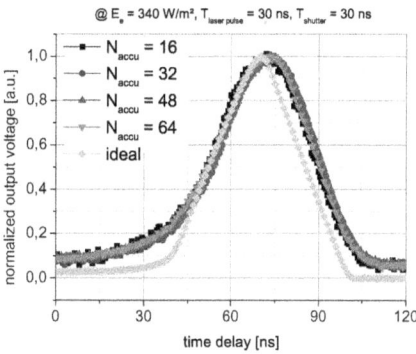

Figure 6.63 Response to the emulated ToF principle at an irradiance level of 340 Wm^{-2}.

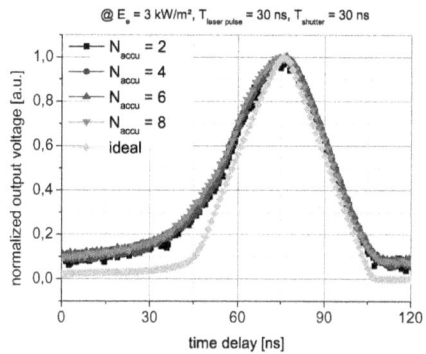

Figure 6.64 Response to the emulated ToF principle at an irradiance level of 3 kWm^{-2}.

gates, the finite transfer speed of the photodetector and a remaining parasitic sensitivity to illumination when the shutter switch is actually OFF.

The proposed ToF-LDPD demonstrates a significantly improved matching performance when compared to the previous approaches. Furthermore, due to the relatively small collection gate a substantial improvement in charge transfer speed was observed allowing for a nearly constant performance over the irradiance range of 3 to 3000 W/m^2. Last but not least, it has to be pointed out that for the first time proper charge transfer was yielded for a shutter width of 30 ns.

6.6.2.2   Camera level verification[13]

The proposed ToF-LDPD was integrated within a 128×96 image sensor which allowed for characterization with the PTM method. The results are given in Table 6.6

---

[13]The author wants to acknowledge Adrian Driewer for the execution of the measurements.

Table 6.6 Extract of PTM measurements of the third generation LDPD based PM-ToF range camera.

extrinsic quantum efficiency	4.3% (@ 525 nm)
linearity error	1.32%
read noise	680 μVrms
sense node capacitance	12.5 fF
linear output voltage range	2.02 V
dark current of the storage nodes	6700 e⁻/s

and are investigated in Section 6.7. The reduced quantum efficiency results from the photoactive region which is approximately three times smaller than the former variants. The linearity error is slightly improved, which probably originates from the way the sense node capacitance is realized. Compared to the previous designs the actual floating diffusion is much smaller. The resulting sense node capacitance, which is slightly higher compared to the previous designs, thus consists of more metal-to-metal capacitances, which are apparently more linear than the reversed-biased junction of the floating diffusion/p-well-diode. This also becomes clear in the improved linear output voltage range. Fortunately, another advantageous effect is that the smaller diffusion results in a reduced dark current.

Employing the approach from Section 6.5.2.3, the extrinsic quantum efficiency is evaluated and displayed in Figure 6.65. Here, the value for an irradiance level of 3 W/m² seems to be too deteriorated, so that the remaining values were taken to calculate the mean extrinsic quantum efficiency, which amounts to 20%. However, as explained in Section 6.5.2.3 the employed shutter length is significantly lower than the laser signal, so that the resulting extrinsic quantum efficiency is underestimated. Thus, the quantum efficiency of the former approach may be employed for further calculations. The parasitic sensitivity to illumination when the shutter switch is actually turned OFF was also observed for this design. Since the separation of the transfer and collection gate sections should actually result in a reduced parasitic sensitivity which, however, was not observed, it can be assumed that it is originated in charge carriers generated deep in the silicon which then propagate to the storage nodes. These can apparently not be controlled by the potential profiles defined by the applied potentials or the built-in gradient in the electrostatic potential profile of the photoactive region. More interesting, is that the parasitic sensitivity is significantly increasing when the left neighboring subpixel is activated. This may be originated in a lateral crosstalk and might be circumvented by improved separation as it can be achieved for instance by slightly increasing the gaps between the subpixels what would, however, come at the expense of increasing pixel pitch. Alternatively p-type implants or trench isolation may be used to introduce a barrier between the subpixels.

6.7   IMPACT OF FINITE CHARGE TRANSFER SPEED AND PARASITIC LIGHT SENSITIVITY ON PM-TOF

The time needed to transfer photogenerated charge carriers into the respective storage nodes is dependent on the irradiance level as it was observed from the evaluation of

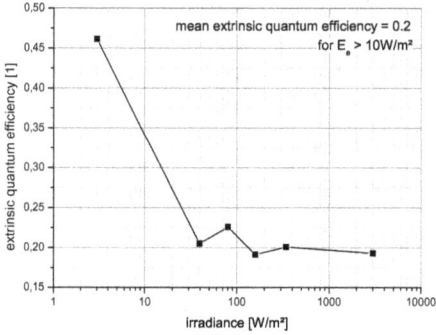

Figure 6.65 Estimation of the extrinsic quantum efficiency.

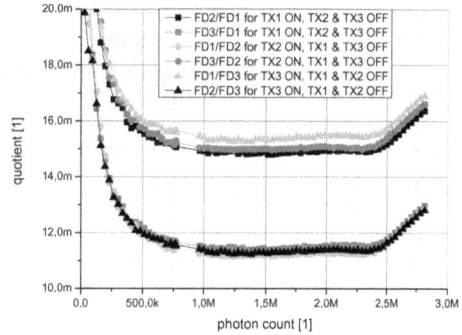

Figure 6.66 Parasitic sensitivity at 525 nm.

the presented devices (c.f. Sections 6.4–6.6). This can be explained by the impact the photogenerated charge carriers have on the electrostatic potential profiles. An increasing amount of photogenerated charge carriers results in a more negative potential in the photoactive region which moreover results in an increased gradient. To properly describe this phenomenon, the semiconductor equations have to be employed (c.f. Equation 3.162). This can be done by the means of finite element simulations. Unfortunately, this results in an extensive calculation effort, since many different scenarios have to be investigated to ensure proper performance of the range camera under all conditions. Alternatively, compact models can be derived which result in a decreased calculation effort. Deriving such compact models, however, is beyond the scope of this work. Here, a more abstract approach is used that is much simpler than the before mentioned variants and thus allows for fast estimation of the impact of finite charge transfer speed on the actual time-of-flight measurement performance.

6.7.1 Concept of the generalized MSI ToF model

The modeling can be subdivided into three tasks. The first one is to estimate the time-dependent irradiance impinging on the photodetector. For the sake of simplicity, the model from Section 6.2 is employed throughout this work. In future, this model may be extended to model e.g. the irradiance at shorter distances more properly. Here, the design of the lenses might also be taken into account. This could be achieved by e.g. use of the Maxwell's equations. The second task is to model the photogenerated charge carrier counts while taking into account effects like finite transfer time and parasitic sensitivity when the shutter should actually be OFF. The last task is to estimate the precision of the range measurement system. In this section the two latter tasks are investigated.

As described throughout this chapter, photodetectors offer only finite charge transfer speed. This phenomenon is modeled, here, by introducing a low-pass filter characteristic through convolution with the photo stream. The resulting affected photo current would then be integrated by the shutter functionality. As the shutter is never entirely OFF instead of a short-time integration an improper integral is used to model the shutter. The varying sensitivities of the shutter are taken into account then by a time-varying function that is multiplied with the photo current within the integrand of the improper integral.

The proposed model for the photogenerated and stored electron count in a storage node for a given irradiance level is

$$N_e = N_{accu} \eta_{ext} \cdot A_{pix} FF \frac{\lambda}{hc} \int_0^{\frac{1}{f_{rep}}} [E_e(t, \tau_{ToF}, \tau_E) * h_{\tau_{LDPD}}(E_e, t)] \cdot S_{LDPD}(E_e, t) dt + \frac{N_{accu}}{f_{rep}} J_{dark}$$

(6.34)

where the symbol $*$ stands for the convolution and $h_{\tau_{LDPD}}$ and S_{LDPD} are functions which are used to model the photodetector. Here, the time of flight $\tau_{ToF} = 2z/c$ and the time instance τ_E at which the laser pulse is emitted are used. The dark current J_{dark} is given in e^-/s. These parameters are discussed in Section 6.7.4. Assuming $h_{\tau_{LDPD}}(E_e, t) = \delta(t)$, $S_{LDPD}(E_e, t) = \text{rect}\left(\frac{t - \tau - T_{SW}/2}{T_{SW}}\right)$ and $J_{dark} = 0$, this model reduces to the plain short time integrator model described in Section 6.2 (cf. Equation 6.10). The proposed model assumes that the irradiance can be separated according to

$$E_e(t) = E_{e-ambient} + \hat{E}_e f_{laser}(t - \tau_{ToF} - \tau_E)$$

(6.35)

where $E_{e-ambient}$ is the ambient level of the irradiance, $\hat{E}_e$ is the peak magnitude of the impinging laser pulse and f_{laser} is the normalized laser function as given in Figure 6.22. Furthermore

$$\int_{-\infty}^{\infty} h_{\tau_{LDPD}}(E_e, t) dt = 1,$$

(6.36)

which results from charge conservation. A very simple function that obeys the above relation is

$$h_{\tau_{LDPD}}(E_e, t) := \begin{cases} \frac{1}{\tau_{LDPD}(E_e)} \exp\left(\frac{-t}{\tau_{LDPD}(E_e)}\right) & \text{for: } t \geq 0 \\ 0 & \text{for: } t < 0, \end{cases}$$

(6.37)

which is defined in analogy to the impulse response function of a single pole low-pass filter. It is assumed, that the dependence on the irradiance level can be modeled by use of a time-independent $\tau_{LDPD}(E_e) = f(\hat{E}_e, E_{e-ambient})$. Lastly

$$S_{LDPD}(E_e, t) = \begin{cases} 1 & \text{for: } \tau \leq t \leq \tau + T_{SW} \\ S_{LDPD-TX-OFF} & \text{elsewhere}, \end{cases}$$

(6.38)

is defined with τ being the time instance at which the respective shutter is activated for the duration T_{SW}. Here, $S_{LDPD-TX-OFF}$ describes a sensitivity to light when the shutter is OFF, because the integrand in Equation 6.34 does not vanish for $t \notin [\tau, \tau + T_{SW}]$. The presented equations were not conducted from an analysis based on the semiconductor equations. They are stated in a heuristic manner and are founded on the observations presented throughout this chapter. As will be presented in this section, the stated functions already allow for a good understanding of fundamental limitations, they are flexible enough to be fitted to the measurement results but of course may fail accurate predictions when the verified parameter range is left.

The above definitions for irradiance, $h_{\tau_{\text{LDPD}}}(t, E_e)$ and S_{LDPD} allow for a simplification of Equation 6.34:

$$N_e = N_{\text{accu}}\eta_{\text{ext}} \cdot A_{\text{pix}}FF\frac{\lambda}{hc}\left[E_{e-\text{ambient}}\left(T_{\text{SW}} + \left(\frac{1}{f_{\text{rep}}} - T_{\text{SW}}\right)\right)\right.$$
$$\left. + \hat{E}_e \int_{-\infty}^{\infty} \left(f_{\text{laser}}\left(t - \tau_{\text{ToF}} - \tau_E\right) * h_{\tau_{\text{LDPD}}}(E_e, t)\right) \cdot S_{\text{LDPD}}(t, E_e)dt\right] + \frac{N_{\text{accu}}}{f_{\text{rep}}}J_{\text{dark}}$$

$$(6.39)$$

which can be solved numerically. Unfortunately, the deviation from the model of the ideal photodetector in Section 6.2 also results in a different mapping between the stored electron count in FD1-3 and the object distance. Assuming an irradiance independent τ_{LDPD}, an ideal rectangularly shaped laser pulse $\hat{E}_e f_{\text{laser}}(t) = \hat{E}_e \cdot \text{rect}\left(\frac{t - \tau_{\text{ToF}} - T_p/2}{T_p}\right)$ with pulse width T_p, that the photodetector is not sensitive to light when the shutter is OFF and that dark current is negligible, yields

$$f_{\text{laser}}(t) * h_{\tau_{\text{LDPD}}}(E_e, t) = \begin{cases} 0 & \text{for: } t \leq \tau_{\text{ToF}} \\ 1 - \exp\left(\frac{-(t - \tau_{\text{ToF}})}{\tau_{\text{LDPD}}}\right) & \text{for: } \tau_{\text{ToF}} \leq t \leq \tau_{\text{ToF}} + T_p \\ \exp\left(\frac{-(t - \tau_{\text{ToF}} - T_p)}{\tau_{\text{LDPD}}}\right) - \exp\left(\frac{-(t - \tau_{\text{ToF}})}{\tau_{\text{LDPD}}}\right) & \text{for: } \tau_{\text{ToF}} + T_p \leq t \end{cases}$$

$$(6.40)$$

for the convolution integral. Assuming that no ambient illumination is present, only the electron count in FD1 and FD2 will differ from zero. The electron count in those storage nodes can be calculated by

$$N_{\text{FD1}} \propto \begin{cases} \tau_{\text{LDPD}} \cdot \exp\left(\frac{\tau_{\text{ToF}}}{\tau_{\text{LDPD}}}\right)\left[\exp\left(\frac{-T_{\text{SW}}}{\tau_{\text{LDPD}}}\right) + \exp\left(\frac{T_p}{\tau_{\text{LDPD}}}\right) - 1 - \exp\left(\frac{T_p - T_{\text{SW}}}{\tau_{\text{LDPD}}}\right)\right] & \text{for } \tau_{\text{ToF}} \leq -T_p \\ T_p + \tau_{\text{LDPD}} + \tau_{\text{ToF}} + \tau_{\text{LDPD}} \cdot \exp\left(\frac{\tau_{\text{ToF}}}{\tau_{\text{LDPD}}}\right)\left[\exp\left(\frac{-T_{\text{SW}}}{\tau_{\text{LDPD}}}\right) - \exp\left(\frac{T_p - T_{\text{SW}}}{\tau_{\text{LDPD}}}\right) - 1\right] & \text{for } -T_p \leq \tau_{\text{ToF}} \leq 0 \\ T_p + \tau_{\text{LDPD}} \cdot \exp\left(\frac{\tau_{\text{ToF}} - T_{\text{SW}}}{\tau_{\text{LDPD}}}\right)\left[1 - \exp\left(\frac{T_p}{\tau_{\text{LDPD}}}\right)\right] & \text{for } 0 \leq \tau_{\text{ToF}} \leq T_{\text{SW}} - T_p \\ T_{\text{SW}} - \tau_{\text{LDPD}} - \tau_{\text{ToF}} + \tau_{\text{LDPD}} \cdot \exp\left(\frac{\tau_{\text{ToF}} - T_{\text{SW}}}{\tau_{\text{LDPD}}}\right) & \text{for } T_{\text{SW}} - T_p \leq \tau_{\text{ToF}} \leq T_{\text{SW}} \\ 0 & \text{for } T_{\text{SW}} \leq \tau_{\text{ToF}} \end{cases}$$

$$(6.41)$$

and

$$
N_{FD2} \propto
\begin{cases}
\tau_{LDPD} \cdot \exp(\tau_{ToF}/\tau_{LDPD}) \left[\exp\left(\frac{-2T_{SW}}{\tau_{LDPD}}\right) + \exp\left(\frac{T_p - T_{SW}}{\tau_{LDPD}}\right) - \exp\left(\frac{T_p - 2T_{SW}}{\tau_{LDPD}}\right) - \exp\left(\frac{-T_{SW}}{\tau_{LDPD}}\right) \right] \\
\hfill \text{for } \tau_{ToF} \leq T_{SW} - T_p \\[2mm]
T_p - T_{SW} + \tau_{LDPD} + \tau_{ToF} + \tau_{LDPD} \cdot \exp\left(\frac{\tau_{ToF}}{\tau_{LDPD}}\right) \left[\exp\left(\frac{-2T_{SW}}{\tau_{LDPD}}\right) - \exp\left(\frac{-T_{SW}}{\tau_{LDPD}}\right) - \exp\left(\frac{T_p - 2T_{SW}}{\tau_{LDPD}}\right) \right] \\
\hfill \text{for } T_{SW} - T_p \leq \tau_{ToF} \leq T_{SW} \\[2mm]
T_p + \tau_{LDPD} \cdot \exp(\tau_{ToF}/\tau_{LDPD}) \left[\exp\left(\frac{-2T_{SW}}{\tau_{LDPD}}\right) - \exp\left(\frac{T_p - 2T_{SW}}{\tau_{LDPD}}\right) \right] \\
\hfill \text{for } T_{SW} \leq \tau_{ToF} \leq 2T_{SW} - T_p \\[2mm]
2T_{SW} - \tau_{LDPD} - \tau_{ToF} + \tau_{LDPD} \cdot \exp\left(\frac{\tau_{ToF} - 2T_{SW}}{\tau_{LDPD}}\right) \\
\hfill \text{for } 2T_{SW} - T_p \leq \tau_{ToF} \leq 2T_{SW} \\[2mm]
0 \hfill \text{for } 2T_{SW} \leq \tau_{ToF}
\end{cases}
$$

$$(6.42)$$

These results demonstrate two important points. Firstly, distinctions of cases have to be made to model the relation between the time-of-flight or distance and the photo-generated electron count. In the above example, however, the number of cases can be reduced by choosing T_p and T_{SW} to be equal and by limiting the used τ_{ToF}-range for evaluation as for instance $0 \leq \tau_{ToF} \leq T_{SW}$. Unfortunately, though the second observation cannot be circumvented that easily. The presented model results in transcendental relations and cannot be rearranged to yield τ_{ToF} in an analytical manner. This has an impact on calibration as well as on the estimation of the precision. It may be circumvented by improvement of the charge transfer speed. The derived equations converge to $z \propto N_{FD2}/(N_{FD1} + N_{FD2})$ for $\tau_{LDPD} \ll \tau_{ToF}$ and $\tau_{LDPD} \ll T_{SW} - \tau_{ToF}$. If the measurement range is chosen small compared to T_{SW} and placed appropriately within $(0, T_{SW})$, the simplifications may be justified resulting in simple analytical equations. However, reducing the measurement range while keeping $T_{SW} = T_p$ constant immediately results in a reduced robustness against ambient light (c.f. discussions in Chapter 2 and 6).

To allow for an estimation of the precision, a unique mapping between the object distance and the stored electron counts has to be expressed to allow for error calculus according to Section 3.2.1. Since that is not in general possible in an analytical manner as described above, a numerical approach is employed in this work. Therefore, supporting data points are calculated with the above model and continuously interpolated a multivariate power series model that is fitted to the supporting dataset. To improve the numerical stability of the fitting procedure, a coordinate transform is firstly carried out. Here, logarithmic functions are employed, since the electron count can vary over several orders of magnitudes. The employed model is

$$
z_{fit} = \sum_{m+n \leq o} \alpha_{m,n} \log^m (N_{FD1} - N_{FD3}) \log^n (N_{FD2} - N_{FD3}), \qquad (6.43)
$$

where $N_{FD1} - N_{FD3} \geq 0$ and $N_{FD2} - N_{FD3} \geq 0$ have to be guaranteed. The above model is fitted in a least-squares manner:

$$
\alpha_{m,n}^* = \operatorname*{argmin}_{\alpha_{m,n}} \left[\sum_{p=1}^{P} (z - z_{fit})^2 \right], \qquad (6.44)
$$

as thus a linear optimization problem with a unique global optimum $\alpha_{m,n}^*$ is yielded. Here, P is the number of calculated supporting data points for the regression and o is the order of the multivariate model. Since the simplified model does yield fast estimates for the electron count, in principle many simulations can be done before the fitting procedure is started. This also becomes necessary, since proper estimates for $\alpha_{m,n}^*$ are fundamental necessities for the prediction of the precision as well. Therefore, the residuals $\sum_{p=1}^{P}(z - z_{\text{fit}})^2$ after the fitting procedures should always be observed. In principle, different fitting approaches may be superior in terms of complexity or residual error, but since complexity was not in focus of this work, residuals can be avoided by increasing o. Throughout this work, a high value of $o = 15$ resulting in 136 coefficients was used[14].

For the sake of simplicity, Gaussian error calculus was employed for this approach. The random error is thus stated as

$$\sigma_z^2 \approx \sum_{k=1}^{3} \left| \frac{\partial z_{\text{fit}}}{\partial N_{\text{FDx}}} \right|^2 \sigma_{N_{\text{FDx}}}^2. \tag{6.45}$$

The partial derivatives can be calculated to

$$\frac{\partial z_{\text{fit}}}{\partial N_{\text{FD1}}} = \sum_{m,n \le o \wedge m \ge 1} \alpha_{m,n}^* \frac{m}{N_{\text{FD1}} - N_{\text{FD3}}} \log^{m-1}(N_{\text{FD1}} - N_{\text{FD3}}) \log^{n}(N_{\text{FD2}} - N_{\text{FD3}}) \tag{6.46}$$

$$\frac{\partial z_{\text{fit}}}{\partial N_{\text{FD2}}} = \sum_{m,n \le o \wedge n \ge 1} \alpha_{m,n}^* \frac{n}{N_{\text{FD2}} - N_{\text{FD3}}} \log^{m}(N_{\text{FD1}} - N_{\text{FD3}}) \log^{n-1}(N_{\text{FD2}} - N_{\text{FD3}}) \tag{6.47}$$

$$\frac{\partial z_{\text{fit}}}{\partial N_{\text{FD3}}} = -\frac{\partial z_{\text{fit}}}{\partial N_{\text{FD1}}} - \frac{\partial z_{\text{fit}}}{\partial N_{\text{FD2}}}. \tag{6.48}$$

The variances at the different storage nodes expressed in electron count can be modeled as

$$\sigma_{N_{\text{FDx}}}^2 = N_{\text{FDx}} + \frac{k_B \theta C_{\text{SN}}}{q^2} + \sigma_{\text{read}}^2 \left| \frac{C_{\text{SN}}}{q \cdot A_{\text{read}}} \right|^2, \tag{6.49}$$

which differs from Equation 6.23, since the dark current is already taken into account during the calculation of N_{FDx} (c.f. Equation 6.34). This is advantageous compared to the former approaches, because it also takes into account the impact of the dark current on the partial derivatives $\frac{\partial z_{\text{fit}}}{\partial N_{\text{FDx}}}$ and not only onto the variance $\sigma_{N_{\text{FDx}}}^2$. It is important to note that the assumption of Poisson distributed electron counts is a simplification. The proposed model shows memory (due to finite charge transfer) and time-varying behavior (due to the shutter function) of the system, which is not entirely captured by the choice of Poisson distributed carrier counts. A more general model could be

---

[14] Overfitting is not a problem here, as the generated data points that are used for the regression are noise free. Nevertheless, a cross-validation is still performed (c.f. Fig. 6.67) in order to ensure that the fitting yields negligible error.

Table 6.7   Parameter set for the verification of the proposed model.

distance range	0.015 to 2.25 m
reflectance range	0.05 to 1
accumulation count	1 to 10 000
ambient illuminance	0 to 150 klx
temperature of the sun	5777 K
optical bandpass filter	820 to 920 nm
lens transmittance	1
field of view	$15° \times 15°$
f-number	0.95
total radiant flux of the laser	75 W
wavelength of the laser	905 nm
laser pulse width	30 ns
shutter width	30 ns
pixel area	$40 \times 40\,\mu m^2$
fill factor	0.38
extrinsic quantum efficiency	4%
τ_{LDPD}	50 fs
$S_{LDPD-TX-OFF}$	0

derived based on linear periodical time-varying systems (c.f. 3.2.3) but is beyond the scope of this work.

6.7.2 Verification

The correctness of the approach presented in the former section can be verified by the statement that the model has to converge into the model from Section 6.2. Therefore, an infinitesimally low value for τ_{LDPD} of 50 fs is chosen and the dark current is set to zero as well as $S_{LDPD-TX-OFF}$. The remaining model parameters used for the verification are gathered in Table 6.7. Since the difference in the two models arises in the way electron counts are determined and how the relation between object distance and electron count is found, read noise and dark current are neglected during the verification. It is important to note that it is advantageous to use different datasets for the fitting process and its evaluation, since this is more likely to demonstrate the robustness of the fitting functions between the actual supporting data points. This is important, because the data points span a three dimensional set with variation of the parameters distance, reflectance and illuminance which may also vary over several orders of magnitude.

Figure 6.67 demonstrates the evaluated residuals after the fitting procedure. Here, the maximum residual is below 1 mm for the given measurement range of 2.25 m. This corresponds to a systematic error below 0.045%, what is likely to yield good estimates for the precision. In general it is important to note, that the residuals should be much smaller than the measurement range. In Figures 6.68–6.70 the comparison of the precision estimates between the two models is given. Except for very low reflectance levels and extremely short distances, the models yield very similar results what proves the correct implementation of the proposed model. It is likely that the remaining differences are caused by the choice of the data points used for the fitting procedure. The minimum

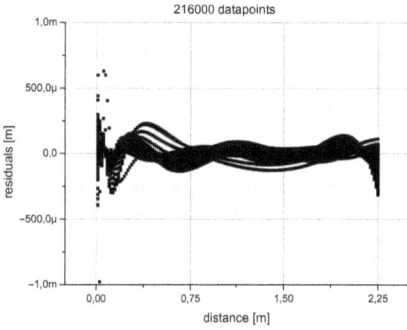

Figure 6.67 Residuals after the fitting procedure of the virtual experiments with the multivariate model from Equation 6.43.

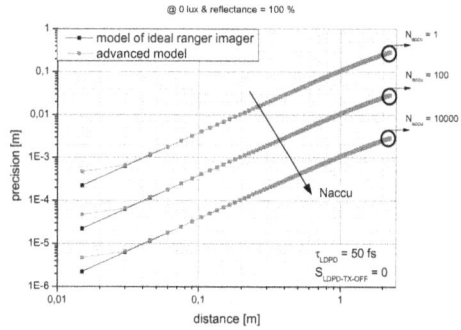

Figure 6.68 Comparison of the advanced model for random error prediction and the basic model from Section 6.2 – 1.

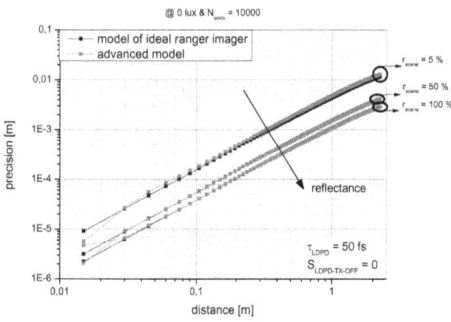

Figure 6.69 Comparison of the advanced model for random error prediction and the basic model from Section 6.2 – 2.

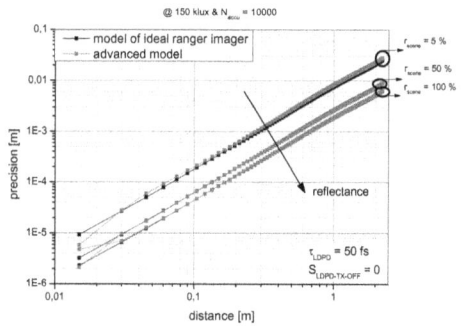

Figure 6.70 Comparison of the advanced model for random error prediction and the basic model from Section 6.2 – 3.

distance and reflectance were the boundaries of the spanned set. Increasing the amount of data points in the vicinity of the extrema and extending them beyond the measurement range can result in smaller residuals and better estimates of the random error.

6.7.3 Fitting and comparison of the ToF-LDPD designs

Figure 6.71 and 6.72 depict the flexibility of the proposed model by output voltage characteristics of the emulated ToF principle. This is different from the presented results of the former section in the way the irradiance is set. Here, the irradiance is not given as a function of reflectance, distance and ambient illuminance but is directly defined at the sensor.

Figures 6.73–6.74 demonstrate fits of the proposed model to measurements of the ToF-LDPD from Section 6.5 at $\hat{E}_e = 3\,\mathrm{W/m^2}$ and $\hat{E}_e = 3\,\mathrm{kW/m^2}$. It can be observed

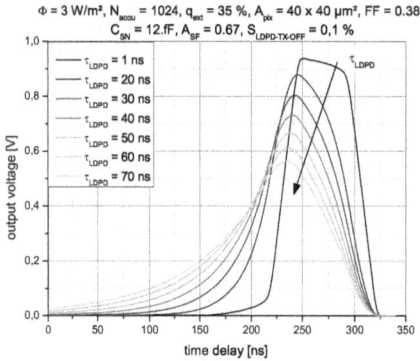

Figure 6.71 Simulated output characteristics of the emulated ToF principle for varying τ_{LDPD}.

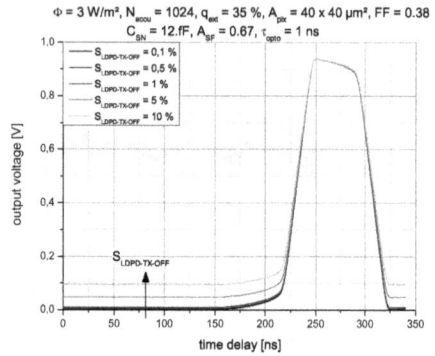

Figure 6.72 Simulated output characteristics of the emulated ToF principle for varying $S_{LDPD-TX-OFF}$.

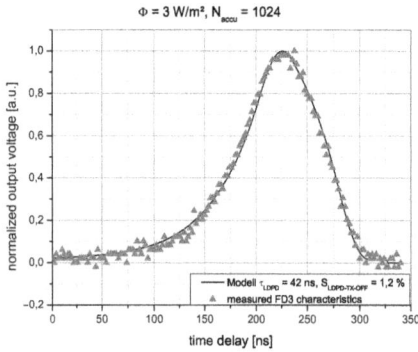

Figure 6.73 Fit of the proposed model to measurements of the ToF-LDPD from Section 6.5 at $\hat{E}_e = 3\,W/m^2$.

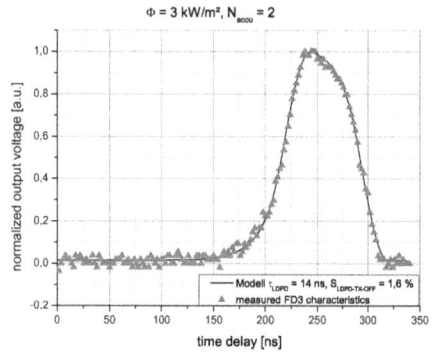

Figure 6.74 Fit of the proposed model to measurements of the ToF-LDPD from Section 6.5 at $\hat{E}_e = 3\,kW/m^2$.

that the flexibility of the proposed model is sufficient to properly adapt the measured shape of the output characteristics of the emulated ToF principle at low and high levels of irradiance.

Here, normalized data is presented, because absolute fits were not satisfying. It is believed, that this may be caused for instance by the uncertainty of the irradiance level and the employed model parameters which resulted from the PTM measurements. Alternatively, differences may also stem from the fact that the proposed model is actually founded on a heuristic description resulting from a limited dataset. From the fitting results, it was again confirmed that the charge transfer is an irradiance dependent phenomenon. The charge transfer characteristic τ_{LDPD} of the ToF-LDPD from Section 6.5 is plotted against the irradiance in Figure 6.75 for TX1 and TX3. In the observed measurement range the data approximately exhibits a logarithmic dependence. Therefore, an LQ-fit was performed for each of the shutters. The $S_{LDPD-TX-OFF}$ term is also given

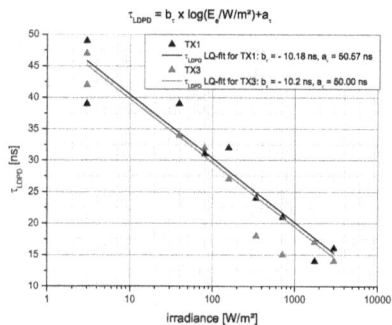

Figure 6.75 τ_{LDPD} fit from measurements of the ToF-LDPD from Section 6.5 for varying irradiance levels.

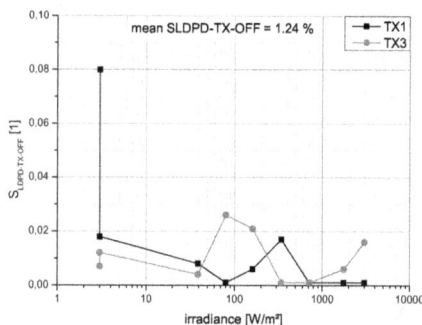

Figure 6.76 $S_{LDPD-TX-OFF}$ fit from measurements of the ToF-LDPD from Section 6.5 for varying irradiance levels.

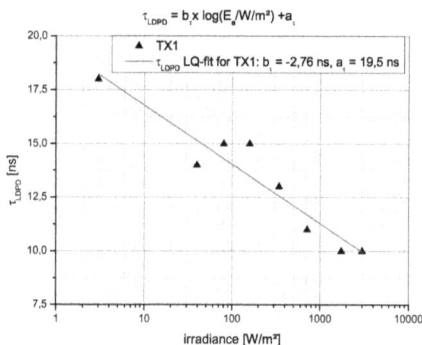

Figure 6.77 τ_{LDPD} fit from measurements of the ToF-LDPD from Section 6.6 for varying irradiance levels.

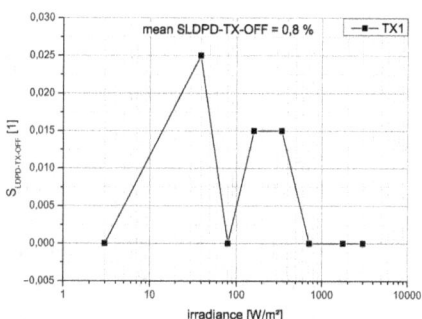

Figure 6.78 $S_{LDPD-TX-OFF}$ fit from measurements of the ToF-LDPD from Section 6.6 for varying irradiance levels.

as function of the irradiance (c.f. Figure 6.76). However, the employed method hardly shows any irradiance dependence. This may be explained by the comparatively high noise level when the shutter is deactivated as it can for instance be observed in Figures 6.73–6.74. Excluding the outlier at 3 W/m², the average $S_{LDPD-TX-OFF}$ amounts to 1.24% which is close to the data from Figure 6.52.

The Figures 6.77 and 6.78 depict the data from the fitting procedure of the ToF-LDPD from Section 6.6. It can be observed that the ToF-LDPD from Section 6.6 is more than a factor of two faster than the ToF-LDPD from Section 6.5. Unfortunately, output characteristics of the emulated ToF principle are too noisy here, to allow for proper estimation of $S_{LDPD-TX-OFF}$. The PTM measurements from the former sections are considered to yield more reliable estimates.

Concluding the latter two sections, it can be stated that the proposed model is flexible enough to resemble the measured characteristics. It is very simple and thus feasible for fast predictions as it can be useful to understand the general characteristics and support the calibration or to enable sensitivity or precision analysis. In the observed dataset, τ_{LDPD} was properly matching a logarithmic plot. This is also a useful relation that can be employed for further analysis. However, the observed relation has to be treated carefully, since it predicts diverging time constants for low irradiance levels and negative values for increasing irradiance levels. Neither of those predictions is physically meaningful. At low irradiance levels, the photogenerated charge carriers will not anymore affect the electrostatic potential distribution, so that the built-in drift field and diffusion processes ultimately limit τ_{LDPD}. The irradiance levels that are necessary to yield negative values are not likely to be observed in the proposed application of time-of-flight imaging. The high electric fields which would result from further deformation of the electrostatic potential profile by increasing the irradiance level, would affect the entire charge transfer process. It will become difficult to properly switch OFF the transfer gates. Thus, further characterization and modeling that should be founded on the semiconductor equations has to be done to yield better estimates of the irradiance dependent charge transfer process.

6.7.4 Impact on precision

The proposed model can be properly adjusted to actual devices as described in the former section. In this section, the impact of the new parameters on the range measurement precision is investigated. For this survey, datasets according to Table 6.8 were computed.

The integral in Equation 6.39 can be examined independently from ambient light and the irradiance level which is affected by spherical broadening and reflectance. The shape of these functions corresponds to the measurement results from the emulated ToF principle. Exemplary numerical solutions of the integral are given in Figure 6.79 for the shutters TX1 and TX2. Ideally, as described in Section 6.1, the laser is emitted at the same time instance at which the shutter TX1 is triggered. This would result in affine linear characteristics for TX1 and TX2 that should intersect at half the measurement range $z_{max} = cT_p/2$ (c.f. Section 2.3.3). The non-ideal laser shape and the finite transfer speed, however, affect this behavior substantially as was observed in the measurements and is verified by the proposed model. These phenomena delay the intersection point which is ideally found at $cT_p/4$ to longer distance values. Since this might shift the range sensitive and bijective section of the response characteristics out of the shutter windows, the time instance τ_E at which the laser pulse is emitted has to be properly adjusted. Thus, Figure 6.79 depicts the characteristics for various τ_E. Since the used measurement range of 2 m is smaller than $cT_p/2$, it might become advantageous to shift the intersection point for instance in the vicinity of 1 m. This can result in reduced calibration error as depicted in Figure 6.80 and moreover in a better precision as it is depicted in Figure 6.81.

Figure 6.80 presents the maximum residuals of the fitting procedure given in Equations 6.43 and 6.44 for various τ_E and τ_{LDPD}. It can clearly be observed, that the residuals diverge once the laser pulse is shifted too far. Comparison of the residuals at $\tau_E \approx 0$ and $\tau_{LDPD} = 1$ ns with the corresponding large-signal characteristics in Figure 6.79 demonstrates that for this value TX2 is rarely increasing for varying distance, whereas TX1 is entirely flat for some distance values. Since for these scenarios

Table 6.8 Parameter set for the evaluation of the expected range measurement precision by means of the proposed model.

distance range	0.105 to 2.1 m
reflectance range	0.02 to 1.04
accumulation count	1 to 1000
ambient illuminance	0 to 110 klx
temperature of the sun	5777 K
optical bandpass filter	820 to 920 nm
lens transmittance	0.9
field of view	$15° \times 15°$
f-number	0.95
total radiant flux of the laser	4×75 W
wavelength of the laser	905 nm
laser pulse width	30 ns
shutter width	30 ns
duty cycle of the laser	1/1000
pixel area	$40 \times 40\ \mu m^2$
fill factor	0.38
extrinsic quantum efficiency	35%
sense node capacitance	not taken into account
read noise	not taken into account
τ_{LDPD}	1 to 50 ns
$S_{LDPD-TX-OFF}$	1×10^{-6} to 0.1
dark current	0 to $1 \times 10^6\ e^-$/s

$\partial z_{fit}/\partial N_{FDx}$ is very high and might even diverge, a very high non-deterministic error results even for low $\sigma^2_{N_{FDx}}$ as it is depicted in Figure 6.81. For the region in which the intersection point is in the vicinity of 1 m, residuals and random error are not that strongly affected by τ_E. Interestingly though, is the phenomenon that variation of τ_E can slightly improve the range measurement precision for a certain set of scenarios, whereas it can simultaneously negatively affect the precision for other scenarios. This can be explained by the adjustment of the signal level for given distance values when τ_E is varied, what affects $\partial z_{fit}/\partial N_{FDx}$ and $\sigma^2_{N_{FDx}}$ and thus the precision. As verified in the former section, τ_{LDPD} was observed to vary in the range of 15 to 50 ns for the design presented in Section 6.5 and 10 to 20 ns for the design proposed in Section 6.6.

To study how the large-signal characteristics are affected by τ_{LDPD}, Figure 6.71 can be evaluated. Since the characteristics' magnitudes shrink for increasing τ_{LDPD} while the curves disperse, $\partial z_{fit}/\partial N_{FDx}$ and $\sigma^2_{N_{FDx}}$ are both negatively affected what results in a decreasing precision. Figure 6.82 depicts the variation of the intersection point for different τ_{LDPD}.

As observed from the measurements in Sections 6.5 and 6.6, the shutters do not properly block photogenerated charge carriers when they should actually remain OFF. This introduces an increased sensitivity to ambient illumination. The ideal PM ToF principle promises improved immunity to background light compared to the CW ToF principle, by reduction of the integration time windows and according increase of the active light source's power. This concentration of the signal power to small time windows reduces the amount of accumulated background related photogenerated charge carriers (c.f. Sections 2.3.3 and 6.1).

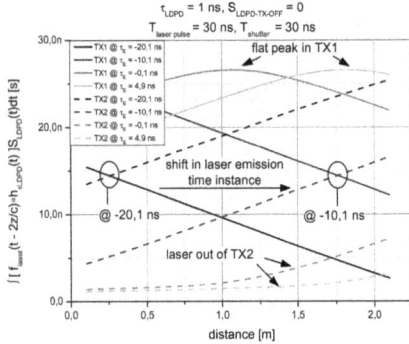

Figure 6.79 Evaluation of the integral in Equation 6.39 for various laser emission time instances τ_E.

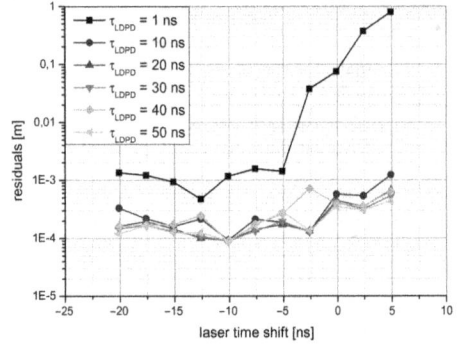

Figure 6.80 Fitting residuals for various τ_E and τ_{LDPD}.

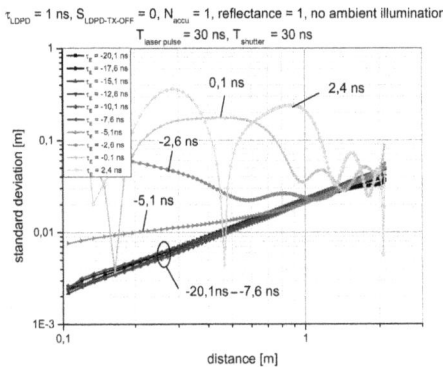

Figure 6.81 Estimated range measurement precision for various τ_E.

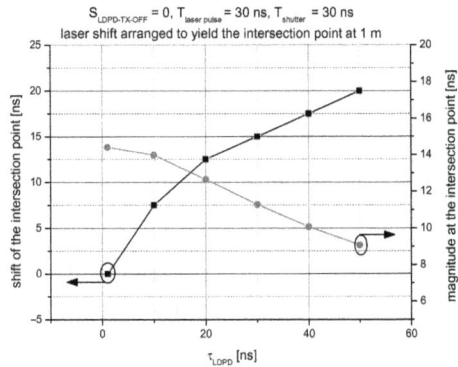

Figure 6.82 Characteristics of the intersection point of TX1 and TX2 from Equation 6.39 for various τ_{LDPD}.

Under harsh conditions the parasitic sensitivity to ambient light when the shutters should actually remain OFF can result in a higher ambient related electron count than signal related electron count. Even when $S_{LDPD-TX-OFF}$ amounts to only 0.1% many charge carriers can be accumulated simply because the time interval in which the shutter is OFF is a factor of 1000 higher than the integration time for the signal related charge carriers. The proposed model allows to evaluate this phenomenon. As depicted in Figure 6.83, the ambient illuminance of 110 klx does not deteriorate the range measurement precision significantly for $S_{LDPD-TX-OFF} \leq 1 \times 10^{-4}$, but almost exponentially increases the random error for $S_{LDPD-TX-OFF} \geq 1 \times 10^{-3}$. Since $S_{LDPD-TX-OFF}$ was evaluated to amount approximately 1 to 2% for the presented

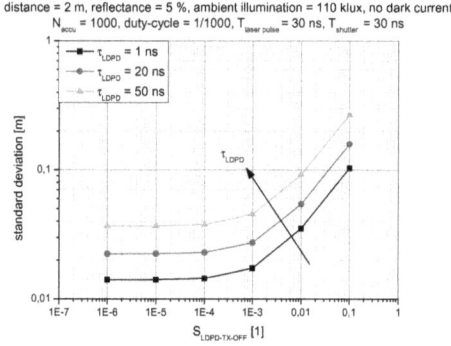

Figure 6.83 Estimated range measurement precision for various $S_{LDPD-TX-OFF}$ and τ_{LDPD}.

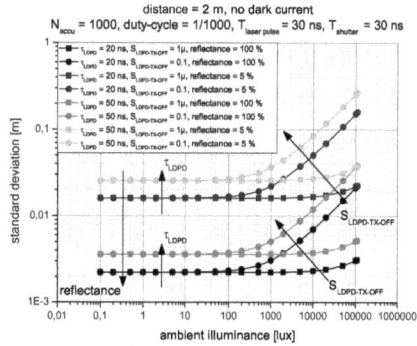

Figure 6.84 Impact of the ambient illumination on the range measurement precision.

designs in Sections 6.5 and 6.6, the performance should be improved by approximately a factor of 10 in future designs, to be sufficiently immune against this phenomenon. In Figure 6.84 the range measurement precision as a function of reflectance, ambient illuminance, and the photodetector's model parameters τ_{LDPD} and $S_{LDPD-TX-OFF}$ is presented. Here, it can clearly be observed, that an increasing $S_{LDPD-TX-OFF}$ shifts the ambient illuminance level at which the precision is deteriorated to lower values. The reflectance of the scenery simultaneously scales the ambient and signal related photon counts and thus appears as a simple shift of the random error values. τ_{LDPD} barely affects the ambient related charge carrier accumulation, since the corresponding integration time is a factor of 1000 longer.

In Figure 6.85 two different methods to take into account the dark current for the estimation of the range measurement precision are compared. One method calculates the amount of stored charge carriers according to Equation 6.39, fits the dataset to Equation 6.43 and determines the precision according to Equation 6.45. The other method does not take into affect the dark current related charge carriers for the calculation of the charge carrier counts and the partial derivatives $\partial z_{fit}/\partial N_{FDx}$ but adds the Poisson distributed dark noise according to Equation 6.23. Interestingly the characteristics for these two approaches don't differ, from which it can be deducted, that the dark current does not affect $\partial z_{fit}/\partial N_{FDx}$ for the given dataset. Moreover it can be observed, that the dark current has negligible effect on the precision for $J_{dark} \leq 100\,\mathrm{ke^-/s}$. Fortunately, the PTM measurements verified values of $J_{dark} \leq 15\,\mathrm{ke^-/s}$ for the design presented in Section 6.5 and $J_{dark} \leq 6.7\,\mathrm{ke^-/s}$ for the design proposed in Section 6.6.

Concluding this survey, it has to be stressed on the fact that the given simulation results are not to be misunderstood as a qualification against the constraints given in Section 6.3. This has to be done in the future by actual measurements. Even more important, it has to be noted that the given results were achieved for a substantially simplified treatment of the existing problem, since τ_{LDPD} and $S_{LDPD-TX-OFF}$

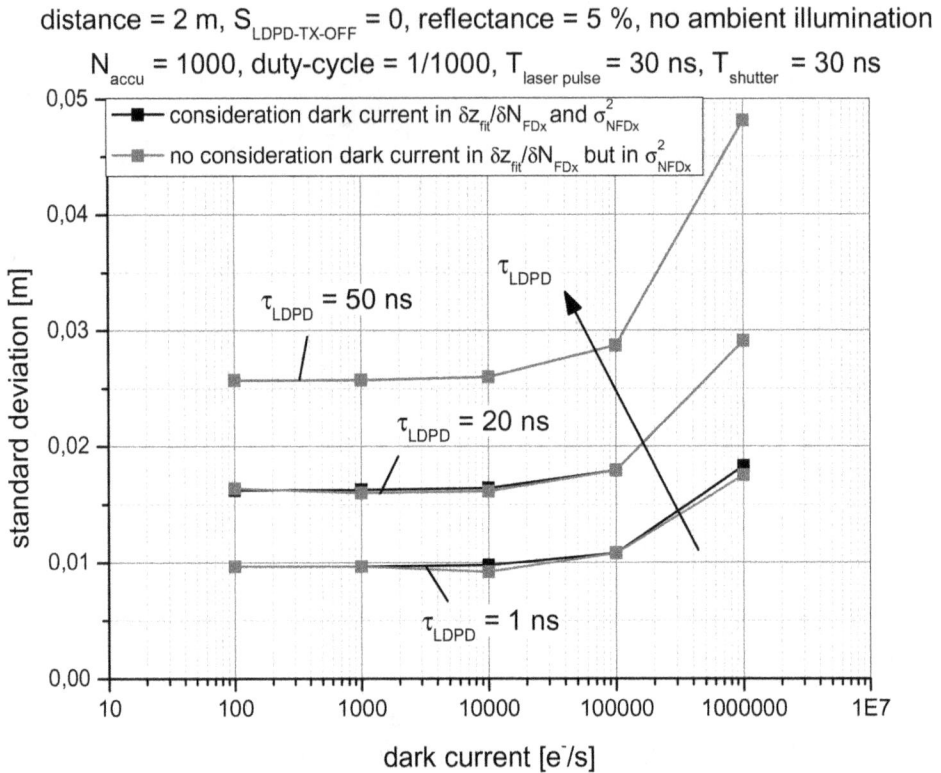

Figure 6.85 Impact of the dark current on the range measurement precision.

were observed to be irradiance dependent. Thus, in the future the proposed model has to evaluate the irradiance level for each scenario followed by the calculation of τ_{LDPD} and $S_{LDPD-TX-OFF}$ from which the charge carrier counts for each scenario can be determined which, unfortunately, significantly increases the calculation effort since the integral in Equation 6.39 has to be calculated for each data point. This has to be followed by the fitting procedure and the estimation of the range measurement precision. For such a model, τ_E will play an even more important role than it did here, since it cannot be simply used to adjust the intersection point to a defined distance because there is not only one intersection point but an entire set. This will result in an even increased effort to actually optimize the performance. Another problem that can arise here is that the values for τ_{LDPD} and $S_{LDPD-TX-OFF}$ might have to be determined for irradiance levels that have not yet been characterized by measurements. As claimed in the former section, the logarithmic fit is neither considered meaningful for low irradiance levels at which at some point diffusion processes and the built-in gradient will govern the charge carrier transport nor at high irradiance levels at which the fit predicts negative transfer times. The impact from the read noise has not yet been determined in combination with the generalized model. Another issue to be investigated is the effect introduced by the multiple accumulation count settings, which can cause dynamic range gaps which basically are reduced signal-to-noise ratios in the vicinity of the irradiance level at which a higher accumulation count setting is used (c.f. [Dar12]). For cases were multiple accumulation count settings are not tolerable, the read noise has a substantially increased impact on the range measurement precision, since the storage node capacitance has to be increased in order to allow measurements at short distances with high reflectance levels (c.f. Sections 6.2 and 6.3). This results in the need for low noise readout circuitry, novel reset noise reduction schemes and proper calculus of the random error by methods as they are explained in Chapter 3.

Chapter 7

Conclusions

This work is dedicated to CMOS imaging with emphasis on noise modeling, characterization and optimization using the example of PM ToF range image sensors. It intends to complement the design of high performance image sensors in general and high performance range imagers in particular.

A comparison of the known range measurement technologies is given in the present work. Different state of the art modulation schemes have been analyzed heuristically, resulting in the fact that ideally the pulse modulated ToF approach can be superior to the competing continuously modulated ToF principle. A comprehensive analysis of the available ToF camera systems and ToF image sensors demonstrated that, so far, there is no ToF range measurement system available that operates satisfactorily in harsh conditions such as when provoked by high ambient illumination, long object distances and low reflectance levels. In this work the PM ToF principle is investigated which, however, requires a high speed photodetector, high dynamic range as well as SNR. For that purpose a so-called lateral drift-field detector was employed.

Temporal noise is one of the most crucial parameters for the design of a sensor, since it ultimately limits the performance and cannot be compensated because it is non-deterministic. This work comprises a comprehensive survey through the theory of noise and covers topics such as noise propagation in linear time-invariant and non-linear time-variant systems, fundamental noise processes and modeling of noise in semiconductor devices and proposes a mathematically non-rigorous but potentially computational efficient algorithm to estimate noise in time-sampled systems. Within the framework of this work a low-frequency noise characterization of various available devices in the $0.35\,\mu$m-CMOS process that was employed for the design of the presented range imagers, was performed. Heuristic explanations for the differing noise performance of the devices was done by support of TCAD simulations and the comparison to the state of research. These fundamental investigations can be understood as a foundation for facilitating low-noise circuit design at the Fraunhofer Institute IMS. The author believes that rigorous calculus based on physically verified component models allows for accurate prediction of the achievable random error in sensor systems or circuits in general. The theory on noise, unfortunately, still has many open questions. In the recent past different methodologies have been published that aim for proper estimation of the noise in non-linear, time-variant systems but are yet to be compared to actual measurements of real systems. Exemplarily, a switched capacitor correlated double sampling stage has been fabricated and characterized within the framework of this work. The presented results proved that non-rigorous estimations can yield huge mismatches to the actual measurements. Many of those methodologies often simplify the observed systems by linearization which is valid only if the random error is negligible compared to the large-signal excitations. In this work this was studied by example of the ideal PM ToF range image sensor. It could be observed that under harsh conditions, noise prediction by Gaussian error analysis yielded significantly smaller noise levels than the more rigorous approach of the transformations of the probability density functions. Noise processes such as flicker noise or RTS noise are gaining importance, but are not yet fully understood and have to be modeled with extensive use of seemingly arbitrary fitting parameters. The same applies for the white noise component of e.g. MOS transistors which is also typically adjusted by fitting parameters. The presented measurements for the low-frequency noise performance characterization of short-channel devices demonstrated significant differences between different samples of a particular device as it is also published in the literature. To allow for better modeling a larger variety of geometries and biasing points have to be characterized. For small devices, several samples should be measured to allow for statistically meaningful estimations. In the presented work different models have already been compared to the observed results. For instance, it was demonstrated that the simple Level-1 to Level-10 flicker noise models do not accurately describe the observed power spectral densities. The quantitative comparison with the extended BSIM3v3/BSIM4 flicker noise model and the EKV models were beyond the scope of this work but should be performed in the future. Unfortunately, the existing measurement equipment does not offer automatized characterization, making the effort for a proper process modeling inadequately large which should be changed in the future. Moreover, there was no measurement equipment available for the characterization beyond the kilohertz frequency range so that observation of the white noise components of MOSFET devices was not possible within the framework of this work.

Temporal noise sources in CMOS APS have been investigated intensively in the past. The widely employed pinned photodiode detector improved the noise performance considerably by reducing dark current and transferring the charges to a small storage capacitor which improves reset noise performance and simultaneously allows to amplify the charge by the ratio of storage node capacitance to the capacitance of the photoactive region. Typically, the remaining dominant noise sources are the reset noise and the noise that is introduced by the source follower. Solutions to reduce the effect of the reset noise are correlated-double sampling and active reset. Unfortunately, the noise arising from the source follower transistor cannot be easily attenuated by signal conditioning circuitry. Thus, within the framework of this work a novel readout structure was developed which aims at a significantly improved low-frequency noise performance. This is gained by the spacial separation of current paths from centres that are known to exhibit high generation/recombination noise and thus result in flicker or RTS noise. As known from the state of the art, this can be achieved by implementation of JFET transistors. In this work it was presented how such a JFET can be integrated and joined with the photodetector itself by directly forming the gate of the JFET transistor with the photodetector's floating diffusion, so that all advantages of the photodetectors – such as the low dark current, low reset noise due to a low sense-node capacitance and the capability of performing multiple accumulations in the photodetector itself which can be understood as a low-noise averaging process – are preserved, whereas the low frequency noise performance is significantly improved. The proposed device was designed with the support of TCAD simulations, manufactured and characterized. Measurements of a JFET transistor with a width of 1.52 μm and a length of 1.2 μm demonstrated similar flicker noise performance as the best available standard $10 \times 10 \, \mu m^2$ n-type MOSFET transistor which is remarkable, since its gate area is approximately by a factor of 100 smaller. It is known that for long-channel devices flicker noise approximately scales inversely proportional to the gate area, but for short channel transistors much higher noise levels are expected due to second order effects or arising phenomena such as RTS noise. This also verifies that this device is superior to standard n-type readout transistors for low-light imaging applications, because those would have to have a comparatively large gate area to meet flicker noise constraints which comes at the expense of an increased reset noise level and a reduced conversion gain. Obviously, this is not that severe for the proposed JFET. Unfortunately, the characterized sample devices only allowed relatively low drain currents at zero gate-source voltage, so that the junctions had to be forward biased resulting in leakage currents through the gates. This phenomenon was analyzed by measurements as well as TCAD simulations, a biasing scheme to prevent leakage currents through the gate was proposed and it was presented how the device will be rearranged in a redesign to avoid this parasitic phenomenon. This has to be evaluated in the future, followed by an implementation of the device in an actual APS configuration.

Within the framework of this work the LDPD based PM ToF image sensor principle was implemented, modeled and optimized. The LDPD can be considered as a modification of a pinned-photodiode which intends to preserve all of its advantages but improves the charge-transfer speed. The author supported the first implementation of this new photodetector for ToF purposes. In this work the physical limitations due to the very fundamental photon noise were modeled by application of Gaussian error analysis and transformation of probability density functions. This model was extended by noise sources that are typically additionally introduced by the image sensor and the camera board, so that a link between the parameters from the sceneries, camera, image sensor and the photodetector is established. It was demonstrated, that the constraints for dynamic range and temporal noise performance can be remarkably harsh. The use of multiple accumulation count settings, however, can substantially relax this, but comes at the expense of lower frame rates, larger complexity of the calibration and the possibility of dynamic range gaps. Measurement results of the first implementation of the LDPD based PM ToF range imager demonstrated a severely insufficient charge transfer speed. It was verified that this was originated by two phenomena. Firstly, the design of the geometry in the vicinity of the transfer gates introduced an electric field in the opposite direction of the intended charge transfer. Secondly, it was verified that the PM ToF approach not only requires a high-speed photodetector but also high-performance shutter generating circuitry. Making extensive use of parasitic modeling, an architecture was developed that yields rise and fall-times below 7 ns for all corners of the employed process and over the automotive temperature range. It provides shutter interlocking functionality which hinders parasitic short circuiting of different storage nodes and operates properly for positive power supply voltages rail levels above 1.7 V. The presented measurement results demonstrated a significantly improved charge-transfer performance. This enabled the implementation of the image sensor in an actual range camera setup, that demonstrated centrimetric resolution for 0.2 to

2 m, with a FOV of 30° × 30° at reflectance levels from 80 to 100%. In the future, the performance at lower irradiance levels and under ambient illumination has to be verified. It is also to be investigated how the calibration process can be simplified without introducing an intolerable systematic error. More important for economical reasons, however, is the task of reducing the amount of data points required by the calibration process to a minimum. Even though the charge-transfer was significantly improved it was still observed that it is deteriorated at low irradiance levels. Furthermore it was observed, that the design of the geometry of the storage nodes, transfer gates and collection gates result in a mismatch between the different shutters. An alternative approach for an LDPD based ToF photodetector was developed, fabricated and characterized within the framework of this work that employs three sub-pixels to demodulate the impinging wave-packets according to the PM-ToF principle. This detector demonstrated superior performance in matching, charge transfer speed, linearity and dark current which is traded against a smaller photoactive area. Since proper functionality was obtained for all three shutters, compared to the former approach where one shutter was not properly operating under low-light conditions, so that background light subtraction would have to be performed by sampling two subsequent images, the reduced photoactive area is not resulting in a worsened precision at comparable frame rate settings. Moreover, this results in a significantly improved background light suppression functionality due to higher sampling frequencies. For the case of using the original LDPD-ToF pixel approach and applying the background light suppression scheme of subtracting two subsequent frames, aliasing frequencies of approximately 5 Hz resulted, whereas the new approach offers aliasing frequencies in the range of 16.7 kHz, due to the capability of sampling the background light with the laser repetition rate. Even though the new approach demonstrated a significantly improved charge transfer speed by more than a factor of two, irradiance dependence was still observed and the transfer time was not guaranteed to be far below the shutter length of 30 ns. Another phenomenon that was observed during the characterization of the different designs is that the shutter is not capable of fully blocking the propagation of photogenerated charge carriers into the storage nodes. To investigate these effects and their impact on the range measurement performance, the model of the photodetector was generalized using a non-linear time-variant approach that allows for observation of large-signal characteristics, sensitivity analysis and estimation of the range measurement precision. The proposed model defined two parameters to model the finite charge transfer speed and the parasitic light sensitivity when the shutters should actually be OFF. The proper implementation was verified by choosing the new defined parameters appropriately, what demonstrated that this more general model converges into the former model of the ideal range image sensor. The model was then fitted to normalized measurements what showed a remarkably good agreement between these two levels. Unfortunately, no calibrated equipment was available so that absolute fits could not be satisfactorily performed. Within the measured irradiance range, it was possible to model the irradiance dependent transfer time by a logarithmic function. This will, however, not be possible beyond the observed irradiance range, because the transfer time is expected to converge for both limits. To analyze this, TCAD simulations or analytical derivations based on the semiconductor equations have to be performed in the future. This could for instance result in compact models for circuit level simulation tools, so that the interdependence of shutter generating and readout circuitry and

the actual photodetector can be investigated. Moreover, this would be desirable from a non-deterministic point of view, too. Such compact models allow to use the tools provided by e.g. the Cadence environment to simulate noise propagation in non-linear time-variant systems more accurately. Analytical derivations demonstrated that the formula for the ideal PM ToF range imager does not properly link the output signals of the sensor with the object distance, if the new defined parameter for the effective transfer time is not negligible compared to the shutter period/laser width, because transcendental equations are yielded. This results in the need for an increased calibration complexity. It was demonstrated by deterministic and non-deterministic analyses that the emission time instances of the laser pulses should not be simply set to the time instance at which the shutter is triggered, but has to be shifted appropriately because of the finite charge transfer speed and the resulting deterioration of the laser pulse shape. The proposed model also allowed to demonstrate the effect of the ambient light which can affect the performance of the range imager significantly. This can be reduced by either increasing the laser repetition rate so that images can be acquired faster or by improvement of the shutters' blocking performance. Dark current is also found to have a severe impact on the precision, which can be reduced by a different design of the storage nodes or by an increase of the laser repetition rate. Apart from the desire to develop a physics based compact model for the photodetector, there is also a need for a more accurate estimation of the impinging irradiance levels. The approach implemented in this work has to be considered as rather rough for short object distances. This could be improved by application of Maxwell's equations. Additionally to the more confident model of the irradiance levels, this would also allow to study imaging artefacts and their impact on calibration and the lateral resolution limitations. Of course, all efforts on the development of such models should be accompanied by actual measurements.

Appendix A

Derivation of the autocorrelation formula of shot noise

As described in Section 3.3.2, the macroscopic current is modeled by superposition of multiple current pulses at random time instances t_k:

$$i(t) = \sum_{k=1}^{K} i_e(t - t_k). \tag{A.1}$$

It is thus a random process of $K + 1$ random variables; namely the t_k time instances and the actual number of current pulses K. To evaluate the autocorrelation function $\mathbb{E}(i(t)i(t + \tau))$ has to be calculated. Firstly, however, $\mathbb{E}(i(t))$ is determined:

$$\mathbb{E}(i(t)) = \int_{-\infty}^{\infty} \int_{-\infty}^{\infty} \cdots \int_{-\infty}^{\infty} \left[\sum_{k=1}^{K} i_e(t - t_k) \right] f_{p-t_n,K}(t_1, t_2 \cdots t_k \cdots t_K, K)$$
$$\cdot \, dt_1 dt_2 \cdots dt_k \cdots dt_K dK \tag{A.2}$$

which can be expressed in terms of conditional probability density functions:

$$\mathbb{E}(i(t)) = \int_{-\infty}^{\infty} \int_{-\infty}^{\infty} \cdots \int_{-\infty}^{\infty} \left[\sum_{k=1}^{K} i_e(t - t_k) \right] f_{p-t_n}(t_1, t_2 \cdots t_k \cdots t_K | K)$$
$$\cdot \, f_{p-K}(K) dt_1 dt_2 \cdots dt_k \cdots dt_K dK. \tag{A.3}$$

The observation time interval is chosen to $[-T, T]$. Later, the limit $T \to \infty$ will be calculated. $p(K)$ is thus limited on $[-T, T]$. Since the time instances t_k were assumed to be independent and equally distributed it follows that

$$f_{p-t_n|K}(t_1, t_2 \cdots t_K | K) = \underbrace{\frac{1}{2T} \cdot \frac{1}{2T} \cdots \frac{1}{2T}}_{K \text{ times}}. \tag{A.4}$$

$\mathbb{E}(i(t))$ can thus be rearranged to

$$\mathbb{E}(i(t)) = \int_{-\infty}^{\infty} \int_{-T}^{T} \cdots \int_{-T}^{T} \left[\sum_{k=1}^{K} i_e(t - t_k) \right] \cdot \frac{1}{2T} dt_1 \frac{1}{2T} dt_2 \cdots \frac{1}{2T} dt_k \cdots \frac{1}{2T} dt_K f_{p-K}(K) dK.$$

$$\tag{A.5}$$

Since the time instances t_k were assumed to be independent, every integral with respect to dt_i will yield $2T$ except for t_k itself:

$$\mathbb{E}(i(t)) = \int_{-\infty}^{\infty} \int_{-T}^{T} \left[\sum_{k=1}^{K} i_e(t - t_k) \right] \frac{1}{2T} dt_k f_{p-K}(K) dK. \tag{A.6}$$

Rearranging the sequence of integration yields

$$\mathbb{E}(i(t)) = \frac{1}{2T} \int_{-\infty}^{\infty} \sum_{k=1}^{K} \int_{-T}^{T} i_e(t - t_k) dt_k f_{p-K}(K) dK. \tag{A.7}$$

The time interval $[-T, T]$ is chosen large enough to fully include the current pulse corresponding to t_k. In that case the integral will yield the elementary charge q:

$$\mathbb{E}(i(t)) = \frac{q}{2T} \int_{-\infty}^{\infty} Kp(K)dK = \frac{q}{2T} \mathbb{E}(K). \tag{A.8}$$

Defining $<n> = \frac{\mathbb{E}(K)}{2T}$ as the mean number of charge carriers emitted in the time interval $[-T, T]$ this becomes

$$\mathbb{E}(i(t)) = <n>q. \tag{A.9}$$

Assuming a DC current, taking the limit $T \to \infty$ does not affect the above result since $\mathbb{E}(K)$ itself is proportional to the observation interval (c.f. Equation 3.110). This result is known as the first part of the so-called *Campbell-Theorems*.

Now, the autocorrelation function is derived:

$$R_{ii}(\tau) = \mathbb{E}\left(\left[\sum_{k=1}^{K} i_e(t - t_k) \right] \left[\sum_{j=1}^{K} i_e(t + \tau - t_j) \right] \right). \tag{A.10}$$

The expectation is to evaluated with respect to all random variables:

$$R_{ii}(\tau) = \int_{-\infty}^{\infty} \int_{-\infty}^{\infty} \cdots \int_{-\infty}^{\infty} \left[\sum_{k=1}^{K} i_e(t - t_k) \right] \left[\sum_{j=1}^{K} i_e(t + \tau - t_j) \right]$$
$$\cdot f_{p-t_n, K}(t_1, t_2 \cdots t_K, K) dt_1 dt_2 \cdots dt_K dK \tag{A.11}$$

which can again be expressed by conditional probability density functions:

$$R_{ii}(\tau) = \int_{-\infty}^{\infty} \int_{-\infty}^{\infty} \cdots \int_{-\infty}^{\infty} \left[\sum_{k=1}^{K} i_e(t - t_k) \right] \left[\sum_{j=1}^{K} i_e(t + \tau - t_j) \right]$$
$$\cdot f_{p-t_n}(t_1, t_2 \cdots t_K | K) f_{p-K}(K) dt_1 dt_2 \cdots dt_K dK. \tag{A.12}$$

Inserting the equally distributed conditional probability density functions for the independent current time instances t_k and rearrangement of the sums yields

$$R_{ii}(\tau) = \int_{-\infty}^{\infty} \int_{-T}^{T} \cdots \int_{-T}^{T} \sum_{k=1}^{K} \sum_{j=1}^{K} i_e(t - t_k) i_e(t + \tau - t_j) \cdot \frac{1}{2T} dt_1 \frac{1}{2T} dt_2 \ldots \frac{1}{2T} dt_K f_{p-K}(K) dK.$$

(A.13)

The sums are now separated into independent components $k \neq j$ and the components $k = j$ for which the single current pulses are autocorrelated. The possible permutations can be visualized by a $K \times K$-matrix:

$$\begin{pmatrix} a_{11} & \cdots & a_{1K} \\ \vdots & \ddots & \vdots \\ a_{K1} & \cdots & a_{KK} \end{pmatrix}.$$

(A.14)

The main diagonal represents the K cases for which $k = j$ applies, whereas the remaining $K^2 - K$ combinations correspond to the independent cases $k \neq j$. This can be used to split the autocorrelation function into two parts:

$$R_{ii}(\tau) = \int_{-\infty}^{\infty} \int_{-T}^{T} \cdots \int_{-T}^{T} \sum_{k=1}^{K} i_e(t - t_k) i_e(t + \tau - t_k) \frac{1}{2T} dt_1 \frac{1}{2T} dt_2 \cdots \frac{1}{2T} dt_K f_{p-K}(K) dK$$

$$+ \int_{-\infty}^{\infty} \int_{-T}^{T} \cdots \int_{-T}^{T} \sum_{\substack{k=1 \\ k \neq j}}^{K} \sum_{j=1}^{K} i_e(t - t_k) i_e(t + \tau - t_j) \frac{1}{2T} dt_1 \frac{1}{2T} dt_2 \cdots \frac{1}{2T} dt_K f_{p-K}(K) dK.$$

(A.15)

Integrating the first term with respect to dt_n, $n \neq k$ yields $2T$ as before. For the second term this applies for the components $n \neq k \neq j$:

$$R_{ii}(\tau) = \int_{-\infty}^{\infty} K \int_{-T}^{T} i_e(t - t_k) i_e(t + \tau - t_k) \frac{1}{2T} dt_k f_{p-K}(K) dK$$

$$+ \int_{-\infty}^{\infty} (K^2 - K) \int_{-T}^{T} \int_{-T}^{T} i_e(t - t_k) i_e(t + \tau - t_j) \frac{1}{2T} dt_k \frac{1}{2T} dt_j f_{p-K}(K) dK.$$

(A.16)

Since integration with respect to dt_n and dK are independent the above equation can be rearranged to

$$R_{ii}(\tau) = \int_{-\infty}^{\infty} K f_{p-K}(K) dK \int_{-T}^{T} i_e(t - t_k) i_e(t + \tau - t_k) \frac{1}{2T} dt_k$$

$$+ \int_{-\infty}^{\infty} (K^2 - K) f_{p-K}(K) dK \int_{-T}^{T} i_e(t - t_k) \frac{1}{2T} dt_k \int_{-T}^{T} i_e(t + \tau - t_j) \frac{1}{2T} dt_j.$$

(A.17)

Calculation of the integrals yields

$$R_{ii}(\tau) = \frac{\mathbb{E}(K)}{2T} \int_{-T}^{T} i_e(t - t_k) i_e(t + \tau - t_k) dt_k + \mathbb{E}(K^2 - K) \frac{q^2}{(2T)^2}. \qquad (A.18)$$

Using for instance the characteristic function, it can be shown that

$$\mathbb{E}(K^2 - K) = \int_{-\infty}^{\infty} K^2 - K f_{p-K}(K) dK = \sum_{0}^{\infty} \frac{(K^2 - K)(<n>2T)^K e^{-<n>2T}}{K!}$$

$$= (<n>2T)^2 \qquad (A.19)$$

applies. Now taking the limit $T \to 0$ finally yields

$$R_{ii}(\tau) = <n>q^2 + <n> \int_{-\infty}^{\infty} i_e(t - t_k) i_e(t + \tau - t_k) dt_k \qquad (A.20)$$

or

$$R_{ii}(\tau) = \mathbb{E}(i(t))^2 + <n> \int_{-\infty}^{\infty} i_e(t - t_k) i_e(t + \tau - t_k) dt_k \qquad (A.21)$$

as there were used in Section 3.3.2.

Appendix B

Measurement setups

B.I NOISE MEASUREMENT SETUP

At the Fraunhofer Institute IMS a low-frequency, low-noise noise measurement setup is available which was developed in [Bro10] and adopted the basic principles from [JT07]. This setup enabled the characterization of the noise performance within a frequency range of approximately 0.1 Hz up to 1–10 kHz depending on the actual noise level, the DC bias and the small-signal output impedance of the device under test (DUT). The DUT is fixed in a μ-metal box in which the biasing batteries are also kept. A transimpedance amplifier (SR570) is introduced in the current loop to act as a current meter. It converts the current to a voltage which can then be directly measured by e.g. a spectrum analyzer. Since the transimpedance amplifier amplifies the DC current as well as the interesting AC noise component a large DC voltage is yielded at the output of the amplifier that could easily push the spectrum analyzer into overload. Apart from that very extreme scenario, the voltage resolution of the spectrum analyzer is also defined by the maximum voltage between signal and ground. To attenuate this effect an offset compensation is realized by usage of an additional battery at the output of the amplifier. Contrary to e.g. a capacitive decoupling scheme, this allows to measure very low frequency components. An *Agilent 35670A* is employed as the spectrum analyzer. It samples a series of voltage levels and computes an FFT to estimate the power spectral density. The DUT, the amplifier and the offset compensation battery are placed inside an aluminum box to avoid pick-up of any surrounding disturbances. It is furthermore advantageous to separate signal ground from the ground of the μ-metal box and the cases of the SR570 and the offset compensation battery and to separate the ground of the DUT, amplifier and μ-metal box from the aluminum box (c.f. e.g. [Sha92]).

The major limitation of this setup is that the measurement procedure can hardly be automatized because the shielding boxes always have to be opened to rearrange biasing points or to switch the DUT. Electrically the system is limited by its noise floor which was thus always measured and compared to the noise level exhibited by the DUT to guarantee meaningful characterization. The uncertainty of the noise measurement is further affected by the linear filtering the amplifier performs due to its limited bandwidth. [JT07] satisfactory modeled the SR570 by employing two amplifying stages. The first stage provides a feedback by means of a resistor R_F so that a current-to-voltage conversion is yielded. The parallel capacitor C_F substantially dominates the pole-frequency of the setup. Due to the Miller effect, the capacitance seen from the

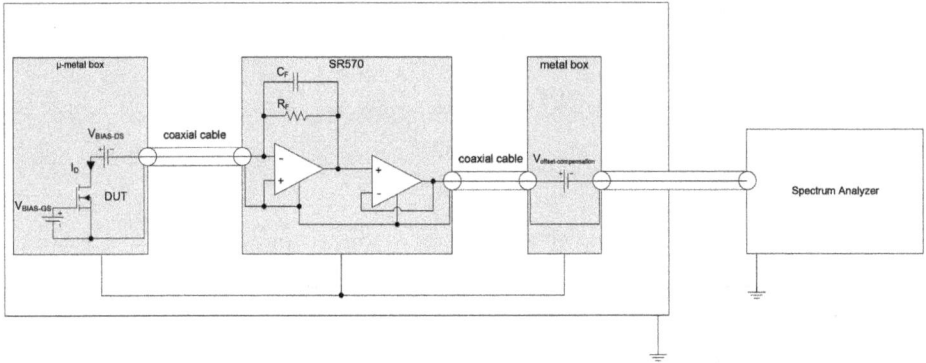

Figure B.1   Schematic of the noise measurement setup.

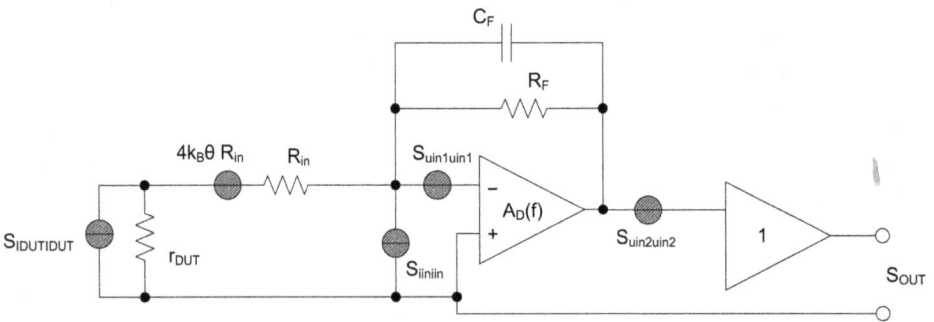

Figure B.2   Schematic of the noise measurement setup's noise model.

DUT is typically dominated by the SR570. Thus, for most scenarios it is not important to precisely evaluate the output capacitance of the DUT and its impact on the measurement accuracy. The bandwidth of the amplifier is increasing for larger amplification. For measurements presented in this work, however, it was typically above 1 kHz. To extend that usable bandwidth slightly, correction of the linear filtering was carried out for all measurements. At frequencies significantly larger than 1 kHz, this resulted in an increasing power spectral density which does not correspond to the theory of noise exhibition from the devices investigated throughout this work.

The noise model that was proposed by [JT07] is depicted in Figure B.2. They modeled the noise of the input resistor R_{in} as thermal noise $4k_B\theta R_{in}$ and added additional noise sources $S_{u_{in1}u_{in1}}$ and $S_{i_{in}i_{in}}$ as input referred noise sources for the first amplifying stage and $S_{u_{in2}u_{in2}}$ for the second stage. For transistors operating in saturation such that their output impedance is significantly larger than R_{in} plus the Miller transformed input impedance of the amplifier, the systems noise floor is dominated by $S_{i_{in}i_{in}}$. The noise sources of the setup were characterized according to [JT07] for all employed amplifier settings. The proposed subtraction of the noise component, however, is not

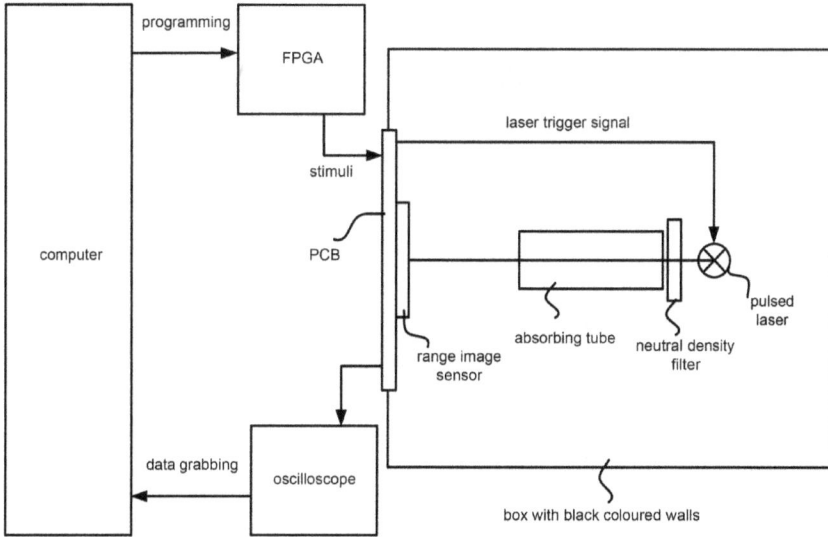

Figure B.3 Schematic of the measurement setup that is employed to carry out the emulated ToF principle.

considered to yield significant improvements of the noise measurement accuracy. For proper measurements the noise floor of the setup should be negligible compared to the noise exhibited by the DUT, which was the case throughout the presented characterizations. If that is not any longer the case, subtraction of these two uncorrelated random processes is rather likely to increase the measurement uncertainty.

B.2 SETUP TO MEASURE ACCORDING TO THE EMULATED TOF PRINCIPLE

In Figure B.3 the setup that was used for the measurements according to the emulated time-of-flight principle is depicted. The center of the setup is the range image sensor which is mounted onto a zero-force injection socket on the PCB. The PCB comprises bypass capacitors for the image sensor's power supply and a delay line that postpones the emission time instance of the laser pulse. The laser module is located 30 to 40 cm from the sensor's surface. The shape of the laser pulse was characterized with a high-speed photoreceiver as explained in Section 6.4.3. The irradiance at the sensor can be varied from 3 to 6000 W/m^2 by usage of neutral density filter and variation of the distance between sensor and laser module. Multiple reflections in the measurement box were omitted by an absorbing tube and a black coloring of the inner walls of the box. The DUT PCB board received stimulating signals from an field-programmable gate array (FPGA) that can be programmed by a conventional computer. The output of the sensor is connected to an oscilloscope that is controlled via the computer so that the setup can automatically rearrange the laser emission time instance and grab the output data.

Appendix C

Photon transfer method

The EMVA1288 standard models the imaging chain as a linear system [Ass10]:

$$\mathbb{E}(N_e) = \eta_{ext}\mathbb{E}(N_p) \tag{C.1}$$

$$\mathbb{E}(y) = [\mathbb{E}(N_e) + \mathbb{E}(N_{dark})] \cdot K_{tot}, \tag{C.2}$$

where $\mathbb{E}(N_e)$ is the average electron count generated by the average photon count $\mathbb{E}(N_p)$ via the extrinsic quantum efficiency η_{ext}. The generated electrons together with the dark signal at the storage nodes are propagated to the cameras output via the total gain K_{tot}. This linear system is assumed to exhibit noise according to

$$\sigma_y^2 = \sigma_q^2 + (\sigma_e^2 + \sigma_{dark}^2) \cdot K_{tot}^2 \tag{C.3}$$

with the output referred variance σ_y^2, the quantization noise component σ_q^2 and the noise components from the dark signal and the generated electrons. According to this definition σ_{dark}^2 is comprised by the read noise of the imager and the camera and a dark current related component. Substitution of $\sigma_e^2 = \mathbb{E}(N_e)$ and usage of Equation C.2 yields

$$\sigma_y^2 = \sigma_q^2 + \sigma_{dark}^2 \cdot K_{tot}^2 + K_{tot}[\mathbb{E}(y) - K_{tot}\mathbb{E}(N_{dark})], \tag{C.4}$$

in which the slope equals K_{tot} and can be found by fitting in an LQ-manner. Substituting this result back into Equation C.2 and fitting again in an LQ-manner yields an estimate for the extrinsic quantum efficiency. This, moreover, allows to plot $\mathbb{E}(N_e)$ against e.g. integration time or temperature enabling investigations of the dark current. σ_e^2 can be plotted against $\mathbb{E}(N_e)$ from which the read noise can be estimated. According to the EMVA1288, the read noise has to be determined at the minimum integration time. Alternatively, one may plot dark noise against the integration time and measure the read noise by extrapolation to integration time zero. The PTM method also provides measures to compare for instance linearity, non-uniformity and full-well capacity. To provide the dataset needed for evaluation of the above relations, according to the EMVA1288, properly calibrated measurement equipment is necessary. The implementation of the described method has some pitfalls which, however, are beyond the scope of this work. The interested reader is referred to [Ass10; Jan07].

Nomenclature

α	angle	in rad
α_f	flicker noise model exponent of the current dependence	in 1
α_{FOV}	plane angle of the field-of-view	in rad
$\alpha^*_{m,n}$	optimal fitting parameters for the fitting model of the object distance	in 1
α_{Hooge}	Hooge constant	in 1
β	angle	in rad
γ_{nD}	thermal noise excess factor	in 1
γ_e	noise excess factor	in 1
$\gamma_{flicker}$	flicker noise model exponent of the frequency dependence	in 1
δ	Dirac delta function	
ε	permittivity	in $A\,s\,V^{-1}\,m$
η_{ext}	extrinsic quantum efficiency	in 1
θ	temperature	in K
θ_p	polar angle	in rad
λ	wavelength	in m
λ_L	laser wavelength	in m
$\mu_{e/h}$	electron/hole mobility	in $m^2\,V^{-1}\,s^{-1}$
ν	frequency	in s^{-1}
ν_s	sampling frequency	in s^{-1}
ν_m	modulation frequency	in s^{-1}
ν_{dem}	demodulation frequency	in s^{-1}
$\xi(t)$	ideal white noise process	
σ	standard variation	
σ_k	k-th central moment	
σ^2_{read}	read noise w/o reset noise	in Vrms
τ	shift in time	in s
τ_E	emission time instance of the laser pulse	in s
τ_{LDPD}	characteristic charge transfer time of the proposed photodetector model	in s
τ_{ToF}	time of flight	in s
ϕ_a	azimuth angle	in rad
ϕ_{FD}	electrostatic potential at the floating diffusion	in V

ϕ_{sg}^{E}	cross-correlation function for energy signals s and g	
Φ_e	radiant flux	W
$\Phi_{e\lambda}$	spectral radiant flux	W m^{-1}
χ^2	fitting residual	
ω	element of the sample space	
Ω	sample space	
Ω_{sa}	solid angle	sr
$\langle ., . \rangle$	inner product	
$A^{\mathcal{T}}$	transpose of matrix A	
a^*	conjugate complex of $a \in \mathbb{C}$	
$x_s(t) * y_s(t)$	convolution of x_s and y_s	
$f \circ g$	composition of f and g	
A	area	in m^2
A_{pa}	photoactive area	in m^2
A_{pix}	total pixel area	in m^2
A_{read}	differential amplification of the readout circuitry from source follower in to analog out	in 1
c	velocity of light	in m s^{-1}
C	capacitance	in A s V^{-1}
C_{ox}	specific gate-oxide capacitance of a MOSFET	in A s/V m^2
C_{SN}	sense-node capacitance	in A s V^{-1}
$d_{aperture}$	diameter of the circular aperture	in m
d_{lens}	diameter of the circular lens	in m
$\mathbb{C}$	set of complex numbers	
CG	conversion gain	in V W^{-1} or V W^{-1} s
E	energy	in W s
E_e	irradiance	in W m^{-2}
$E_{e\lambda}$	spectral irradiance	in W m^{-3}
$\mathbb{E}$	expectation	
$f_\#$	f-number	in 1
f_{foc}	focal length	in m
f_{p-X_r}	probability distribution function	
$f_{p-X_r \mid Y_r}$	conditional probability distribution function	
f_{laser}	normalized laser pulse shape	in 1
F_{p-X_r}	probability density function	
f_{rep}	laser repetition rate	in Hz
FF	fill factor	in 1
$\mathcal{F}$	Fourier transform	
$\mathcal{F}^{-1}$	inverse Fourier transform	
g_m	transconductance $\partial I_D / \partial V_{GS}$ of a MOSFET	A V^{-1}
g_{mb}	transconductance $\partial I_D / \partial V_{BS}$ of a MOSFET	A V^{-1}
G	conductance	A V^{-1}
G_{nD}	thermal noise conductance at the drain	A V^{-1}
h	Planck constant	in W s^2
h_s	impulse response function	

$h_{\tau_{LDPD}}$	function to model the finite charge transfer process of the photodetector	
$H_{\mathscr{F}-s}$	Fourier transformed of the transfer function	
$H_{\mathscr{L}-s}$	z-Transformed of the transfer function	
i	electric current	in A
I_B	base current	in A
I_C	collector current	in A
I_D	drain current	in A
I_E	emitter current	in A
I_G	gate current	in A
I_{IS}	current of a JFET at $V_{GS} = 0V$	in A
I_S	source current	in A
I	intensity of a wave function	in $W\,m^{-1}$
j	imaginary unit	in 1
J_{dark}	dark current	in e^-/s
$\mathscr{L}$	Laplace transform	
L	width of a MOSFET	
L_e	emitting radiance	in $W\,s\,r^{-1}\,m^{-2}$
L_{X_r}	noise power	in W
k_B	Boltzmann constant	in $W\,s\,K^{-1}$
K_{tot}	total camera gain	
$K_{X_r Y_r}$	cross-covariance	
m	mass	in kg
m_k	k-th moment of a random variable	
m_{mod}	modulation factor of a amplitude modulated signal	in 1
M_{X_r}	characteristic function	
$\mathbb{N}$	set of natural numbers	
n	electron density	m^{-3}
n_i	intrinsic charge carrier concentration	m^{-3}
n_{MB}	number of degrees of freedom according to the equipartition theorem	1
$<n>$	mean rate of emitted carrier per time interval	Hz
N_A	acceptor concentration	m^{-3}
N_{accu}	accumulation count	m^{-3}
N_D	donor concentration	m^{-3}
N_{e^-}	photgenerated electron count	1
N_{FDi}	stored electron count at the diffusion node i	in 1
P	measure of probability	
p	hole density	m^{-3}
PDP	photon detecting probability	in %
q	elementary charge	in A s
Q_F	quality factor	
Q_n^2	electric noise charge	in A s
r_{BE}	small-signal base-emitter resistance	in $V\,A^{-1}$
r_{CE}	small-signal collector-emitter resistance	in $V\,A^{-1}$
r_{DS}	small-signal drain-source resistance	in $V\,A^{-1}$
R	resistance	in $V\,A^{-1}$
R_B	extrinsic base resistance	in VA^{-1}

R_{source}	source resistance	in VA^{-1}
R_{rec}	recombination rate	in s^{-1}m^{-3}
$\mathscr{R}$	responsivity	in V/W/m^2
$\mathbb{R}$	set of real numbers	
$\mathfrak{R}$	real part of a complex number	
$R_{X_r X_r}$	autocorrelation function	
$R_{X_r Y_r}$	cross-correlation function	
s	complex frequency from the Laplace space	in s^{-1}
$S_{I_B I_B}$	noise current power spectral density referred of the base current	in A^2/Hz
$S_{I_C I_C}$	noise current power spectral density referred of the collector current	in A^2/Hz
$S_{I_D I_D}$	noise current power spectral density referred of the drain current	in A^2/Hz
S_{LDPD}	function to model the imperfect shutter of the photodetector	
$S_{\text{LDPD}-\text{TX}-\text{OFF}}$	characteristic shutter blocking parameter of the proposed photodetector model	in 1
$S_{V_{BE} V_{BE}}$	noise current power spectral density referred of the collector current referred to the base-emitter voltage	in V^2/Hz
$S_{V_{GS} V_{GS}}$	noise current power spectral density referred of the drain current referred to the gate-source voltage	in V^2/Hz
$S_{X_r X_r}$	noise power spectral density	in W s
t	time	in s
t_{ox}	gate-oxide thickness of a MOSFET	in m
T	time interval	in s
T_{int}	integration time	in s
T_{p}	pulse width	in s
T_{per}	period of a periodic function	in s
T_{SW}	shutter window	in s
u	real-valued wave function	in W$^{1/2}$m$^{-1/2}$
$u_{T_{\text{SW}},\tau}$	short time integration function starting τ for the duration of T_{SW}	
$u_{\text{demoulated}}$	demodulated signal of a time-of-flight sensor	in a.u. (e.g. in V or A s)
U	complex-valued wave function	in W$^{1/2}$m$^{-1/2}$
v	electric voltage	in V
v_{x}	velocity	in m s^{-1}
V_{BE}	electric base-emitter voltage between	in V
V_{BS}	electric bulk-source voltage between	in V
V_{CE}	electric collector-emitter voltage between	in V
V_{DS}	electric drain-source voltage between	in V
V_{GB}	electric gate-bulk voltage between	in V
V_{GS}	electric gate-source voltage between	in V
V_{FDi}	electric voltage between the floating diffusion node i and ground potential	in V

V_P	pinch-off voltage	in V
V_T	thermal voltage	in V
W	width of a MOSFET	
$W(t)$	Wiener process	
x	space coordinates	in m
x_r	boundary of random variable $X_r(\omega)$	
x_s	signal	
x_{s-n}	time discrete signal	
$x_{s-sampled}$	sampled signal	
$x_{triangulation}$	length of the base of a triangulation based range measurement setup	in m
X_r	random variable	
$X_r(t)$	stochastic process	
$X_{\mathscr{F}-s}$	Fourier transformed of the signal x_s	
$X_{\mathscr{L}-s}$	Laplace transformed of the signal x_s	
$X_{\mathscr{Z}-s}$	z-Transformed of the signal x_s	
y_r	boundary of random variable $X(\omega)$	
y_s	signal	
Y_r	random variable	
$Y_r(t)$	stochastic process	
$Y_{\mathscr{F}-s}$	Fourier transformed of the signal y_s	
$Y_{\mathscr{L}-s}$	Laplace transformed of the signal y_s	
$Y_{\mathscr{Z}-s}$	z-Transformed of the signal y_s	
z	distance from object to range camera or more general a measurement system	in m
z_{fit}	fitted object distance	in m
z_z	complex, discrete frequency of z-Transform	
$\mathbb{Z}$	set of integers	
$\mathscr{Z}$	z-Transform	

Abbreviations

APS	active pixel sensors
BiCMOS	CMOS process that is enhanced to offer bipolar devices
BSIM	Berkeley Short-channel insulated gate FET (IGFET) model
CAD	computer aided design
CCD	charge-coupled device
CDF	cumulative distribution function
CDS	correlated double sampling
CMOS	complementary metal-oxide-semiconductor
CW	continuously modulated waves
CX	collection gate
DAE	differential-algebraic equation
DD	draining diffusion
DOV	depth-of-view
DUT	device under test
ESD	electrostatic discharge
EKV	Enz-Krummenacher-Vittoz model
FD	floating diffusion
FET	field-effect transistor
FFT	fast Fourier transform
FPN	fixed pattern noise
FOV	field-of-view
FPGA	field-programmable gate array
GOX	gate oxide
IC	integrated circuit
IR	infrared
JFET	junction FET
MSI	multiple short time integration
MOS	metal-oxide-semiconductor
MOSFET	metal-oxide-semiconductor field-effect transistor
MTF	modulation transfer function
NMOS	n-type MOSFET
LADAR	laser detection and ranging
Laser	light amplification by stimulated emission of radiation
LDPD	lateral drift-field photodetector
LED	light emitting diode

LIDAR   light detection and ranging
LNA   low-noise amplifier
LOCOS   local oxidation of silicon
LPTV   linear periodical time-variant
LTI   linear time-invariant
ODE   ordinary differential equation
OTA   operational transconductance amplifier
PCB   printed circuit board
PDE   partial differential equation
PDF   probability density equation
PM   pulsed modulated
PMOS   p-type MOSFET
PPD   pinned photodiode
PTM   photon transfer method
RADAR   radio detection and ranging
RGBZ   image sensor capable of detecting color and range information
RTS   random telegraph signal
SDE   stochastic differential equation
SONAR   sound navigation and ranging
SPAD   single photon avalanche diode
STI   shallow trench isolation
TCAD   technology CAD
TCSPC   time-correlated single photon counting
ToF   time-of-flight
TX   transfer gate
WSS   wide-sense stationary

Bibliography

[Add97]   Paul S. Addison. *Fractals and Chaos – An illustrated course*. 2nd ed. Institute of Physics Publishing, 1997.

[Adva]   Advanced Scientific Concepts, ed. *DragonEye 3D Flash LIDAR Space Camera™*. URL: http://advancedscientificconcepts.com/products/drag oneye.html (visited on 05/03/2012).

[Advb]   Advanced Scientific Concepts, ed. *Portable 3D Flash LIDAR Camera Kit™*. URL: http://advancedscientificconcepts.com/products/portable. html (visited on 05/03/2012).

[Advc]   Advanced Scientific Concepts, ed. *TigerEye 3D Flash LIDAR Camera Kit™*. URL: http://advancedscientificconcepts.com/products/tigereye. html (visited on 05/03/2012).

[AH02]   Phillip E. Allen and Douglas R. Holberg. *CMOS Analog Circuit Design* 2nd ed. Oxford University Press, 2002.

[Amb82]   András Ambrózy. *Electronic Noise*. 1st ed. McGraw-Hill International Book Company, 1982.

[Ass10]   European Machine Vision Association. *EMVA Standard 1288*. Nov. 2010.

[Bak10]   R. Jacob Baker. *CMOS: Circuit Design, Layout, and Simulation*. 3rd ed. John Wiley & Sons Inc., 2010.

[Bel85]   D.A. Bell. *Noise and the Solid State*. 1st. Pentech Press, 1985.

[Ber07]   Berufgenossenschaft Energie Textil Elektro Medizinerzeugnisse. *Unfal-lverhütungsvorschrift* Laserstrahlung. 2007.

[BS97]   Heinz Bittel and Leo Storm. *Rauschen*. 1st ed. Springer-Verlag Berlin Heidelberg New York, 1997.

[Bla04]   F. Blais. "Review of 20 years of range sensor development". In: *Journal of Electronic Imaging* 13(1) (Jan. 2004), pp. 231–240.

[Blu96]   Alfons Blum. *Elektronisches Rauschen*. B.G. Teubner Stuttgart, 1996.

[BG01]   Fabrizio Bonani and Giovanni Ghione. *Noise in Semiconductor Devices – Modeling and Simulation*. 1st ed. Springer-Verlag Wien GmbH, 2001.

[Bo11]   Fabrizio Bonani, Simona Donati Guerrieri, Giovanni Ghione, and Riccardo Tisseur. "Modeling and simulation of noise in transistors under large-signal condition". In: *Noise and Fluctuations (ICNF), 2011 21st Conference on*. June 2011, pp. 10–15.

[BMY]   Bong Ki Mheen, Mi Jin Kim, and Young Joo Song. Pat. US 2007 0158710 A1. Laid Open.

[BS70]   Willard S. Boyle and Georg E. Smith. "Charge Coupled Semicon-
             ductor Devices". In: *Bell System Technical Journal* 49.4 (Apr. 1970),
             pp. 597–593.

[Braa]   Brainvision Inc., ed. *TOF-based Range Image Camera DISTANZA®
             series*. 2011th ed. URL: http://www.brainvision.co.jp/xoops/contents/
             product/tof/CEATEC2011_pamphlet_en.pdf (visited on 05/03/2012).

[Brab]   Brainvision Inc, ed. *TOFCam Stanley*. URL: http://www.brainvision.
             co.jp/xoops/modules/tinyd4/index.php?id=5 (visited on 05/03/2012).

[Bre78]   R. J. Brewer. "A low noise CCD output amplifier". In: *Electron Devices
             Meeting, 1978 International*. Vol. 24. IEEE, 1978, pp. 610–612.

[Bre80]   R. J. Brewer. "The low light level potential of a CCD imaging array". In:
             IEEE Transactions on Electron Devices 27.2 (Feb. 1980), pp. 401–405.

[Bro10]   Christian Brockners. "Entwicklung und Charakterisierung eines nieder-
             frequenten Rauschmessplatzes zur Bestimmung des Rauschverhaltens
             von Halbleiterbauelementen". diploma thesis. Universität Duisburg-
             Essen – Fraunhofer IMS, 2010.

[Bü07]   B. Büttgen, M'H.-A. El Mechat, F. Lustenberger, and P. Seitz.
             "Pseudonoise Optical Modulation for Real-Time 3-D Imaging With Min-
             imum Interference". In: *IEEE Transactions on Circuits and Systems I:
             Regular Papers* 54.10 (Oct. 2007), pp. 2109–2119.

[Bü08]   B. Büttgen and P. Seitz. "Robust Optical Time-of-Flight Range Imaging
             Based on Smart Pixel Structures". In: *IEEE Transactions on Circuits and
             Systems I: Regular Papers* 55.6 (July 2008), pp. 1512–1525.

[Bü06]   Bernhard Büttgen, Felix Lustenberger, and Peter Seitz. "Demodulation
             Pixel Based on Static Drift Fields". In: *IEEE Transactions on Electron
             Devices* 53.11 (Nov. 2006), pp. 2741–2747.

[Buc83]   M.J. Buckingham. *Noise in Electronic Devices and Systems*. 1st ed. Ellis
             Horwood Limited, 1983.

[CDS06]   Inc. Cadence Design Systems, ed. *Application Notes on Direct Time-
             Domain Noise Analysis using Virtuoso Spectre*. July 2006.

[CDS11a]   Inc. Cadence Design Systems, ed. *Virtuoso®Spectre®Circuit Simulator
             and Accelerated Parallel Simulator RF Analysis User Guide*. 10.1.1. June
             2011.

[CDS11b]   Inc. Cadence Design Systems, ed. *Virtuoso®Spectre®Circuit Simulator
             Components and Device Models Manual*. 10.1.1. June 2011.

[CDS11c]   Inc. Cadence Design Systems, ed. *Virtuoso®Spectre®Circuit Simulator
             RF Analysis Theory*. 10.1.1. June 2011.

[CWW01]   Ja-Hao Chen, Shyh-Chyi Wong Wong, and Yeong-Her Wang. "An
             analytic three-terminal band-to-band tunneling model on GIDL in
             MOSFET". In: *IEEE Transaction on Electron Devices* 48.7 (July 2001),
             pp. 1400–1405.

[CWM09]   Yue Chen, Xinyang Wang, A.J. Mierop, and A.J.P. Theuwissen. "A
             CMOS Image Sensor With In-Pixel Buried-Channel Source Follower and
             Optimized Row Selector". In: *IEEE Transactions on Electron Devices*
             56.11 (Nov. 2009), pp. 2390–2397.

[CH99]   Yuhua Cheng and Chenming Hu. *MOSFET Modeling and BSIM3 User's
             Guide*. Kluwer Academic Publishers, 1999.

[CDK09]   Richard M. Conroy, Adrian A. Dorrington, Rainer Künnemeyer, and Michael J. Cree. "Range imager performance comparison in homodyne and heterodyne operating modes". In: *Proceedings of SPIE* 7239.1 (Jan. 2009), pp. 723905–723905–10.

[Czy10]   Andreas Czylwik. *Theorie statistischer Signale – Communications 2*. scriptum. 2010.

[Dar12]   Arnaud Darmont. *High Dynamic Range Imaging*. 1st ed. SPIE Press, 2012.

[DR58]   Wilbur B. Davenport, Jr. and William L. Root. *An Introduction to the Theory of Random Signals and Noise*. International Student Edition. Mcgraw-Hill Book Company, INC., 1958.

[DM02]   M. Jamal Deen and Ognian Marinov. "Effect of Forward and Reverse Substrate Biasing on Low-Frequency Noise in Silicon PMOS-FETs". In: *IEEE Transactions on Electron Devices* 49.3 (Mar. 2002), pp. 409–413.

[DLSV96]   A. Demir, E.W.Y. Liu, and A.L. Sangiovanni-Vincentelli. "Time-domain non-Monte Carlo noise simulation for nonlinear dynamic circuits with arbitrary excitations". In: *IEEE Transactions on Computer-Aided Design of Integrated Circuits and Systems* 15.5 (May 1996), pp. 493–505.

[DR99]   A. Demir and J. Roychowdhury. "Modeling and simulation of noise in analog/mixed-signal communication systems". In: *Custom Integrated Circuits, 1999. Proceedings of the IEEE 1999*. 1999, pp. 385–393.

[DSV00]   Alper Demir and Alberto Sangiovanni-Vincentelli. *Analysis and Simulation of Noise in Nonlinear Electronic Circuits and Systems*. Second printing. Kluwer, 2000.

[DCC07]   Adrian Dorrington, Michael Cree, Dale A. Carnegie, Andrew Payne, and Richard Conroy. "Heterodyne range imaging as an alternative to photogrammetry". In: *Proceedings of SPIE* 6491.1 (Jan. 2007), pp. 64910D–64910D–9.

[DCC06]   Adrian A. Dorrington, Dale A. Carnegie, and Michael J. Cree. "Toward 1-mm depth precision with a solid state full-field range imaging system". In: *Proceedings of SPIE* 6068.1 (Feb. 2006), 60680–60680K–10.

[DJ06]   Smitha Dronavalli and Renuka P. Jindal. "CMOS Device Noise Considerations for Terabit Lightwave Systems". In: *IEEE Transactions on Electron Devices* 53.4 (Apr. 2006), pp. 623–630.

[Du11]   D. Durini, A. Spickermann, J. Fink, W. Brockherde, A. Grabmaier, and B. Hosticka. "Experimental Comparison of Four Different CMOS Pixel Architectures Used in Indirect Time-of-Flight Distance Measurement Sensors". In: *2011 International Image Sensor Workshop (IISW)*. June 2011.

[Du10]   Daniel Durini, Andreas Spickermann, Rana Mahdi, Werner Brockherde, Holger Vogt, Anton Grabmaier, and Bedrich J. Hosticka. "Lateral drift-field photodiode for low noise, high-speed, large photo-active-area CMOS imaging applications". In: *Nuclear Instruments and Methods in Physics Research Section A: Accelerators, Spectrometers, Detectors and Associated Equipment* 624.2 (Dec. 2010), pp. 470–475.

[Du09a]   Daniel Romero Durini. "Solid-State Imaging in Standard CMOS Processes". PhD thesis. Fraunhofer IMS/Universität Duisburg-Essen, 2009.

[Du10]   Daniel Romero Durini, Werner Brockherde, and Bedrich Hosticka. Pat. US 2010/0308213 A1. Laid Open.

[Du09]   Daniel Romero Durini, Werner Brockherde, and Bedrich Hosticka. Pat. DE 10 2009 020 218 B3. Laid Open.

[Ein05]   A. Einstein. "Über die von der molekularkinetischen Theorie der Wärme geforderte Bewegung von in ruhenden Flüssigkeiten suspendierten Teilchen". In: *Annalen der Physik* 322 (1905), pp. 549–560.

[EEC]   Department of Electrical Engineering and Berkeley Computer Sciences University of California, eds. *BSIM3v3.2.2 MOSFET Model – Users' Manual*. 1999th ed. URL: http://www-device.eecs.berkeley.edu/bsim/?page=BSIM3_LR (visited on 08/16/2012).

[EKC92]   Tarik Elewa, Bendik Kleveland, Sorin Cristoloveanu, Boubaker Boukriss, and Alain Chovet. "Detailed Analysis of Edge Effects in SIMOX-MOS Transistors". In: *IEEE Transactions on Electron Devices* 39.4 (Apr. 1992), pp. 874–882.

[ESM04]   O. Elkhalili, O.M. Schrey, P. Mengel, M. Petermann, W. Brockherde, and B.J. Hosticka. "A 4 times 64 pixel CMOS image sensor for 3-D measurement applications". In: *Solid-State Circuits, IEEE Journal of* 39.7 (July 2004), pp. 1208–1212.

[Elk05]   Omar Elkhalili. "Entwicklung von optischen 3D CMOS-Bildsensoren auf der Basis der Pulslaufzeitmessung". PhD thesis. Fraunhofer IMS/Universität Duisburg-Essen, 2005.

[EV06]   Christian C. Enz and Eric A. Vittoz. *Charge-based MOS Transistor Modeling – The EKV model for low-power and RF IC design*. John Wiley & Sons, 2006.

[Fal03]   Kenneth Falconer. *Fractal Geometry – Mathematical Foundations and Applications*. 2nd ed. John Wiley & Sons, Inc., 2003.

[FXL02]   H.D. Fleetwood, D.M. Xiong, Z.-Y. Lu, C.J. Nicklaw, J.A. Felix, R.D. Schrimpf, and S.T. Pantelides. "Unified Model of Hole Trapping, 1/f Noise, and Thermally Stimulated Current in MOS Devices". In: *IEEE Transactions on Nuclear Science* 49.6 (Dec. 2002), pp. 2674–2683.

[Fos93]   Eric R. Fossum. "Active pixel sensors: are CCDs dinosaurs?" In: *Proc. SPIE 1900*. Vol. 2. SPIE, Feb. 2, 1993.

[Fos97]   Eric R. Fossum. "CMOS image sensors: electronic camera-on-a-chip". In: *Electron Devices, IEEE Transactions on* 44.10 (Oct. 1997), pp. 1689–1698.

[Fota]   Fotonic, ed. *Fotonic D40 – Technical Information*. URL: http://www.fotonic.com/assets/documents/fotonic_d40_highres.pdf (visited on 05/02/2012).

[Fotb]   Fotonic, ed. *Fotonic D70 – Technical Information*. URL: http://www.fotonic.com/assets/documents/fotonic_d70_highres.pdf (visited on 05/02/2012).

[Fotc]   Fotonic, ed. *Fotonic T300 – Technical Information*. URL: http://www.fotonic.com/assets/documents/fotonic_t300_highres.pdf (visited on 05/02/2012).

[FGM06]   Boyd Fowler, Michael D. Godfrey, and Steve Mims. "Reset noise reduction in capacitive sensors". In: *Circuits and Systems I: Regular Papers, IEEE Transactions on* 53.8 (Aug. 2006), pp. 1658–1669.

[FPR09]   Mario Frank, Matthias Plaus, Holger Rapp, Ullrich Köthe, Bernd Jähne, and Fred A. Hamprecht. "Theoretical and experimental error analysis of continuouswave time-of-flight range camers". In: *Optical Engineering* 48.1 (Jan. 2009), pp. 013602–013602–16.

[Fro08]   Frost & Sullivan. *World Image Sensors Market.* 2008.

[Gar85]   C.W. Gardiner. *Handbook of Stochastic Methods.* Second Edition. Springer Verlag, 1985.

[Gar90]   W.A. Gardner. *Introduction to Random Processes.* Second Edition. McGraw-Hill, 1990.

[GS01]   Geoffrey Grimmett and David Stirzaker. *Probability and Random Processes.* Third Edition. Oxford University Press, 2001.

[Has05]   Alan Hastings. *The Art of Analog Layout.* 2nd ed. Prentice Hall, 2005.

[HS93]   Harold, M. Hastings and George Sugihara. *Fractals – A User's Guide for the Natural Sciences.* 1st ed. Oxford University Press, 1993.

[HP91]   H. Heywang and H.W. Pötzl. *Bänderstruktur und Stromtransport.* 2nd ed. Springer-Verlag Berlin Heidelberg New York, 1991.

[Hol94]   William Timothy Holman. "A Low Noise CMOS Voltage Reference". PhD thesis. Georgia Institute of Technology, 1994.

[HPJ11]   Sung-Min Hong, Anh-Tuan Pham, and Christoph Jungemann. *Deterministic Solvers for the Boltzmann Transport Equation.* 1st ed. Springer-Verlag Wien GmbH, 2011.

[HSS06]   Bedrich Hosticka, Peter Seitz, and Andrea Simoni. "Optical time-of-flight sensors for solid-state 3D-vision". In: *Encyclopedia of sensors.* Ed. by C.A. Grimes, E.C. Dickley, and M.V. Pishko. Vol. 7. 2006, pp. 259–289.

[HKH90]   Kwok K. Hung, Ping K. Ko, Chenming Hu, and Yiu, C. Cheng. "A Unified Model for the Fliker Noise in Metal-Oxide-Semiconductor Field-Effect Transistors". In: *IEEE Transactions on Electron Devices* 37.3 (Mar. 1990), pp. 654–665.

[Hä97]   G. Häusler. "About the scaling behaviour of optical range sensors." In: *Proc. 3rd Int. Workshop on Automatic Processing of Fringe Patterns.* (Sept. 1997), pp. 147–155.

[HE11]   Gerd Häusler and Svenja Ettl. "Limitations of Optical 3D Sensors". In: *Optical Measurement of Surface Topography.* Ed. by Richard Leach. 2nd. Springer-Verlag, 2011, pp. 23–48.

[IC 10]   IC Insights. *O-S-D Report.* 2010.

[IC 12]   IC Insights. *O-S-D Report.* 2012.

[IEE]   IEE, ed. *3D-MLI – Technology Fact Sheet.* April 2008. URL: http://www.steadlands.com/data/iee/mli.pdf (visited on 05/03/2012).

[INY99]   I. Inoue, H. Nozaki, H. Yamashita, T. Yamaguchi, H. Ishiwata, H. Ihara, R. Miyagawa, H. Miura, N. Nakamura, Y. Egawa, and Y. Matsunaga. "New LVBPD (low voltage buried photo-diode) for CMOS imager". In: *Electron Devices Meeting, 1999. IEDM Technical Digest. International.* IEEE, 1999, pp. 883–886.

[ITY03]   I. Inoue, N. Tanaka, H. Yamashita, T. Yamaguchi, H. Ishiwata, and H. Ihara. "Low-leakage-current and low-operating-voltage buried photodiode for a CMOS imager". In: *IEEE Transactions on Electron Devices* 50.1 (Jan. 2003), pp. 43–47.

[ITR11a]   ITRS. *International Technology Roadmap for Semiconductors – Emerging Research Devices*. 2011.

[ITR11b]   ITRS. *International Technology Roadmap for Semiconductors – Modeling and Simulation*. 2011.

[IUS05]   A.H. Izhal, T. Ushinaga, T. Sawada, M. Homma, Y. Maeda, and S. Kawahito. "A CMOS time-of-flight range image sensor with gates on field oxide structure". In: *Sensors, 2005 IEEE*. Nov. 2005, 4 pp.

[Jan07]   James R. Janesick. *Photon Transfer*. 1st ed. SPIE-Press, 2007.

[JT07]   M. Jankovec and M. Topic. "Development and Characterization of a Lowfrequency Noise Measurement System for Optoelectronic Devices". In: *Informacije Midem – Ljubljana* 37.2 (2007), pp. 80–86.

[Jer01]   R. Jeremias, W. Brockherde, G. Doemens, B. Hosticka, L. Listl, and P. Mengel. "A CMOS photosensor array for 3D imaging using pulsed laser". In: *Solid-State Circuits Conference, 2001. Digest of Technical Papers. ISSCC. 2001 IEEE International*. 2001, pp. 252–253, 452.

[Jer09]   Ralf Friedrich Jeremias. "CMOS Bildsensoren mit Kurzzeitverschluß zur Tiefenerfassung nach dem Lichtlaufzeit-Meßprinzip". PhD thesis. Universität Duisburg-Essen, 2009.

[Joh28]   J.B. Johnson. "Thermal Agitation of Electricity in Conductors". In: *Physical Review* 32 (July 1928), pp. 97–109.

[JM12]   Christoph Jungemann and Bernd Meinerzhagen. *Hierarchical Device Simulation: The Monte-Carlo Perspective*. 1st ed. Springer-Verlag Wien GmbH, 2012.

[KSW08]   Uwe Kiencke, Michael Schwarz, and Thomas Weickert. *Signalverarbeitung: Zeit-Frequenz-Analyse und Schätzverfahren*. Oldenbourg Wissenschaftsverlag, 2008.

[KHL05]   J.Y. Kim, S.I. Hwang, J.J. Lee, J.H. Ko, Y. Kim, J.C. Ahn, T. Asaba, and Y.H. Lee. "Characterization and Improvement of Random Noise in 1/3.2" UXGA CMOS Image Sensor with 2.8 µm Pixel using 0.13 µm-Technology". In: 2005.

[KHK10]   Seong-Jin Kim, Sang-Wook Han, Byongmin Kang, Keechang Lee, J.D.K. Kim, and Chang-Yeong Kim. "A Three-Dimensional Time-of-Flight CMOS Image Sensor With Pinned-Photodiode Pixel Structure". In: *IEEE Electron Device Letters* 31.11 (Nov. 2010), pp. 1272–1274.

[KKK12]   Seong-Jin Kim, Byongmin Kang, J.D.K Kim, Keechang Lee, Chang-Yeong Kim, and Kinam Kim. "A 1920x1080 3.65 µm-pixel 2D/3D image sensor with split and binning pixel structure in 0.11 µm standard CMOS". In: *Solid-State Circuits Conference Digest of Technical Papers (ISSCC), 2012 IEEE International*. Feb. 2012, pp. 396–398.

[KYO12]   Wonjoo Kim, Wang Yibing, I. Ovsiannikov, SeungHoon Lee, Yoondong Park, Chilhee Chung, and E. Fossum. "A 1.5Mpixel RGBZ CMOS image sensor for simultaneous color and range image capture". In: *Solid-State Circuits Conference Digest of Technical Papers (ISSCC), 2012 IEEE International*. Feb. 2012, pp. 392–394.

[Koe08]   Bernhard Koenig. "Optimized Distance Measurement with 3D-CMOS Image Sensor and Real-Time Processing of the 3D Data for Applications in Automotive and Safety Engineering". PhD thesis. Universität Duisburg-Essen, 2008.

[Kos97]   Walter F. Kosonocky, Guang Yang, Rakesh K. Kabra, Chao Ye, Zeynep Pektas, John L. Lawrence, Vincent Mastrocolla, Frank V. Shallcross, and Vipulkumar Patel. "360×360 element three-phase very high frame rate burst image sensor: design operation and performance". In: *Electron Devices, IEEE Transactions on* 44.10 (Oct. 1997), pp. 1617–1624.

[Kos96]   Walter F. Kosonocky, Guang Yang, Chao Ye, Rakesh K. Kabra, Liansheng Xie, John L. Lawrence, Vincent Mastrocolla, Frank V. Shallcross, and Vipulkumar Patel. "360×360-Element Very-High Frame-Rate Burst-Image Sensor". In: *Solid-State Circuits Conference Digest of Technical Papers (ISSCC), 1996 IEEE International.* Feb. 1996, pp. 182–183.

[Kun95]   Kenneth S. Kundert. *The Designer's Guide to SPICE and Spectre®.* Kluwer Academic Publishers, 1995.

[Kun97]   Kenneth S. Kundert. "Simulation methods for RF integrated circuits". In: *1997 IEEE/ACM International Conference on Computer-Aided Design, 1997. Digest of Technical Papers.* Nov. 1997, pp. 752–765.

[Kun90]   Kenneth S. Kundert, Jacob K. White, and Alberto Sangiovanni-Vincentelli. *Steady-State Methods for Simulating Analog and Microwave Circuits.* Kluwer Academic Publishers, 1990.

[Kun04]   Kenneth S. Kundert and Olaf Zinke. *The Designer's Guide to Verilog-AMS.* Kluwer Academic Publishers, 2004.

[Kun99]   K.S. Kundert. "Introduction to RF simulation and its application". In: *IEEE Journal of Solid-State Circuits* 34.9 (Sept. 1999), pp. 1298–1319.

[LVFS07]   Assaf Lahav, Dmitry Veigner, Amos Fenigstein, and Amit Shiwalkar. "Optimization of Random Telegraph Noise Non Uniformity in a CMOS Pixel with a pinned-photodiode". In: *Proc. Int. Image Sens. Workshop 2007* 2007, pp. 2019–223.

[Lan00]   Robert Lange. "3D Time-of-Flight Distance Measurement with Custom Solid-State Image Sensors in CMOS/CCD-Technology". PhD thesis. University of Siegen, 2000.

[LLM11]   T.Y. Lee, Y.J. Lee, D.K. Min, S.H. Lee, W.H. Kim, S.H. Kim, J.K. Jung, I. Ovsiannikov, Y.G. Jin, Y.D. Park, E.R. Fossum, and C.H. Chung. "A 192×108 pixel ToF-3D image sensor with single-tap concentric gate demodulation pixels in 0.13 μm technology". In: *Electron Devices Meeting (IEDM), 2011 IEEE International.* Dec. 2011, pp. 8.7.1–8.7.4.

[LN05]   F. Li and A. Nathan. *CCD Image Sensors in Deep-Ultraviolet.* 1st ed. Springer-Verlag Berlin Heidelberg New York, 2005.

[Li11]   Xiang Li. "Rauschcharakterisierung fortschrittlicher CMOS Bauelemente in einem 0.35 μm – Prozess". MA thesis. Universität Siegen – Fraunhofer IMS, 2011.

[Lin08]   David Lindley. *Uncertainty.* Reprint. Anchor, 2008.

[Loh]   Jan Lohstrom. Pat. US 4,245,233. Laid Open.

[Lou03]   N.V. Loukianova, H.O. Folkerts, J.P.V. Maas, D.W.E. Verbugt, A.J. Mierop, W. Hoekstra, E. Roks, and Theurissen A.J.P. "Leakage current modeling of test structures for characterization of dark current in CMOS image sensors". In: *IEEE Transaction on Electron Devices* 50.1 (Jan. 2003), pp. 77–83.

[Maa03]   Stephen A. Maas. *Nonlinear Microwave and RF Circuits*. Artech House Inc., 2003.

[Mac54]   Stefan Machlup. "Noise in Semiconductors: Spectrum of a Two-Parameter Random Signal." In: *Journal of Applied Physics* 25.3 (Mar. 1954), pp. 341–343.

[MML97]   Stéphan Maëstre, Pierre Magnan, Francis Lavernhe, and Franck Corbière. "Hot carriers effects and electroluminescence in the CMOS photodiode active pixel sensors". In: *Sensors and Camera Systems for Scientific, Industrial, and Digital Photography Applications IV* 5017.59 (May 2003), pp. 59–67.

[MFH10]   R. Mahdi, J. Fink, and B.J. Hosticka. "Lateral Drift-Field Photodetector for high speed 0.35 μm CMOS imaging sensors based on non-uniform lateral doping profile: Design, theoretical concepts, and TCAD simulations". In: *Research in Microelectronics and Electronics (PRIME), 2010 Conference on Ph.D* IEEE, July 2010, pp. 1–4.

[Man67]   Benoit Mandelbrot. "Some Noises with 1/f Spectrum, a Bridge Between Direct Current and White Noise". In: *IEEE Transactions on Information Theory* 13.2 (Apr. 1967), pp. 289–298.

[MHM10]   P. Martin-Gonthier, E. Havard, and P. Magnan. "Custom transistor layout design techniques for random telegraph signal noise reduction in CMOS image sensors". In: *Electronics Letters* 46.19 (Sept. 2010), pp. 1323–1324.

[MM09]   P. Martin-Gonthier and P. Magnan. "RTS noise impact in CMOS image sensors readout circuit". In: *16th IEEE International Conference on Electronics, Circuits, and Systems, 2009. ICECS 2009.* IEEE, Dec. 2009, pp. 928–931.

[MM11]   P. Martin-Gonthier and P. Magnan. "Novel Readout Circuit Architecture for CMOS Image Sensors Minimizing RTS Noise". In: *IEEE Electron Device Letters* 32.6 (June 2011), pp. 776–778.

[MBG11]   F. Matheis, W. Brockherde, A. Grabmaier, and B.J. Hosticka. "Modeling and calibration of 3D-Time-of-Flight pulse-modulated image sensors". In: *2011 20th European Conference on Circuit Theory and Design (ECCTD)*. Aug. 2011, pp. 417–420.

[MOI87]   Y. Matsunaga, S. Oosawa, M. Iesaka, S. Manabe, N. Harada, and N. Suzuki. "A high sensitivity output amplifier for CCD image sensor". In: *Electron Devices Meeting, 1987 International*. Vol. 33. IEEE, 1987, pp. 116–119.

[McK93]   John McKelvey. *Solid State Physics*. 1st ed. Krieger Publishing Company, 1993.

[McW55]   A.L. McWorther. "1/f noise and related surface effects in germanium". PhD thesis. Boston: MIT, 1955.

[Meh02]   A. Mehrotra. "Noise analysis of phase-locked loops". In: *IEEE Transactions on Circuits and Systems I: Fundamental Theory and Applications* 49.9 (Sept. 2002), pp. 1309–1316.

[MSV05]   Amit Mehrotra and Sangiovanni-Vincentelli. *Noise Analysis of Radio Frequency Circuits*. Kluwer Academic Publishers, 2005.

[Mer95]   S. Merchant. "Arbitrary lateral diffusion profiles". In: *IEEE Transactions on Electron Devices* 42.12 (Dec. 1995), pp. 2226–2230.

[MES]   MESA Imaging AG, ed. *SR4000 Data Sheet*. August 2011. URL: http://www.mesa-imaging.ch/dlm.php?fname=pdf/SR4000_Data_Sheet.pdf (visited on 05/03/2012).

[MST08]   Bongki Mheen, Young-Joo Song, and A.J.P. Theuwissen. "Negative Offset Operation of Four-Transistor CMOS Image Pixels for IncreasedWell Capacity and Suppressed Dark Current". In: *IEEE Electron Device Letters* 29.4 (Mar. 2008), pp. 347–349.

[Mil11]   Drake Miller. "Random Dopants and Low-Frequency Noise Reduction in Deep-Submicron MOSFET Technology". PhD thesis. Oregon State University, 2011.

[Mü90]   Rudolf Müller. *Rauschen*. 2nd ed. Springer-Verlag Berlin Heidelberg New York, 1990.

[Mü95]   Rudolf Müller. *Grundlagen der Halbleiter-Elektronik*. 7th ed. Springer-Verlag Berlin Heidelberg New York, 1995.

[MC93]   C.D. Motchenbacher and J.A. Connelly. *Low-Noise Electronic System Design*. John Wiley & Sons Inc, 1993.

[MF73]   C.D. Motchenbacher and F.C. Fitchen. *Low-Noise Electronic Design*. John Wiley & Sons Inc, 1973.

[NTK07]   O. Nastov, R. Telichevesky, K. Kundert, and J. White. "Fundamentals of Fast Simulation Algorithms for RF Circuits". In: *Proceedings of the IEEE* 95.3 (Mar. 2007), pp. 600–621.

[NFK08]   C. Niclass, C. Favi, T. Kluter, M. Gersbach, and E. Charbon. "A 128×128 Single-Photon Image Sensor With Column-Level 10-Bit Time-to-Digital Converter Array". In: *Solid-State Circuits, IEEE Journal of* 43.12 (Dec. 2008), pp. 2977–2989.

[Nyq28]   Harry Nyquist. "Thermal Agitation of Electric Charge in Conductors". In: *Physical Review* 32 (July 1928), pp. 110–113.

[odo]   odos imaging ltd., ed. *real.IZ high resolution time-of-flight 2+3D™ camera development kit – product brief*. March 2012. URL: http://www.odos-imaging.com/uploads/pages/products/PBOI3DCAM1DVK01_V2.pdf (visited on 05/02/2012).

[Ogg95]   Thierry Oggier, Rolf Kaufmann, Michael Lehmann, Bernhard Büttgen, Simon Neukom, Michael Richter, Matthias Schweizer, Peter Metzler, Felix Lustenberger, and Nicolas Blanc. "Novel pixel architecture with inherent background suppression for 3D time-of-flight imaging". In: *Proceedings of SPIE* 5665.1 (Jan. 2005), pp. 1–8.

[OL07]   Rainer Ohm and Hans Dieter Lüke. *Signalübertragung*. 12th ed. Springer-Verlag, 2007.

[PCH05]   Bedabrata Pain, Thomas Cunningham, Bruce Hancock, Chris Wrigley, and Chao Sun. "Excess noise and dark current mechanisms in CMOS imagers". In: *2005 International Image Sensor Workshop (IISW)*. June 2005, pp. 144–148.

[Pan]   Panasonic, ed. *3D Image Sensor Product Lineup*. URL: http://pewa.panasonic.com/components/built-in-sensors/3d-image-sensors/ (visited on 05/02/2012).

[PMP12] L. Pancheri, N. Massari, M. Perenzoni, M. Malfatti, and D. Stoppa. "A QVGA-range image sensor based on buried-channel demodulator pixels in 0.18 μm CMOS with extended dynamic range". In: *Solid-State Circuits Conference Digest of Technical Papers (ISSCC), 2012 IEEE International*. Feb. 2012, pp. 394–396.

[PSM10] Lucio Pancheri, David Stoppa, Nicola Massari, Mattia Malfatti, Lorenzo Gonzo, Quanzi D. Hossain, and Gian-Franco Dalla Betta. "A 120 × 160 pixel CMOS range image sensor based on current assisted photonic demodulators". In: *Proceedings of SPIE* 7726.1 (Apr. 2010), pp. 772615–772615–9.

[PU02] Athanasios Papoulis and S. Unnikrishna Pillai. *Probability, Random Variables, and Stochastic Processes*. 4th ed. Mcgraw-Hill Higher Education, 2002.

[PBK10] Seong-Hyung Park, Jung-Deuk Bok Bok, Hyuk-Min Kwon, Woon-Il Choi, Man-Lyun Ha, Ju-Il Lee, and Hi-Deok Lee. "Decrease of Dark Current by Reducing Transfer Transistor Induced Partition Noise With Localized Channel Implantation". In: *Electrin Device Letter, IEEE* 31.11 (Nov. 2010), pp. 1278–1280.

[PCDC06] A.D. Payne, D.A. Carnegie, A.A. Dorrington, and M.J. Cree. "Full field image ranger hardware". In: *Third IEEE International Workshop on Electronic Design, Test and Applications, 2006. DELTA 2006*. Jan. 2006, 6 pp.

[PPBS08] F. Pedrotti, L. Pedrotti, W. Bausch, and H. Schmidt. *Optik für Ingenieure*. Vol. 4. Springer Verlag, 2008.

[PMS11] M. Perenzoni, N. Massari, D. Stoppa, L. Pancheri, M. Malfatti, and L. Gonzo. "A 160×120-Pixels Range Camera With In-Pixel Correlated Double Sampling and Fixed-Pattern Noise Correction". In: *Solid-State Circuits, IEEE Journal of* 46.7 (July 2011), pp. 1672–1681.

[PN86] Robert F. Pierret and Gerold W. Neudeck. *Advanced Semiconductor Fundamentals*. 1st ed. Addison-Wesley Publishing Company, 1986.

[PMDa] PMD Technologies, ed. *PMD PhotonICs® 19k-S3 – Preliminary datasheet*. URL: http://www.pmdtec.com/fileadmin/pmdtec/media/PMD-PhotonICs-19k-S3.pdf (visited on 05/02/2012).

[PMDb] PMD Technologies, ed. *PMD[vision]® CamCube 3.0 – spec sheet*. URL: http://www.pmdtec.com/fileadmin/pmdtec/downloads/documentation/datenblatt_ camcube3.pdf (visited on 05/02/2012).

[Raz03] Behzad Razavi. *Design of Analog CMOS Integrated Circuits*. 1st ed. McGraw-Hill, 2003.

[Res11] Research in China. *Global and China CMOS Camera Module Industry Report*. Sept. 2011.

[RM01] Gabriel Alfonso Rincón-Mora. *Voltage References*. 1st ed. John Wiley & Sons Inc., 2001.

[RCS92] Edwin Roks, P. G. Centen, L. Sankaranarayanan, Jan W. Slotboom, Jan T. Bosiers, and Wim F. Huinink. "A bipolar floating base detector (FBD) for CCD image sensors". In: *IEDM '92*. 1992, pp. 109–112.

[RCB96]   Edwin Roks, Peter Centen, Jan Bosiers, and Wim Huinink. "The Double-Sided Floating-Surface Detector: An Enhanced Charge-Detection Architecture for CCD Image Sensors". In: Vol. 43.No. 9 (Sept. 1996), pp. 1583–1591.

[RCB95]   Edwin Roks, Peter G. Centen, Jan T. Bosiers, and Wim F. Huinink. "The Double-Sided Floating-Surface Detector: An Enhanced Charge-Detection Architecture for CCD Image Sensors". In: *Solid State Device Research Conference, 1995. ESSDERC '95. Proceedings of the 25th European.* IEEE, Sept. 1995, pp. 327–330.

[Ros96]   Sheldon M. Ross. *Stochastic Processes.* Second Edition. John Wiley & Sons, Inc., 1996.

[RLF98]   J. Roychowdhury, D. Long, and P. Feldmann. "Cyclostationary noise analysis of large RF circuits with multitone excitations". In: *IEEE Journal of Solid-State Circuits* 33.3 (Mar. 1998), pp. 324–336.

[SA90]   Nelson S. Saks and Mario G. Ancona. "Spatial Uniformity of Interface Trap Distribution in MOSFET's". In: *IEEE Transaction on Electron Devices* 37.4 (Apr. 1990), pp. 1057–1063.

[ST07]   B.E.A. Saleh and M.C. Teich. *Fundamentals of photonics.* 2nd ed. John Wiley & Sons, Inc., 2007.

[SIK10]   Tomonari Sawada, Kana Ito, and Shoji Kawahito. "Empirical Verification of Range Resolution for a TOF Range Image Sensor with Periodiacl Charge Draining Operation Under Influence of Ambient Light". In: *J. Inst. Image Inf. TV Eng. (Academic Journal 2010)* 64.3 (2010), pp. 373–380.

[SHM00]   Rudolf Schwarte, Gerd Häusler, and Reinhard W. Malz. "Three-Dimensional Imaging Techniques". In: *Computer Vision and Applications.* Ed. by Bernd Jähne and Horst Haußecker. Vol. 1. Academic Press, 2000, pp. 177–208.

[Sei08]   P. Seitz. "Quantum-Noise Limited Distance Resolution of Optical Range Imaging Techniques". In: *IEEE Transactions on Circuits and Systems I: Regular Papers* 55.8 (Sept. 2008), pp. 2368–2377.

[Sei07]   Peter Seitz. "Photon-noise limited distance resolution of optical metrology methods". In: *Proceedings of SPIE* 6616.1 (June 2007), pp. 66160D–66160D–10.

[SSV95]   Peter Seitz, Thomas Spirig, Oliver Vietze, and Peter Metzler. "Lock-in CCD and the convolver CCD: applications of exposure-concurrent photocharge transfer in optical metrology and machine vision". In: *Proceedings of SPIE* 2415.1 (Apr. 1995), pp. 276–284.

[Sha92]   Syed Jaffar Shah. *Application Note 025 – Field Wiring and Noise Considerations for Analog Signals.* Ed. by National Instruments. Apr. 1992.

[SCJ66]   W. Shockley, John A. Copeland, and R.P. James. "The Impedance Field Method of Noise Calculation in Active Semiconductor Devices". In: *Quantum theory of atoms, molecules, and the solid-state.* Ed. by Per-Olov Lowdin. 1966, pp. 573–563.

[Sic08]   Thorsten R. Sickenberger. "Efficient Transient Noise Analysis in Circuit Simulation". PhD thesis. Humboldt-Universität zu Berlin, 2008.

[Sin01]   Jasprit Singh. *Semiconductor Devices*. 1st ed. John Wiley and Sons, Inc., 2001.

[Sof]   SoftKinetic, ed. */400 Family DS410 Datasheet*. V1. URL: http://www. softkinetic.com/Portals/0/Download/DS410_Datasheet_V1_050411.pdf (visited on 05/02/2012).

[Sp09]   A. Spickermann, D. Durini, S. Bröcker, W. Brockherde, B.J. Hosticka, and A. Grabmaier. "Pulsed time-of-flight 3D-CMOS imaging using photogate-based active pixel sensors". In: *ESSCIRC, 2009. ESSCIRC '09. Proceedings of*. Sept. 2009, pp. 200–203.

[Sp10]   Andreas Spickermann. "Photodetektoren und Auslesekonzepte für die 3DTime-of-Flight Bildsensoren in 0.35 µm-Standard-CMOS-Technologie". PhD thesis. Fraunhofer IMS/Universität Duisburg-Essen, 2010.

[Sp11b]   Andreas Spickermann, Werner Brockherde, and Bedrich Hosticka. Pat. DE 102009 037 596 A1. Laid Open.

[Sp11a]   Andreas Spickermann,Werner Brockherde, and Bedrich Hosticka Hosticka. Pat. EP 2 290 393 A2. Laid Open.

[SDS11]   Andreas Spickermann, Daniel Durini, Andreas Suss, Wiebke Ulfig, Werner Brockherde, Bedrich J. Hosticka, Stefan Schwope, and Anton Grabmaier. "CMOS 3D image sensor based on pulse modulated time-of-flight principle and intrinsic lateral drift-field photodiode pixels". In: *ESSCIRC, 2011. ESSCIRC '11. Proceedings of*. IEEE, Sept. 2011, pp. 111–114.

[Sü13a]   Andreas Süss, Werner Brockherde, and Bedrich J. Hosticka Hosticka. Pat. US 2013 0270610 A1. 2013. Laid Open.

[Sü12]   Andreas Süss and Bedrich J. Hosticka. "A novel JFET readout structure applicable for pinned and lateral drift-field photodiodes". In: *Proc. SPIE 8439, Optical Sensing and Detection II*. SPIE, Apr. 2012, pp. 84391–1–84391–8.

[Sü13]   Andreas Süss, Christian Nitta, Andreas Spickermann, Daniel Durini, Gabor Varga, Melanie Jung, Werner Brockherde, Bedrich J. Hosticka, Holger Vogt, and Stefan Schwope. "Speed Considerations for LDPD Based Time-of-Flight CMOS 3D Image Sensors". In: *ESSCIRC, 2013. ESSCIRC '13. Proceedings of* 2013.

[Sta94]   International Standard, ed. *ISO 5725-1 – Accuracy (trueness and precision) of measurement methods and results – Part 1: General principles and definitions*. 1st ed. Dec. 1994.

[SMP11]   David Stoppa, Nicola Massari, Lucio Pancheri, Mattia Malfatti, Matteo Perenzoni, and Lorenzo Gonzo. "A Range Image Sensor Based on 10-µm Lock-In Pixels in 0.18-µm CMOS Imaging Technology". In: *IEEE Journal of Solid-State Circuits* 46.1 (Jan. 2011), pp. 248–258.

[Syn09a]   Synopsys, ed. *Sentaurus Device User Guide*. C-2009.06. June 2009.

[Syn09b]   Synopsys, ed. *Sentaurus Process User Guide*. C-2009.06. June 2009.

[Sys09]   Cadence Design Systems, ed. *Accellera – Verilog-AMS Language Reference Manual*. 2.3.1. June 2009.

[SN07]   S.M. Sze and Kwok K. Ng. *Physics of Semiconductor Devices*. John Wiley & Sons Inc., 2007.

[TSI10]   Hiroaki Takeshita, Tomonari Sawada, Tetsuya Iida, Keita Yasutomi, and Shoji Kawahito. "High-speed charge transfer pinned-photodiode for a CMOS time-of-flight range image sensor". In: *Proc. SPIE7536, Sensors, Cameras, and Systems for Industrial/Scientific Applications XI.* SPIE, Jan. 2010, 75360R–1–75360R–9.

[TKI82]   N. Teranishi, A. Kohono, Y. Ishihara, E. Oda, and K. Arai. "No image lag photodiode structure in the interline CCD image sensor". In: *Electron Devices Meeting, 1982 International.* Vol. 28. IEEE, 1982, pp. 324–327.

[TK99]   Manolis Terrovitis and Ken Kundert. *Device Noise Simulation of ΔΣ Modulators.* 1999. URL: http://cktsim.org/Analysis/delta-sigma.pdf (visited on 08/09/2012).

[The95]   Albert J.P. Theuwissen. *Solid-State Imaging With Charge-Coupled Devices.* 1st ed. Kluwer Academic Publishers, 1995.

[Tia00]   Hui Tian. "Noise Analysis in CMOS Image Sensors". PhD thesis. Stanford University, 2000.

[TFG01]   Hui Tian, Boyd Fowler, and Abbas El Gamal. "Analysis of temporal noise in CMOS photodiode active pixel sensor". In: *Solid-State Circuits, IEEE Journal of* 36.1 (Jan. 2001), pp. 92–101.

[THK13]   Yasuhisa Tochigi, Katsuiko Hanzawa, Yuri Kato, Rihhito Kuroda, Hideki Mutoh, Ryuta Hirose, Hideki Tominaga, Kenji Takubo, Yasushi Kondo, and Shigetoshi Sugawa. "A Global-Shutter CMOS Image Sensor with Readout Speed of 1Tpixel/s Burst and 780Mpixel/s Continuous". In: *IEEE Journal of Solid-State Circuits* 48.1 (Jan. 2013), pp. 329–338.

[Tri12]   Dipl. Ing. Stefan Schwope TriDiCam GmbH. *E-mail.* May 7, 2012.

[TM10]   Yannis Tsividis and McAndrew. *Operation and Modeling of the MOS Transistor.* Oxford University Press, 2010.

[Unb97]   Rolf Unbehauen. *Systemtheorie* 1. 7th Edition. Oldenbourg Verlag, 1997.

[Van78]   A. Van der Ziel. "Flicker noise in semiconductors: Not a true bulk effect". In: *Applied Physics Letters* 33.10 (Nov. 1978), pp. 883–884.

[Van88]   A. Van der Ziel. "Unified Presentation of 1/f Noise in Electronic Devices: Fundamental 1/f Noise Sources". In: *Proceedings of the IEEE* 76.3 (Mar. 1988), pp. 233–258.

[van70]   Aldert van der Ziel. *Noise – Sources, Characterization, Measurement.* 1st ed. Prentice-Hall, Inc., 1970.

[van86]   Aldert van der Ziel. *Noise in solid state devices and circuits.* 1st ed. John Wiley and Sons, Inc., 1986.

[NTK05]   D. Van Nieuvenhove, W. van der Tempel, and M. Kuijk. "Novel Standard CMOS Detector using Minority Current for guiding Photo-Generated Electrons towards Detecting Junctions". In: *Proceedings Symposium IEEE/LEOS Benelux Chapter.* 2005, pp. 229–232.

[VV00]   Ewout P. Vandamme and Lode K.J. Vandamme. "Critical Discussion on Unified 1/f Noise Models for MOSFETs". In: *IEEE Transactions on Electron Devices* 47.11 (Nov. 2000), pp. 2146–2152.

[VSH11]   G. Varga, A. Süss, and B.J. Hosticka. "A sequential method for noise estimation in switched-capacitor systems using a switching time-frequency domain". In: *2011 20th European Conference on Circuit Theory and Design (ECCTD).* Aug. 2011, pp. 500–503.

[Vit83]   Eric A. Vittoz. "MOS transistors operated in the lateral bipolar mode and their application in CMOS technology". In: *Solid-State Circuits, IEEE Journal of* 18.3 (June 1983), pp. 273–279.

[WRH11]   R.J. Walker, J.A. Richardson, and R.K. Henderson. "A 128 *times* 96 pixel eventdriven phase-domain $\Sigma\Delta$ fully digital 3D camera in 0.13 µCMOS imaging technology". In: *Solid-State Circuits Conference Digest of Technical Papers (ISSCC), 2011 IEEE International.* Feb. 2011, pp. 410–412.

[WS01]   Ching-Chun Wang and Charles G. Sodini. "The Effect of Hot Carriers on the Operation of CMOS Active Pixel Sensors". In: *Electron Devices Meeting, 2001, IEDM '01. Technical Digest.* IEEE, Dec. 2001, pp. 24.5.1–24.5.4.

[WRM06]   Xinyang Wang, P. R. Rao, A. Mierop, and A. J. P. Theuwissen. "Random Telegraph Signal in CMOS Image Sensor Pixels". In: *Electron Devices Meeting, 2006. IEDM '06. International.* IEEE, Dec. 2006, pp. 1–4.

[WRT06]   Xinyang Wang, Padmakumar R. Rao, and Albert T. Theuwissen. "Fixed-Pattern Noise Induced by Transmission Gate in Pinned 4T CMOS Image Sensor Pixels". In: *Solid-State Device Research Conference, 2006, ESSDERC 2006.* IEEE, Sept. 2006, pp. 331–334.

[WSR08]   Xinyang Wang, M. F. Snoeij, P. R. Rao, A. Mierop, and A.J.P. Theuwissen. "A CMOS Image Sensor with a Buried-Channel Source Follower". In: *Solid-State Circuits Conference, 2008. ISSCC 2008. Digest of Technical Papers. IEEE International.* IEEE, Feb. 2008, pp. 62–595.

[Web92]   Hubert Weber. *Einführung in die Wahrscheinlichkeitsrechnung und Statistik für Ingenieure.* 3rd ed. B.G. Teubner Stuttgart, 1992.

[Whi74]   M. H. White, D. R. Lampe, F. C. Blaha, and I. A. Mack. "Characterization of surface channel CCD image arrays at low light levels". In: *IEEE Journal of Solid-State Circuits* 9.1 (Feb. 1974), pp. 1–12.

[WC99]   John L., Jr. Wyatt and Geoffrey J. Coram. "Nonlinear Device Noise Models: Satisfying the Thermodynamic Requirements". In: *IEEE Transactions on Electron Devices* 46.1 (Jan. 1999), pp. 184–193.

[YMIMH88]   H. Yamashita, Y. Matsunaga, M. Iesaka, S. Manabe, and N. Harada. "A New Hgh Sensitivity Photo-transistor for Area Image Sensors". In: *Electron Devices Meeting, 1988 International.* 1988, pp. 78–81.

[Yol10]   Yole Dévelopement. CMOS *Image Sensors – Technologies & Markets.* 2010.

[ZDZ10]   G. Zach, M. Davidovic, and H. Zimmermann. "A 16×16 Pixel Distance Sensor With In-Pixel Circuitry That Tolerates 150 klx of Ambient Light". In: *Solid-State Circuits, IEEE Journal of* 45.7 (July 2010), pp. 1345–1353.

[Zad50]   L.A. Zadeh. "Frequency Analysis of Variable Networks". In: *Proceedings of the IRE* 38.3 (Mar. 1950), pp. 291–299.

Index

For Product Safety Concerns and Information please contact our EU
representative GPSR@taylorandfrancis.com
Taylor & Francis Verlag GmbH, Kaufingerstraße 24, 80331 München, Germany

www.ingramcontent.com/pod-product-compliance
Lightning Source LLC
Chambersburg PA
CBHW061400210326
41598CB00035B/6042